350 Solved Electrical Engineering Problems

For the FE/PE Exams in Electrical Engineering

Edward Karalis, Ph.D.

© Copyright 2000, Engineering PRess
All rights reserved. Except as permitted under the United States Copyright Act of 1976, no part of this publication may be reproduced or distributed in any form or by any means, or stored in a data base or retrieval system, without the prior written permission of the copyright owner.

ISBN 1-57645-049-X

Printed in the United States of America

Engineering Press
P.O. Box 200129
Austin, TX 78720-0129

PREFACE

The purpose of this text is to provide a concise review of the fundamental concepts of electrical engineering. This is accomplished with selected example problems to illustrate each basic principle. A collection of 350 different problems are included and the relevant equations or formulas are used to derive their corresponding solution.

The illustrative problems are chosen from the following seven areas of electrical engineering, circuit theory, electronics, control systems, communications, fields, power, and computers. These include direct and alternating current circuits, linear and nonlinear electronics, classical and modern control, communications signals and systems, static and dynamic electromagnetics, power transmission and devices, and digital logic both combinational and sequential. The problems are of varying difficulty, and a mix of multi-part essay and multiple-choice formats with step-by-step detailed solutions.

This book is useful to both engineering students and practicing engineers. It can be used for college reference, graduate comprehensive and qualifying, Engineer-In-Training (EIT), and Professional Engineering (PE) examination reviews. Also, it can serve as a quick desktop reference for engineering professionals.

BRIEF CONTENTS

1.0 NETWORKS — 1-1
    Direct current
    Alternating Current

2.0 ELECTRONICS — 2-1
    Linear
    Nonlinear

3.0 CONTROL — 3-1
    Classical
    Modern

4.0 COMMUNICATIONS — 4-1
    Signals
    Systems

5.0 ELECTROMAGNETICS — 5-1
    Static
    Dynamic

6.0 POWER — 6-1
    Transmission
    Devices

7.0 DIGITAL — 7-1
    Combinational
    Sequential

BIBLIOGRAPHY — B-1

INDEX — I-1

CONTENTS

1.0 NETWORKS 1-1
1.1 Voltage 1-1
1.2 Current 1-2
1.3 Resistance 1-3
1.4 Inductance 1-4
1.5 Capacitance 1-5
1.6 Ohms's Law 1-6
1.7 Voltage divider 1-7
1.8 Current divider 1-8
1.9 RC charge circuit 1-9
1.10 RC decay circuit 1-10
1.11 RL charge circuit 1-11
1.12 RL decay circuit 1-12
1.13 Kirchhoff's voltage law 1-13
1.14 Kirchhoff's current law 1-14
1.15 Thevenin's theorem 1-15
1.16 Norton's theorem 1-16
1.17 Millman's theorem 1-17
1.18 Duality theorem 1-18
1.19 Superposition theorem 1-19
1.20 Reciprocity theorem 1-20
1.21 Substitution theorem 1-21
1.22 Sinusoidal function 1-22
1.23 Reactance 1-24
1.24 Impedance 1-25
1.25 Admittance 1-26
1.26 Series to parallel 1-27
1.27 Parallel to series 1-28
1.28 Series resonance 1-29
1.29 Parallel resonance 1-31
1.30 RC natural response 1-33
1.31 RC forced response 1-34
1.32 RL natural response 1-35
1.33 RL forced response 1-36
1.34 RLC natural response 1-37
1.35 RLC forced response 1-38
1.36 Power 1-39
1.37 AC voltage divider 1-40
1.38 Complex plane 1-41
1.39 Laplace transform 1-43
1.40 Impulse function 1-44
1.41 Step Function 1-45
1.42 Ramp function 1-46
1.43 Square wave 1-47
1.44 Impulse response 1-49
1.45 Phasor 1-50
1.46 Frequency response 1-51
1.47 One port network 1-52
1.48 Two port network 1-53
1.49 Insertion loss 1-54
1.50 Switched capacitor network 1-55
1.51 Waveform analysis 1-56
1.52 Transmission constants 1-58
1.53 Hybrid parameters 1-60
1.54 Coupled circuit 1-62
1.55 Attenuator 1-64

2.0 ELECTRONICS 2-1
2.1 Diode 2-1
2.2 Zener diode 2-2
2.3 Bipolar junction transistor 2-3
2.4 Field effect transistor 2-4
2.5 MOSFET depletion mode 2-5
2.6 MOSFET enhancement mode 2-6
2.7 Common emitter amplifier 2-7
2.8 Common base amplifier 2-8
2.9 Common collector amplifier 2-9
2.10 Common source amplifier 2-10
2.11 Common gate amplifier 2-11
2.12 Common drain amplifier 2-12
2.13 RC coupled amplifier 2-13
2.14 Low frequency model 2-15
2.15 High frequency model 2-16
2.16 Tuned amplifier 2-17
2.17 Wideband amplifier 2-19
2.18 Small signal amplifier 2-21
2.19 Large signal amplifier 2-22
2.20 Voltage amplifier 2-24
2.21 Current amplifier 2-26
2.22 Power amplifier 2-28
2.23 Feedback amplifier 2-30
2.24 Differential amplifier 2-32
2.25 Operational amplifier 2-33
2.26 Inverting op amp 2-34
2.27 Noninverting op amp 2-35
2.28 Summing op amp 2-36
2.29 Integrating op amp 2-37
2.30 Differentiating op amp 2-38
2.31 Transfer function op amp 2-39
2.32 Comparator 2-40
2.33 Clipper 2-41
2.34 Clamper 2-42
2.35 Limiter 2-43
2.36 Miller's theorem 2-44
2.37 Voltage sweep generator 2-45
2.38 Current sweep generator 2-46

2.39 Phase shift oscillator	2-47
2.40 Hartley oscillator	2-48
2.41 Collpitts oscillator	2-49
2.42 Pierce oscillator	2-50
2.43 UJT oscillator	2-52
2.44 Astable multivibrator	2-53
2.45 Monostable multivibrator	2-54
2.46 Bistable multivibrator	2-55
2.47 Schmitt trigger	2-56
2.48 Log attenuator	2-57
2.49 Multiplier	2-58
2.50 Hall effect	2-59
2.51 Darlington amplifier	2-60
2.52 Cascode amplifier	2-62
2.53 Hybrid amplifier	2-64
2.54 Buffer amplifier	2-66
2.55 Op amp filter	2-68

3.0 CONTROL — 3-1

3.1 Open-loop control	3-1
3.2 Closed-loop control	3-2
3.3 Block diagrams	3-3
3.4 Signal flow graphs	3-4
3.5 Mason's rule	3-5
3.6 First order loop	3-6
3.7 Second order loop	3-7
3.8 Type 0 error	3-8
3.9 Type 1 error	3-9
3.10 Type 2 error	3-10
3.11 Sensitivity	3-11
3.12 Frequency domain	3-12
3.13 Bode plot	3-13
3.14 Nichols chart	3-14
3.15 Nyquist plot	3-15
3.16 Stability	3-16
3.17 Gain and phase margins	3-17
3.18 Lead compensation	3-18
3.19 Lag compensation	3-19
3.20 Lead-lag compensation	3-20
3.21 Time domain	3-21
3.22 Complex plane	3-22
3.23 Partial fractions	3-23
3.24 Transfer function	3-24
3.25 Ruth-Hurwitz	3-25
3.26 Root locus	3-26
3.27 PD controller	3-27
3.28 PI controller	3-28
3.29 PID controller	3-29
3.30 State space model	3-30
3.31 State variable controller	3-31
3.32 Eigenvalues and eigenvectors	3-32
3.33 Diagonal matrix	3-33
3.34 Transition matrix	3-34
3.35 Controllability	3-35
3.36 Observability	3-36
3.37 Lyaponov criterion	3-37
3.38 Nonlinear control	3-38
3.39 Phase plane	3-39
3.40 Limit cycles	3-40
3.41 Describing function	3-41
3.42 Stability criterion	3-42
3.43 Stochastic control	3-43
3.44 Optimal control	3-44
3.45 Maximum principle	3-45
3.46 Kalman filter	3-46
3.47 Adaptive control	3-47
3.48 z Transform	3-48
3.49 Discrete-time control	3-49
3.50 Direct digital control	3-50
3.51 Control loop ratios	3-51
3.52 Time analysis	3-53
3.53 S-plane analysis	3-55
3.54 Frequency analysis	3-57
3.55 Transfer function analysis	3-59

4.0 COMMUNICATIONS — 4-1

4.1 Fourier series	4-1
4.2 Fourier integral	4-2
4.3 Convolution	4-3
4.4 Correlation	4-4
4.5 Spectral density	4-5
4.6 Cross correlation	4-6
4.7 Cross spectral density	4-7
4.8 Probability	4-8
4.9 PDF function	4-10
4.10 CDF function	4-11
4.11 Joint probability	4-12
4.12 Conditional probability	4-13
4.13 Mean	4-14
4.14 Mean squared value	4-15
4.15 Variance	4-16
4.16 Standard deviation	4-17
4.17 Noise figure	4-18
4.18 Sensitivity	4-19
4.19 Dynamic range	4-20
4.20 Intermodulation distortion	4-21
4.21 Intercept point	4-22
4.22 Lowpass filter	4-23
4.23 Highpass filter	4-24
4.24 Bandpass filter	4-25
4.25 Bandreject filter	4-26
4.26 Delay line LPF	4-27
4.27 Delay line HPF	4-28
4.28 Matched filter	4-29
4.29 AM modulation	4-30
4.30 FM modulation	4-31

4.31 PM modulation	4-32	
4.32 nFM modulation	4-33	
4.33 wFM modulation	4-34	
4.34 AM/FM modulation	4-35	
4.35 FM/AM modulation	4-36	
4.36 SSB modulation	4-37	
4.37 QAM modulation	4-38	
4.38 PAM modulation	4-39	
4.39 PDM modulation	4-40	
4.40 PPM modulation	4-41	
4.41 PCM modulation	4-42	
4.42 FSK modulation	4-43	
4.43 PSK modulation	4-44	
4.44 Balanced modulator	4-45	
4.45 Envelope detector	4-46	
4.46 Square law detector	4-47	
4.47 Synchronous detector	4-48	
4.48 Ratio detector	4-49	
4.49 AM receiver	4-50	
4.50 FM receiver	4-51	
4.51 Sample and Hold	4-52	
4.52 Noise Power	4-54	
4.53 Channel capacity	4-56	
4.54 DAC	4-58	
4.55 ADC	4-60	

5.0 ELECTROMAGNETICS 5-1

5.1 Resistance	5-1
5.2 Inductance	5-2
5.3 Capacitance	5-3
5.4 Parallel plate capacitor	5-4
5.5 Spherical capacitor	5-5
5.6 Cylindrical capacitor	5-6
5.7 Electric force	5-7
5.8 Electric field	5-8
5.9 Electric field strength	5-9
5.10 Electric potential	5-10
5.11 Electric flux density	5-11
5.12 Electric energy	5-12
5.13 Electric deflection	5-13
5.14 Magnetic force	5-14
5.15 Magnetic field	5-15
5.16 Magnetic field strength	5-16
5.17 Magnetic potential	5-17
5.18 Magnetic flux density	5-18
5.19 Magnetic energy	5-19
5.20 Magnetic deflection	5-20
5.21 Laplace's equation	5-21
5.22 Ampere's law	5-22
5.23 Faraday's law	5-23
5.24 Gauss's law	5-24
5.25 Maxwell's law	5-25
5.26 Toroid	5-26
5.27 Solenoid	5-27
5.28 Electric dipole	5-28
5.29 Electric quadrupole	5-29
5.30 Magnetic dipole	5-30
5.31 Magnetic quadrupole	5-31
5.32 Antenna radiation	5-32
5.33 Near field	5-33
5.34 Far field	5-34
5.35 Vertical antenna	5-35
5.36 Dipole antenna	5-36
5.37 Loop antenna	5-37
5.38 Transmission line	5-38
5.39 Coaxial cable	5-39
5.40 Parallel conductors	5-40
5.41 Quarter wave stub	5-41
5.42 Reflections (R<Z)	5-42
5.43 Reflections (R>Z)	5-43
5.44 Rectangular waveguide	5-44
5.45 Circular wave guide	5-45
5.46 TM wave	5-46
5.47 TE wave	5-47
5.48 TEM wave	5-48
5.49 Rectangular cavity	5-49
5.50 Cylindrical cavity	5-50
5.51 Line parameters	5-51
5.52 Impedance matching	5-53
5.53 Electric interference	5-55
5.54 Magnetic interference	5-57
5.55 Radar	5-59

6.0 POWER 6-1

6.1 Line resistance	6-1
6.2 Line inductance	6-2
6.3 Line capacitance	6-3
6.4 Short line	6-4
6.5 Medium line	6-5
6.6 Long line	6-6
6.7 Power factor	6-7
6.8 Maximum power transfer	6-8
6.9 Self inductance	6-10
6.10 Mutual inductance	6-11
6.11 Transformer	6-12
6.12 Delta to Wye conversion	6-14
6.13 Wye to Delta conversion	6-15
6.14 Four wire Wye:Wye	6-16
6.15 Three wire Wye:Wye	6-17
6.16 Three wire Wye:Delta	6-18
6.17 Delta:Delta	6-19
6.18 Delta:Wye	6-20
6.19 Unbalanced Wye load	6-21
6.20 Unbalanced Delta load	6-22
6.21 Generator effect	6-23
6.22 Motor effect	6-24

6.23	Back EMF	6-25	
6.24	Torque	6-26	
6.25	Separate winding DC generator	6-27	
6.26	Series DC generator	6-28	
6.27	Shunt DC generator	6-29	
6.28	Compound DC generator	6-30	
6.29	Separate winding DC motor	6-31	
6.30	Series DC motor	6-32	
6.31	Shunt DC motor	6-33	
6.32	Compound DC motor	6-34	
6.33	AC Delta generator	6-35	
6.34	AC Wye generator	6-36	
6.35	AC Delta motor	6-37	
6.36	AC Wye motor	6-39	
6.37	AC induction motor	6-41	
6.38	AC synchronous motor	6-42	
6.39	Resolver	6-43	
6.40	Synchro	6-44	
6.41	Servo	6-45	
6.42	Stepper motor	6-46	
6.43	Half wave rectifier	6-47	
6.44	Full wave rectifier	6-49	
6.45	Bridge rectifier	6-50	
6.46	Voltage regulator	6-51	
6.47	SCR	6-52	
6.48	TRIAC	6-53	
6.49	Batteries in series	6-54	
6.50	Batteries in parallel	6-55	
6.51	Per unit notation	6-56	
6.52	Ground faults	6-58	

6.53	Breaker ratings	6-60	
6.54	Power factor correction	6-62	
6.55	Motor rating	6-64	

7.0 DIGITAL 7-1

7.1	Minterms	7-1
7.2	Maxterms	7-2
7.3	DeMorgan's Theorem 1	7-3
7.4	DeMorgan's Theorem 2	7-4
7.5	Logic analysis	7-5
7.6	NAND logic	7-6
7.7	NOR logic	7-7
7.8	SOP Standard form	7-8
7.9	POS Standard form	7-9
7.10	Multiplexer logic	7-10
7.11	Sequencer	7-11
7.12	Moore machine	7-12
7.13	Mealy machine	7-13
7.14	Decoded states	7-14
7.15	Encoded states	7-15
7.16	Flip-flop logic	7-16
7.17	D Flip-flop counter	7-18
7.18	JK Flip-flop counter	7-20
7.19	Shift register logic	7-22
7.20	DDS logic	7-24

BIBLIOGRAPHY B-1
INDEX I-1

1.0 NETWORKS

1.1 Voltage

A storage battery has a charging profile as shown in Figure 1.1. After 300 seconds a total charge of fifty coulombs is transferred to the battery. Determine the battery voltage, assuming it was completely discharged.

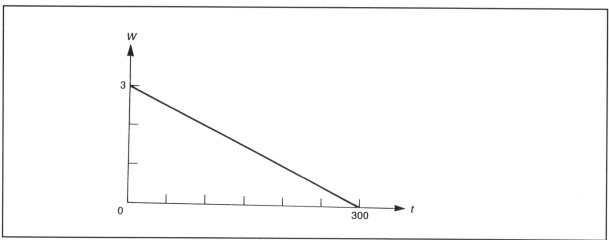

Figure 1.1 Battery charging profile

Solution:

Q = 50 coulombs

V = J / Q volts

W = J / t watts

The charging profile can be expressed as

y = mx + b

where m = - 0.01,   b = 3,   and x = t

W = - 0.01t + 3 watts

$$J = \int_a^b W\,dt = \int_0^{300} (-0.01t + 3)\,dt \quad \text{joules}$$

$$J = [-0.005t^2 + 3t]_0^{300} = 450 \quad \text{joules}$$

V = J / Q = 450 / 50 = 9 volts

1.2 Current

A storage battery has five coulombs of charge remaining. It is recharged at a uniform rate for 300 seconds as shown in Figure 1.2. Determine the charging current.

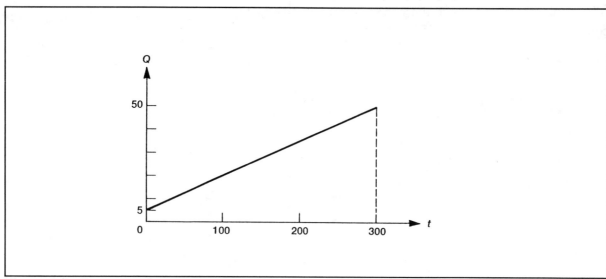

Figure 1.2 Battery recharge rate

Solution:

$$I = Q/t \quad \text{amperes}$$

The recharge rate can be expressed as

$$y = mx + b$$

where m = 0.15, b = 5, and x = t

$$Q = 0.15t + 5 \quad \text{coulombs}$$

$$I = \frac{dQ}{dt} = \frac{d}{dt}(0.15t + 5)$$

$$I = 0.15 \text{ amperes}$$

1.3 Resistance

A cylindrical copper pipe section has an outside diameter of 1.0 cm and a wall thickness of 0.025 cm. Its overall length is 250 cm as shown in Figure 1.3. Determine its resistance assuming the resistivity constant, ρ, equals 1.72×10^{-6} ohms-cm.

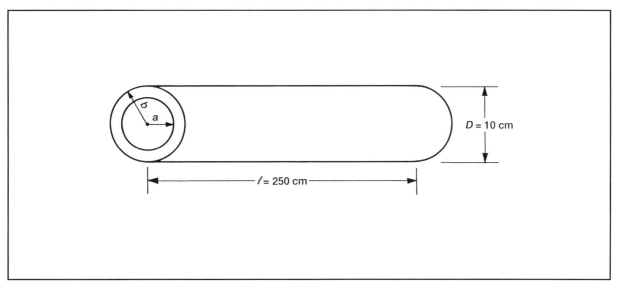

Figure 1.3 Copper pipe section

Solution:

$$R = \rho \frac{l}{A} \quad \text{ohms}$$

$$A = \int_a^b 2\pi r \, dr = [\pi r^2]_a^b = \pi[b^2 - a^2]$$

where $r = D/2$, $\quad b = 0.5$, $\quad a = 0.475$

$$A = \pi(b^2 - a^2) = 0.0765 \text{ cm}^2$$

$$R = \frac{1.72 \times 10^{-6} \times 250}{0.0765}$$

$$R = 5.620 \times 10^{-3} \quad \text{ohms}$$

1.4 Inductance

The applied current through the inductor shown in Figure 1.4 is $i(t) = 1000t$. Determine (a) the voltage across the inductor for $L = 2\times 10^{-6}$ henries, and (b) the current through the inductor at $t = 2\times 10^{-3}$ seconds.

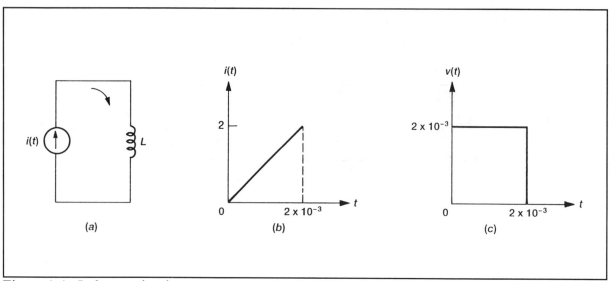

Figure 1.4 Inductor circuit

Solution:

a) $v(t) = L\dfrac{di}{dt} = 2\times 10^{-6}\dfrac{d}{dt}(1000t)$

$v(t) = 2\times 10^{-3}$ volts

b) $i(t) = \dfrac{1}{L}\displaystyle\int_a^b v(t)dt = \dfrac{1}{2\times 10^{-6}}\int_0^{2\times 10^{-3}}(2\times 10^{-3})dt$

$i(t) = [1\times 10^3 t]_0^{2\times 10^{-3}} = 2$ amperes

1.5 Capacitance

The applied voltage across the capacitor shown in Figure 1.5 is $v(t) = 1000t$. Determine (a) the current flow for $C = 2 \times 10^{-6}$ farads, and (b) the voltage across the capacitor at $t = 2 \times 10^{-3}$ seconds.

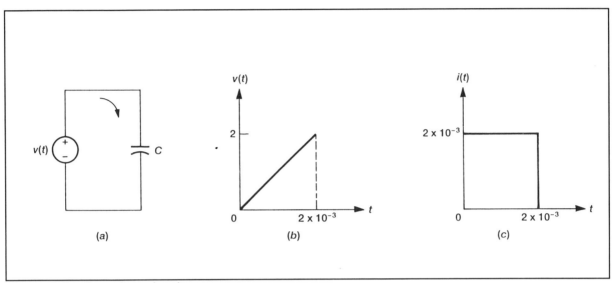

Figure 1.5 Capacitor circuit

Solution:

a) $i(t) = \dfrac{dQ}{dt} = C\dfrac{dv}{dt} = 2 \times 10^{-6} \dfrac{d}{dt}(1000t)$

$i(t) = 2 \times 10^{-3}$ amperes

b) $v(t) = \dfrac{1}{C}\displaystyle\int_a^b i(t)dt = \dfrac{1}{2 \times 10^{-6}}\displaystyle\int_0^{2 \times 10^{-3}}(2 \times 10^{-3})dt$

$v(t) = [1 \times 10^3 t]_0^{2 \times 10^{-3}} = 2$ volts

1.6 Ohms's Law

A three volt flashlight bulb has a turn on resistance characteristic function as shown in Figure 1.6. Calculate the average current and power dissipation during the first three seconds.

Hint: $R = -t^2 + 6t + 9$ ohms.

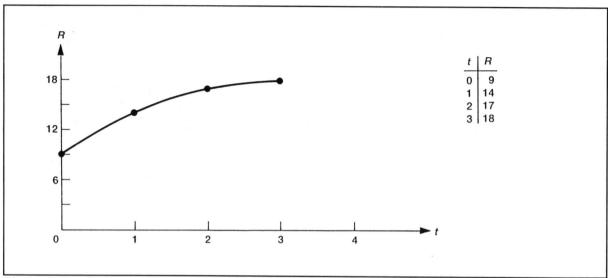

Figure 1.6 Light bulb resistance function

Solution:

$$V = I \times R$$

$$R_{avg} = \frac{\int_a^b R\,dt}{(b-a)}$$

$$R_{avg} = \frac{\int_0^3 (-t^2 + 6t + 9)\,dt}{(3-0)} = \frac{[-\frac{t^3}{3} + 3t^2 + 9t]_0^3}{3}$$

$$R_{avg} = \frac{(-9 + 27 + 27)}{3} = \frac{45}{3} = 15 \text{ ohms}$$

$$I_{avg} = V / R_{avg} = 3/15 = 0.2 \text{ amperes}$$

$$P_{avg} = I^2_{avg} \times R_{avg} = (0.2)^2 \times 15 = 0.6 \text{ watts}$$

1.7 Voltage divider

A three resistor voltage divider circuit is shown in Figure 1.7. Determine the voltage drop developed across resistor R_3 for an applied voltage of 12 volts.

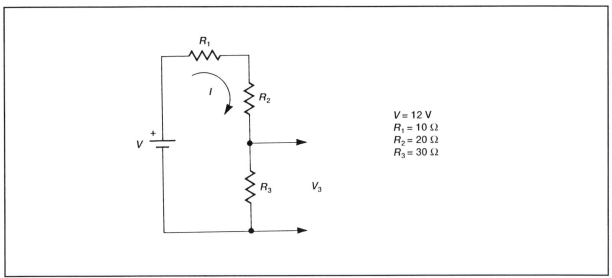

Figure 1.7 Voltage divider circuit

Solution:

$$V = IR_1 + IR_2 + IR_3$$

$$V = I(R_1 + R_2 + R_3)$$

$$V_3 = IR_3$$

$$\frac{V_3}{V} = \frac{IR_3}{I(R_1 + R_2 + R_3)}$$

$$V_3 = \frac{VR_3}{(R_1 + R_2 + R_3)}$$

$$V_3 = 12\left(\frac{30}{10 + 20 + 30}\right)$$

$$V_3 = 12(1/2) = 6 \text{ volts}$$

1.8 Current divider

A two resistor current divider circuit is shown in Figure 1.8. Determine the current passing through resistor R_1 for an applied voltage of 12 volts.

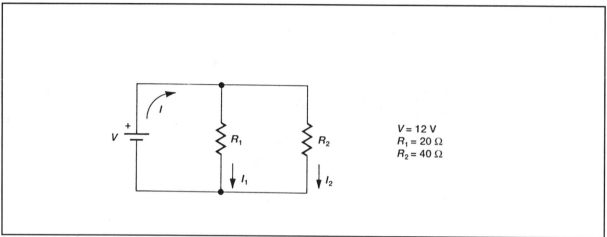

Figure 1.8 Current divider circuit

Solution:

$$V = V_1 = V_2 = I_1 R_1 = I_2 R_2$$

$$I = V/R_1 + V/R_2 = V(1/R_1 + 1/R_2) = 0.9 \text{ amperes}$$

$$I_1 = V/R_1$$

$$\frac{V}{I} = \frac{1}{\frac{1}{R_1} + \frac{1}{R_2}} = \frac{R_1 R_2}{R_1 + R_2} = R_p$$

$$\frac{I_1}{I} = \frac{V/R_1}{V/R_p}$$

$$\frac{I_1}{I} = \left(\frac{V}{R_1}\right)\left(\frac{R_p}{V}\right) = \left(\frac{R_2}{R_1 + R_2}\right)$$

$$I_1 = I\left(\frac{R_2}{R_1 + R_2}\right) = (0.9)\left(\frac{40}{20 + 40}\right)$$

$$I_1 = 0.6 \text{ amperes}$$

1.9 RC charge circuit

Determine the voltage drop across the capacitor shown in Figure 1.9 at t = RC. Assume that the capacitor is completely discharged to zero volts at t = 0.

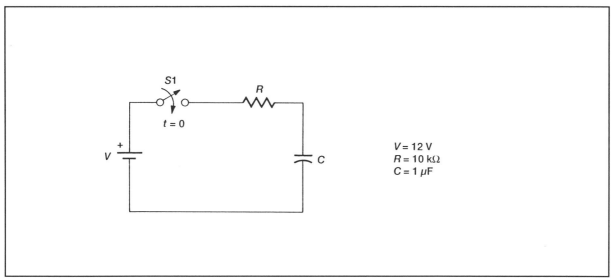

Figure 1.9 RC charge circuit

Solution:

$$V = Ri(t) + \frac{1}{C}\int i(t)dt$$

$i(t) = (V/R)e^{-t/RC}$

$V_R = Ri(t) = Ve^{-t/RC}$

$V_C = V - V_R = V - Ve^{-t/RC} = V(1 - e^{-t/RC}) = 12(1 - e^{-t/RC})$

Note: RC = t = (volts/amperes) × ((amperes × seconds)/volts) = seconds

At t = RC = $1 \times 10^4 \times 1 \times 10^{-6}$ = 1×10^{-2} seconds

$V_C = 12(1 - e^{-t/RC}) = 12(1 - e^{-1}) = 12(0.632)$

$V_C = 7.585$ volts

1.10 RC decay circuit

Determine the voltage drop across the capacitor shown in Figure 1.10 at t = RC. Assume that the capacitor is completely charged to 12 volts at t = 0.

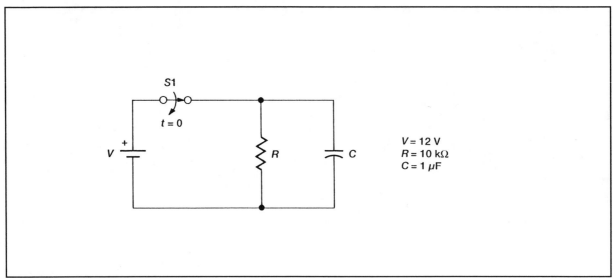

Figure 1.10   RC decay circuit

Solution:

$$0 = Ri(t) + \frac{1}{C}\int i(t)dt$$

$i(t) = (V/R)e^{-t/RC}$

$V_R = V_C = Ri(t)$

$V_C = Ve^{-t/RC}$

Note: RC = t = (volts/amperes)×((amperes×seconds)/volts) = seconds

At t = RC = $1 \times 10^4 \times 1 \times 10^{-6}$ = 1×10^{-2} seconds

$V_C = Ve^{-t/RC} = 12(e^{-1}) = 12(0.368)$

V_C = 4.414 volts

1.11 RL charge circuit

Determine the voltage drop across the resistor shown in Figure 1.11 at $t = L/R$. Assume that the circuit current equals zero at $t = 0$.

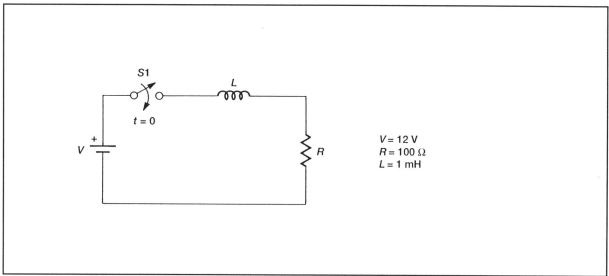

Figure 1.11 RL charge circuit

Solution:

$$V = Ri(t) + L\frac{d}{dt}i(t)$$

$$i(t) = (V/R)(1 - e^{-tR/L})$$

$$V_R = Ri(t) = V(1 - e^{-tR/L})$$

$$V_L = V - V_R = V - V(1 - e^{-tR/L}) = Ve^{-tR/L}$$

Note: $L/R = t = (\text{volts} \times \text{seconds} / \text{amperes}) / (\text{volts} / \text{amperes}) = \text{seconds}$

At $t = L/R = (1 \times 10^{-3}) / (1 \times 10^{2}) = 1 \times 10^{-5}$ seconds

$V_R = 12(1 - e^{-tR/L}) = 12(1 - e^{-1}) = 12(0.632) = 7.585$ volts

1.12 RL decay circuit

Determine the voltage drop across the resistor shown in Figure 1.12 at t = L/R. Assume that the circuit current equals V/R at t = 0.

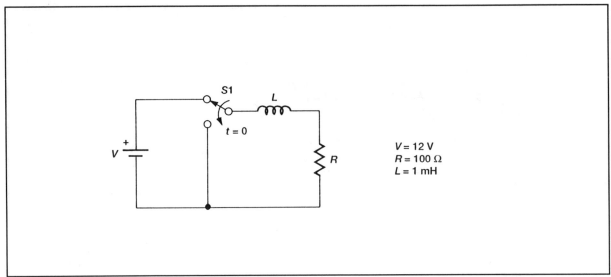

Figure 1.12 RL decay circuit

Solution:

$$0 = Ri(t) + L\frac{d}{dt}i(t)$$

$$i(t) = (V/R)e^{-tR/L}$$

$$V_R = V_L = Ri(t)$$

$$V_R = Ve^{-tR/L}$$

Note: L/R = t = (volts×seconds / amperes) / (volts / amperes) = seconds

At t = L/R = $(1\times10^{-3}) / (1\times10^{2})$ = 1×10^{-5} seconds

V_R = $12(e^{-tR/L})$ = $12(e^{-1})$ = $12(0.368)$ = 4.414 volts

1.13 Kirchhoff's voltage law

Determine the loop currents (a) I_1, and (b) I_2, of Figure 1.13 using Kirchhoff's voltage law. Kirchhoff's voltage law states that the sum of the voltage drops around a closed loop equals zero.

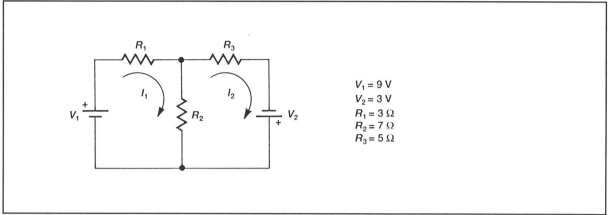

Figure 1.13 Kirchhoff's voltage law

Solution:

$$V_1 + R_1 I_1 + R_2 I_1 - R_2 I_2 = -9 + 3I_1 + 7I_1 - 7I_2 = 0 \quad (1)$$

$$V_2 + R_3 I_2 + R_2 I_2 - R_2 I_1 = +3 + 5I_2 + 7I_2 - 7I_1 = 0 \quad (2)$$

$$10 I_1 - 7 I_2 = 9 \quad (1)$$

$$-7 I_1 + 12 I_2 = -3 \quad (2)$$

$$I_i = \Delta_i / \Delta$$

a) $I_1 = \dfrac{\Delta_1}{\Delta} = \dfrac{\begin{vmatrix} 9 & -7 \\ -3 & 12 \end{vmatrix}}{\begin{vmatrix} 10 & -7 \\ -7 & 12 \end{vmatrix}} = \dfrac{108 - 21}{120 - 49} = \dfrac{87}{71} = 1.52$ amperes

b) $I_2 = \dfrac{\Delta_2}{\Delta} = \dfrac{\begin{vmatrix} 10 & 9 \\ -7 & -3 \end{vmatrix}}{\begin{vmatrix} 10 & -7 \\ -7 & 12 \end{vmatrix}} = \dfrac{-30 + 63}{120 - 49} = \dfrac{33}{71} = 0.42$ amperes

1.14 Kirchhoff's current law

Determine the node voltages (a) V_1 and (b) V_2, of Figure 1.14 using Kirchhoff's current law. Kirchhoff's current law states that the sum of the currents at a given circuit node equals zero.

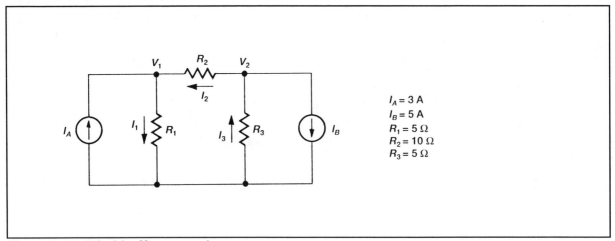

Figure 1.14 Kirchhoff's current law

Solution:

Let $I_1 = V_1/5$, $I_2 = (V_1 - V_2)/10$, and $I_3 = V_2/5$

$$-3 + I_1 + I_2 = 0 \quad (1)$$

$$+5 - I_2 - I_3 = 0 \quad (2)$$

$$V_1/5 + (V_1 - V_2)/10 = 0.3V_1 - 0.1V_2 = 3 \quad (1)$$

$$-(V_1 - V_2)/10 - V_2/5 = -0.1V_1 - 0.1V_2 = -5 \quad (2)$$

a) $V_1 = \dfrac{\Delta_1}{\Delta} = \dfrac{\begin{vmatrix} 3 & -0.1 \\ -5 & -0.1 \end{vmatrix}}{\begin{vmatrix} 0.3 & -0.1 \\ -0.1 & -0.1 \end{vmatrix}} = \dfrac{-0.8}{-0.04} = 20 \text{ volts}$

b) $V_2 = \dfrac{\Delta_2}{\Delta} = \dfrac{\begin{vmatrix} 0.3 & 3 \\ -0.1 & -5 \end{vmatrix}}{\begin{vmatrix} 0.3 & -0.1 \\ -0.1 & -0.1 \end{vmatrix}} = \dfrac{-1.2}{-0.04} = 30 \text{ volts}$

1.15 Thevenin's theorem

Find the Thevenin's equivalent circuit, as seen by the load resistor, R_L, for the circuit shown in Figure 1.15. Thevenin's theorem states that any two terminal network may be replaced by its equivalent open circuit voltage in series with a resistor. The resistance value is found by replacing all internal voltage sources with a short circuit and all internal current sources with an open circuit.

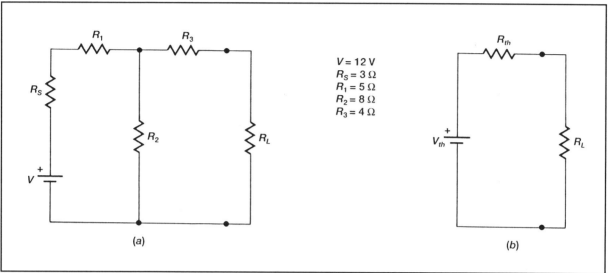

Figure 1.15 Thevenin's theorem

Solution:

a) $R_{th} = R_3 + (R_1 + R_s) || R_2 = R_3 + \dfrac{(R_1 + R_s)(R_2)}{(R_1 + R_s + R_2)}$

$R_{th} = 4 + \dfrac{(5+3)(8)}{(5+3+8)} = 4 + 4 = 8$ ohms

b) $V_{th} = 12\left(\dfrac{R_2}{R_2 + R_1 + R_s}\right) = 12\left(\dfrac{8}{8+5+3}\right)$

$V_{th} = 12(8/16) = 6$ volts

1.16 Norton's theorem

Find the Norton's equivalent circuit, as seen by the load resistor, R_L, for the circuit shown in Figure 1.16. Norton's theorem states that any two terminal network may be replaced by its equivalent short circuit current in parallel with a resistor. The resistance value is found by replacing all internal voltage sources with a short circuit and all internal current sources with an open circuit.

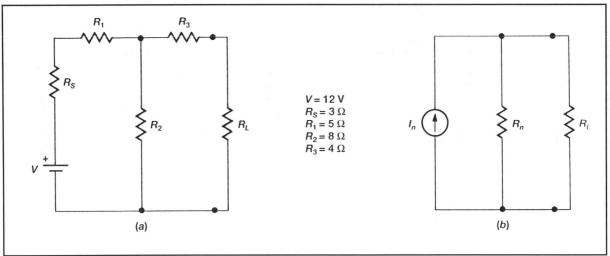

Figure 1.16 Norton's theorem

Solution:

$$R_{sh} = R_s + R_1 + R_2 \| R_3$$

$$R_{sh} = R_s + R_1 + \frac{R_2 R_3}{R_2 + R_3}$$

$$R_{sh} = 3 + 5 + \frac{8 \times 4}{8 + 4} = 10.66 \text{ ohms}$$

$$I_{sh} = V/R_{sh} = 12/10.66 = 1.125 \text{ amperes}$$

a) $I_n = I_{sh}\left(\dfrac{R_2}{R_2 + R_3}\right) = 1.125\left(\dfrac{8}{8 + 4}\right) = 0.75$ amperes

b) $R_n = R_3 + (R_1 + R_s) \| R_2 = R_3 + \dfrac{(R_1 + R_s)(R_2)}{(R_1 + R_s + R_2)}$

$$R_n = 4 + \frac{(5 + 3)(8)}{(5 + 3 + 8)} = 4 + 4 = 8 \text{ ohms}$$

1.17 Millman's theorem

Simplify the circuit shown in Figure 1.17 using Millman's theorem, find (a) the equivalent output voltage, V_m, and (b) the source resistance, R_s.

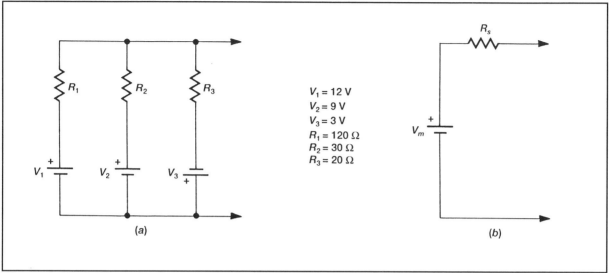

Figure 1.17 Millman's theorem

Solution:

a) $V_m = \sum\limits_{i}^{N} \dfrac{\frac{V_i}{R_i}}{\frac{1}{R_i}} = I/Y$

$$V_m = \frac{12/120 + 9/30 - 3/20}{1/120 + 1/30 + 1/20} = \frac{0.25}{0.091} = 2.7 \text{ volts}$$

b) $R_s = \dfrac{1}{1/R_1 + 1/R_2 + 1/R_3}$

$$R_s = \frac{1}{1/120 + 1/30 + 1/20} = \frac{1}{0.091} = 10.9 \text{ ohms}$$

1.18 Duality theorem

Determine the dual of the circuit shown in Figure 1.18a and write (a) the loop and (b) the node equations of the respective circuits.

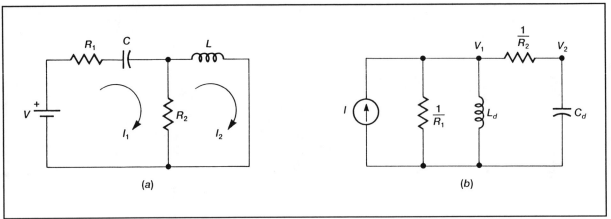

Figure 1.18 Duality theorem

Solution:

The dual circuit of Figure 1.18a is provided in Figure 1.18b and the following relationships hold.

$$I = V, \quad 1/R_1 = R_1, \quad 1/R_2 = R_2, \quad L_d = C, \quad C_d = L$$

a) Loop equations for Figure 1.18a

$$I_1(R_1 + R_2) + \frac{1}{C}\int I_1 dt - I_2 R_2 = V \quad (1)$$

$$-I_1 R_2 + L\frac{d}{dt}I_2 + I_2 R_2 = 0 \quad (2)$$

b) Node equations for Figure 1.18b

$$V_1\left(\frac{1}{R_1} + \frac{1}{R_2}\right) + \frac{1}{L_d}\int V_1 dt - \frac{V_2}{R_2} = I \quad (1)$$

$$-\frac{V_1}{R_2} + C\frac{d}{dt}V_2 + \frac{V_2}{R_2} = 0 \quad (2)$$

1.19 Superposition theorem

Determine the output voltage for the circuit shown in Figure 1.19a using the superposition theorem, and find the equivalent circuit.

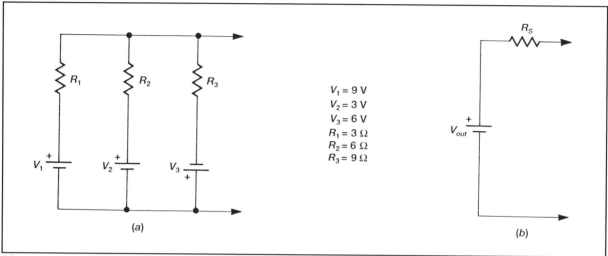

Figure 1.19 Superposition theorem

Solution:

a) For $V_2 = V_3 = 0$

$$V_{out} = V_1\left(\frac{R_2||R_3}{R_1 + R_2||R_3}\right) = 9\left(\frac{3.6}{6.6}\right) = 4.9 \text{ volts}$$

b) For $V_1 = V_3 = 0$

$$V_{out} = V_2\left(\frac{R_1||R_3}{R_2 + R_1||R_3}\right) = 3\left(\frac{2.25}{8.25}\right) = 0.82 \text{ volts}$$

c) For $V_1 = V_2 = 0$

$$V_{out} = V_3\left(\frac{R_1||R_2}{R_3 + R_1||R_2}\right) = -6\left(\frac{2}{11}\right) = -1.09 \text{ volts}$$

$$V_{out} = 4.9 + 0.82 - 1.09 = 4.63 \text{ volts}$$

$$R_s = \frac{1}{\frac{1}{R_1} + \frac{1}{R_2} + \frac{1}{R_3}} = 1.63 \text{ ohms}$$

1.20 Reciprocity theorem

The reciprocity theorem states that if a voltage in one branch produces a current in another branch, then the same voltage in the second branch will produce the same current in the first. Using the circuit in Figure 1.20a determine the current flowing through R_5. Then use the circuit in Figure 1.20b to calculate the current flowing through R_1.

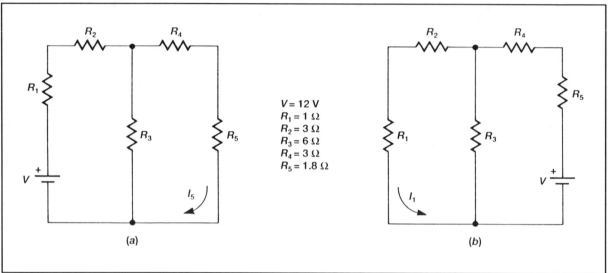

Figure 1.20 Reciprocity theorem

Solution:

a) $V_{th} = V\left(\dfrac{R_3}{R_1 + R_2 + R_3}\right) = 12\left(\dfrac{6}{10}\right) = 7.2$ volts

$R_{th} = R_4 + (R_1 + R_2) \| (R_3) = 3 + 2.4 = 5.4$ ohms

$I_5 = \dfrac{V_{th}}{R_{th} + R_5} = \dfrac{7.2}{5.4 + 1.8} = 1.0$ ampere

b) $V_{th} = V\left(\dfrac{R_3}{R_3 + R_4 + R_5}\right) = 12\left(\dfrac{6}{10.8}\right) = 6.66$ volts

$R_{th} = R_2 + (R_4 + R_5) \| (R_3) = 3 + 2.66 = 5.66$ ohms

$I_1 = \dfrac{V_{th}}{R_{th} + R_1} = \dfrac{6.66}{5.66 + 1} = 1.0$ ampere

1.21 Substitution theorem

Determine the voltage drops across (a) R_2, and (b) across R_4 in Figure 1.21. Substitute R_4 with an equivalent voltage source, and (c) verify that the voltage drop across R_2 remains the same.

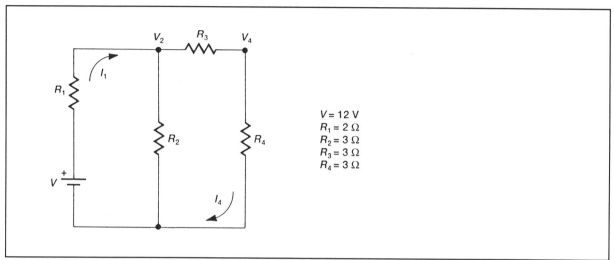

$V = 12$ V
$R_1 = 2\,\Omega$
$R_2 = 3\,\Omega$
$R_3 = 3\,\Omega$
$R_4 = 3\,\Omega$

Figure 1.21 Substitution theorem

Solution:

a) $V_2 = V\left(\dfrac{R_2||(R_3+R_4)}{R_1+R_2||(R_3+R_4)}\right) = 12\left(\dfrac{2}{4}\right) = 6$ volts

$I = \dfrac{V}{R} = \left(\dfrac{V}{R_1+R_2||(R_3+R_4)}\right) = \dfrac{12}{4} = 3$ amperes

$I_4 = I\left(\dfrac{R_2}{R_2+R_3+R_4}\right) = 3\left(\dfrac{3}{9}\right) = 1$ ampere

b) $V_4 = I_4 R_4 = 1 \times 3 = 3$ volts

c) Substituting $V_4 = 3$ volts for R_4 and using Millmans's theorem,

$V_2 = \dfrac{\dfrac{V}{R_1}+\dfrac{0}{R_2}+\dfrac{V_4}{R_3}}{\dfrac{1}{R_1}+\dfrac{1}{R_2}+\dfrac{1}{R_3}} = \dfrac{6+0+1}{\dfrac{1}{2}+\dfrac{1}{3}+\dfrac{1}{3}} = \dfrac{7}{1.166} = 6$ volts

1.22 Sinusoidal function

A sinewave periodic signal waveform, v(t) = 10sin377t, is shown in Figure 1.22.

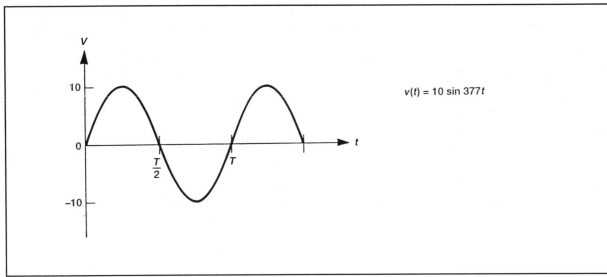

Figure 1.22 Sinusoidal function

1. The average value over the period is most nearly:

    a) 3.18　　　b) 5.0　　　c) 6.36　　　d) 0

2. The effective value over the period is most nearly:

    a) 5.0　　　b) 0　　　c) 7.07　　　d) 6.36

3. The signal frequency equals:

    a) 120　　　b) 60　　　c) 30　　　d) 6.28

4. The signal period is most nearly:

    a) 16.6×10^{-3}　　b) 33×10^{-3}　　c) 08.3×10^{-3}　　d) 0.159

1.22 Solution:

1. $V_{avg} = \dfrac{1}{T}\int_0^T v(t)dt = \dfrac{1}{T}\int_0^T V_{pk}\sin(\omega t)dt = \dfrac{1}{2\pi}\int_0^{2\pi} 10\sin(\theta)d\theta$

   $V_{avg} = \dfrac{10}{2\pi}[-\cos\theta]_0^{2\pi} = \dfrac{10}{2\pi}[(-1)-(-1)] = 0$ volts

   The correct answer is (d).

2. $V_{eff}^2 = \dfrac{1}{T}\int_0^T v(t)^2 dt = \dfrac{1}{T}\int_0^T [V_{pk}\sin(\omega t)]^2 dt = \dfrac{100}{4\pi}\int_0^{2\pi}(1-\cos 2\theta)d\theta$

   $V_{eff}^2 = \dfrac{400}{4\pi}[\int_0^{2\pi} d\theta - \int_0^{2\pi}\cos 2\theta d\theta] = \dfrac{400}{4\pi}[\theta - 0.5\sin 2\theta]_0^{2\pi}$

   $V_{eff}^2 = 400/4\pi[(2\pi - 0) - (0 - 0)] = (100/2)$

   $V_{eff} = (100/2)^{½} = (10/1.414) = 7.07$ volts

   The correct answer is (c).

3. $\omega = 2\pi f$

   $f = \omega/2\pi = 377/2\pi = 60$ Hz,

   The correct answer is (b).

4. $T = 1/f = 16.6 \times 10^{-3}$ seconds

   The correct answer is (a).

1.23 Reactance

Determine the values for (a) the inductive reactance and (b) capacitive reactances, X_L and X_C, in Figures 1.23a and 1.23b respectively.

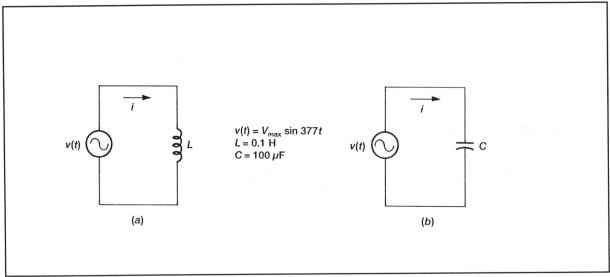

Figure 1.23 Reactance

Solution:

a) $i = \dfrac{1}{L}\int v(t)dt = \dfrac{1}{L}\int V_{max}\sin(\omega t)dt$

$i = \dfrac{V_{max}}{\omega L}(-\cos\omega t) = \dfrac{V_{max}}{\omega L}(\sin\omega t - 90°)$

$X_L = \omega L = 2\pi f L = 377L = 37.7$ ohms

b) $i = C\dfrac{d}{dt}v(t) = C\dfrac{d}{dt}(V_{max}\sin\omega t)$

$i = \omega C V_{max}(\cos\omega t) = \omega C V_{max}(\sin\omega t + 90°)$

$X_C = 1/(\omega C) = 1/(2\pi f C) = 1/(377C) = 26.5$ ohms

1.24 Impedance

Calculate (a) the impedance values for the two circuits in Figure 1.24, and (b) determine their corresponding loop currents.

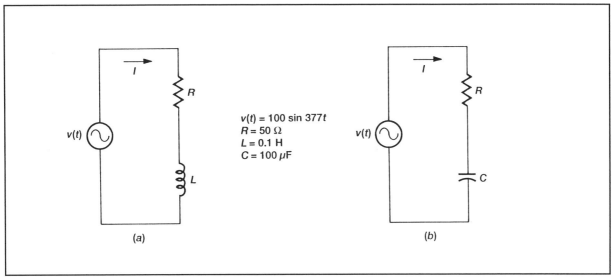

Figure 1.24 Impedance

Solution:

a) $V_{max} = I_{max} Z = I_{max}(R + jX_L) = I_{max}(R + j\omega L) = I_{max}(50 + j2\pi f L)$

$|Z| = (ZZ^*)^{1/2} = (R^2 + X_L^2)^{1/2} = [(50)^2 + (37.7)^2]^{1/2} = 62.62$ ohms

$\theta = \tan^{-1}(X_L / R) = \tan^{-1}(37.7/50) = 37°$

$I_{max} = \dfrac{V_{max}}{Z} = \dfrac{100}{62.62\angle 37°} = 1.6\angle -37°$ amperes

b) $V_{max} = I_{max} Z = I_{max}(R - jX_C) = I_{max}[R - j/(\omega C)] = I_{max}[50 - j/(2\pi f C)]$

$|Z| = (ZZ^*)^{1/2} = (R^2 + X_C^2)^{1/2} = [(50)^2 + (26.5)^2]^{1/2} = 56.68$ ohms

$\theta = \tan^{-1}(-X_C / R) = \tan^{-1}(-26.5/50) = -27.9°$

$I_{max} = \dfrac{V_{max}}{Z} = \dfrac{100}{56.68\angle -27.9°} = 1.76\angle 27.9°$ amperes

1.25 Admittance

Calculate (a) the admittance values for the two circuits shown in Figure 1.25, and (b) determine their corresponding node currents.

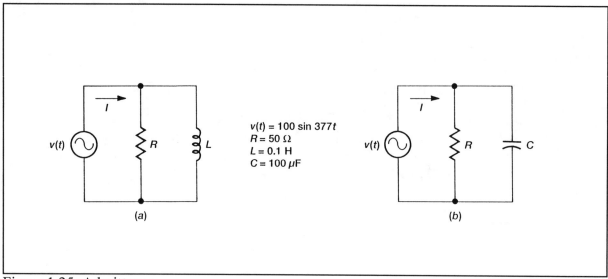

Figure 1.25 Admittance

Solution:

a) $I_{max} = V_{max}Y = V_{max}[1/R + 1/(jX_L)] = V_{max}(G - jB) = V_{max}[1/50 - j(1/37.7)]$

$|Y| = (YY^*)^{½} = (G^2 + B^2)^{½} = [(0.02)^2 + (0.0265)^2]^{½} = 0.0332$ siemens

$\theta = \tan^{-1}(-B/G) = \tan^{-1}(0.0265/0.02) = -52.95°$

$I_{max} = V_{max}Y = 100(0.0332)\angle -52.95° = 3.32\angle -52.95°$ amperes

b) $I_{max} = V_{max}Y = V_{max}[1/R - 1/(jX_C)] = V_{max}(G + jB) = V_{max}[1/50 + j(1/26.5)]$

$|Y| = (YY^*)^{½} = (G^2 + B^2)^{½} = [(0.02)^2 + (0.0377)^2]^{½} = 0.0427$ siemens

$\theta = \tan^{-1}(B/G) = \tan^{-1}(0.0377/0.02) = +62.05°$

$I_{max} = V_{max}Y = 100(0.0427)\angle 62.05° = 4.27\angle 62.05°$ amperes

1.26 Series to parallel

Given the series connected circuit in Figure 1.26a, determine (a) the conductance, G, and (b) the susceptance, B, of the equivalent parallel connected circuit shown in Figure 1.26b.

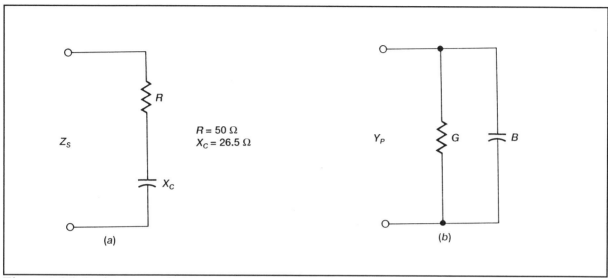

Figure 1.26 Series to parallel

Solution:

$$Z_s = R - jX_C = R - j/(\omega C) = 50 - j/(2\pi f C) = 50 - j26.5 \text{ ohms}$$

$$|Z_s| = (ZZ^*)^{1/2} = (R^2 + X_C^2)^{1/2} = [(50)^2 + (26.5)^2]^{1/2} = 56.68 \text{ ohms}$$

$$\theta = \tan^{-1}(-X_C / R) = \tan^{-1}(-26.5/50) = -27.9°$$

$$Y_s = \frac{1}{Z_s} = \frac{1}{56.68 \angle -27.9°} = 0.176 \angle +27.9° \text{ siemens}$$

$$Y_p = Y_s \cos\theta + Y_s \sin\theta = G + jB$$

$$Y_p = (0.0176)(0.88) + (0.0176)(0.46) = 0.0155 + j0.008$$

a) G = 0.0155 siemens

b) B = 0.008 siemens

1.27 Parallel to series

Given the parallel connected circuit of Figure 1.27a, determine the equivalent series connected circuit shown in Figure 1.27b.

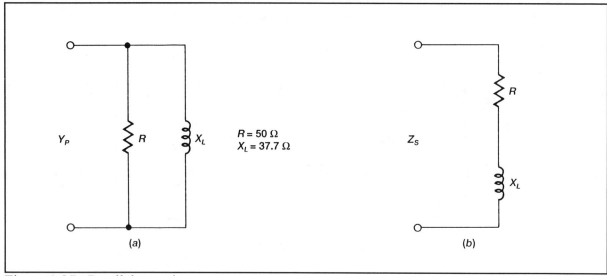

Figure 1.27 Parallel to series

Solution:

$Y_p = 1/R + 1/(jX_L) = 1/R - j/X_L = G - jB = 1/50 - j/(2\pi f L) = 0.02 - j0.0265$ siemens

$|Y_p| = (Y_p Y_p^*)^{1/2} = (G^2 + B^2)^{1/2} = [(0.02)^2 + (0.0265)^2]^{1/2} = 0.033$ ohms

$\theta = \tan^{-1}(-B/G) = \tan^{-1}(-0.0265/0.02) = -52.9°$

$Y_p = 0.033 \angle -52.9°$ siemens

$Z_p = \dfrac{1}{Y_p} = \dfrac{1}{0.033 \angle -52.9°} = 30.30 \angle +52.9°$ ohms

$Z_s = Z_p \cos\theta + Z_p \sin\theta = R + jX_L$

$Z_s = (30.30)(0.60) + (30.30)(0.79) = 18.18 + j23.93$ ohms

$R = 18.18$ ohms

$X_L = 23.93$ ohms

1.28 Series resonance

Figure 1.28 illustrates a series resonant circuit driven with a voltage source, v(t) = 100sin2πf₀t.

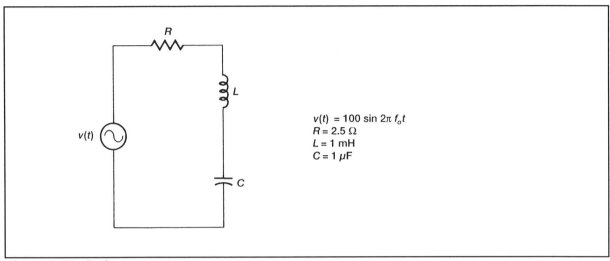

Figure 1.28 Series resonance

1. The resonance frequency is most nearly:

   a) 1×10^4 b) 3×10^4 c) 6.4×10^3 d) 5×10^3

2. The value of Q at resonance is most nearly:

   a) 13   b) 79   c) 32   d) 156

3. The voltage across the reactive elements at resonance is most nearly:

   a) 920   b) 1,000   c) 1,264   d) 827

4. The circuit impedance at resonance is most nearly:

   a) 32   b) 2.5   c) 5.2   d) 13

5. The circuit -3dB bandwidth is most nearly:

   a) 398   b) 2×10^3 c) 962   d) 156

1.28 Solution:

1. At resonance $X_L = X_C$

   $2\pi f_0 L = 1/(2\pi f_0 C)$

   $$f_0 = \frac{1}{2\pi\sqrt{LC}} = \frac{1}{2\pi\sqrt{1 \times 10^{-9}}} = \frac{1}{2\pi \times 3.16 \times 10^{-5}} = 5{,}033 \text{ Hz}$$

   The correct answer is (d).

2. $Q = X_L / R = 31.6 / 2.5 = 12.64$

   The correct answer is (a).

3. $X_L = 2\pi f_0 L = 31.6$ ohms

   $X_C = 1/(2\pi f_0 C) = 31.6$ ohms

   $V_L = V_C = QV = 1{,}264$ volts

   The correct answer is (c).

4. $|Z_0| = [(R^2 + (X_L - X_C)^2]^{\frac{1}{2}} = R = 2.5$ ohms

   $Z_0 = R + j[2\pi f_0 L - 1/(2\pi f_0 C)] = R = 2.5$ ohms

   The correct answer is (b).

5. $BW = f_0 / Q = 5{,}033 / 12.64 = 398$ Hz

   The correct answer is (a).

1.29 Parallel resonance

Figure 1.29a represents a parallel resonant frequency circuit, driven with a voltage source, $v(t) = 100\sin 2\pi f_0 t$. Figure 1.29b represents the equivalent circuit with a series resistance, R_s.

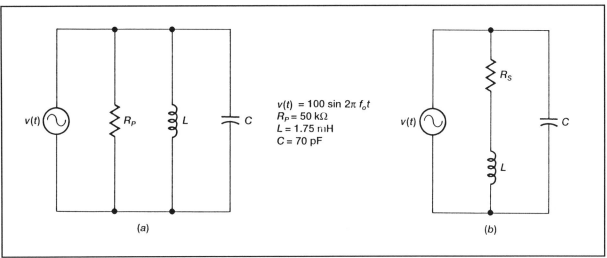

Figure 1.29 Parallel resonance

1. The resonance frequency is most nearly:

   a) 2.85×10^6 b) 4.55×10^5 c) 2.86×10^4 d) 9.1×10^5

2. The value of Q at resonance is most nearly:

   a) 0.1   b) 5   c) 20   d) 10

3. The circuit impedance at resonance is most nearly:

   a) 5×10^4 b) 5×10^3 c) 2.5×10^3 d) 500

4. The value of R_s is most nearly:

   a) 5×10^3 b) 50   c) 500   d) 100

5. The circuit -3dB bandwidth is most nearly:

   a) 4.55×10^3 b) 9.1×10^3 c) 2.27×10^4 d) 4.55×10^4

1.29 Solution:

1. At resonance $X_L = X_C$

   $2\pi f_0 L = 1/(2\pi f_0 C)$

   $f_0 = \dfrac{1}{2\pi\sqrt{LC}} = \dfrac{1}{2\pi\sqrt{1.225 \times 10^{-13}}} = 455 \times 10^3$ Hz

   The correct answer is (b).

2. $Q = \dfrac{R_P}{X_L} = \dfrac{R_P}{2\pi f_0 L} = \dfrac{50 \times 10^4}{2\pi \times 455 \times 10^3 \times 1.75 \times 10^{-3}} = 10$

   The correct answer (d).

3. $|Y_0| = [(1/R_P)^2 + (1/X_C - 1/X_L)^2]^{1/2} = 1/R_P$ siemens

   $Y_0 = 1/R_P + j[2\pi f_0 C - 1/(2\pi f_0 L)] = 1/R_P$ siemens

   $Z_0 = 1/Y_0 = R_P = 5 \times 10^4$ ohms

   The correct answer is (a).

4. $R_s = \dfrac{L}{R_P C} = \dfrac{1.75 \times 10^{-3}}{5 \times 10^4 \times 70 \times 10^{-12}} = 500$ ohms

   $Q = \dfrac{X_L}{R_s} = \dfrac{2\pi \times 455 \times 10^3 \times 1.75 \times 10^{-3}}{500} = \dfrac{5 \times 10^3}{5 \times 10^2} = 10$

   The correct answer is (c).

5. BW $= R_s/(2\pi L) = 500/(2\pi \times 1.75 \times 10^{-3}) = 4.55 \times 10^4$ Hz

   The correct answer is (d).

1.30 RC natural response

Determine the value of angle, θ, at which the switch in Figure 1.30 should be closed such that the natural response current, i_n, is minimum.

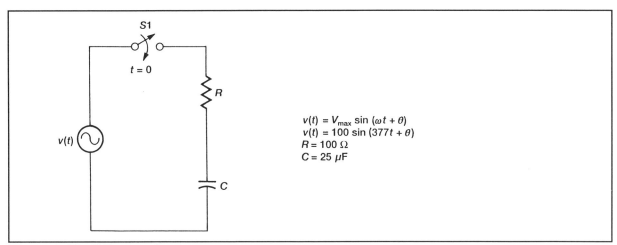

Figure 1.30 RC natural response

Solution:

$$i(t) = i_n + i_s$$

$$Ri(t) + \frac{1}{C}\int i(t)dt = V_{max}\sin(\omega t + \theta)$$

$$i(t) = k_1 e^{-t/RC} + k_2 \sin[\omega t + \theta + \tan^{-1}(X_C/R)]$$

At t = 0

$I(0) = [V_{max}/R](\sin\theta)$ and

$[V_{max}/R](\sin\theta) = k_1(1) + k_2\sin[\theta + \tan^{-1}(X_C/R)]$

if $\theta = -\tan^{-1}(X_C/R)$, then $k_1 = [V_{max}/R](\sin\theta)$

$X_C = 1/(377 \times 25 \times 10^{-6}) = 106.1$ ohms

$\theta = -\tan^{-1}(X_C/R) = -\tan^{-1}(106.1/100) = -46.7°$

$i_n = [V_{max}/R](\sin\theta)e^{-t/RC} = -0.73 e^{-t/RC}$

$RC = 100 \times 25 \times 10^{-6} = 2.5 \times 10^{-3}$

1.31 RC forced response

Determine the value of the current, i_s, in the circuit shown in Figure 1.31. Assume that the natural response current, i_n, equals zero.

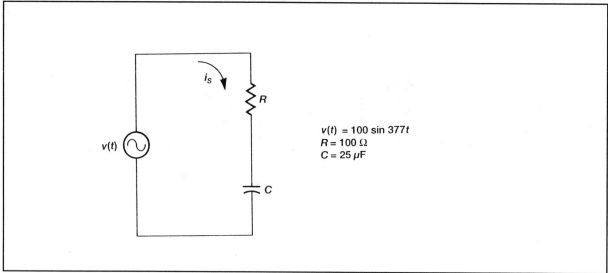

Figure 1.31 RC forced response

Solution:

$i_s = V/Z$

$|Z| = [(R)^2 + (X_C)^2]^{1/2}$

$|Z| = [(100)^2 + (106.1)^2]^{1/2} = 145.8$ ohms

$\theta = \tan^{-1}(-X_C/R) = \tan^{-1}(-106.1/100) = -46.7°$

$$i_s = \frac{100 \sin 377t}{145.8 \angle -46.7°} = \frac{100}{145.8 \angle -46.7°} = 0.68 \angle 46.7° \text{ amperes}$$

$i_s = 0.68 \sin(377t + 46.7°)$ amperes

1.32 RL natural response

Determine the value of angle, θ, at which the switch in Figure 1.32 should be closed such that the natural response current, i_n, is minimum.

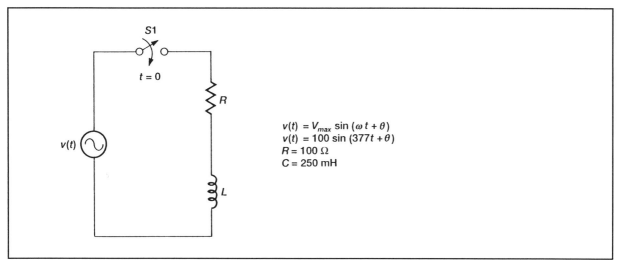

Figure 1.32 RL natural response

Solution:

$i(t) = i_n + i_s$

$$Ri(t) + L\frac{d}{dt}i(t) = V_{max}\sin(\omega t + \theta)$$

$i(t) = k_1 e^{-(R/L)t} + k_2 \sin[\omega t + \theta - \tan^{-1}(X_L/R)]$

At t = 0

I(0) = 0 and

$0 = k_1(1) + k_2 \sin[\theta - \tan^{-1}(X_L/R)]$

if $\theta = \tan^{-1}(X_L/R)$, then $k_1 = 0$

$X_L = 377 \times 250 \times 10^{-3} = 94.25$ ohms

$\theta = \tan^{-1}(X_L/R) = \tan^{-1}(94.25/100) = 43.3°$

$i_n = k_1 e^{-(R/L)t} = 0 \times e^{-(R/L)t} = 0$

1.33 RL forced response

Determine the value of the current, i_s, in the circuit shown in Figure 1.33. Assume that the natural response current, i_n, equals zero.

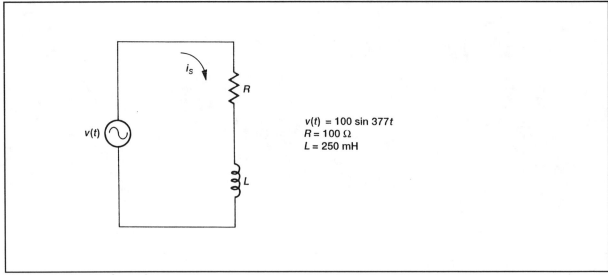

Figure 1.33 RL forced response

Solution:

$i_s = V/Z$

$|Z| = [(R)^2 + (X_L)^2]^{1/2}$

$|Z| = [(100)^2 + (94.25)^2]^{1/2} = 137.42$ ohms

$\theta = \tan^{-1}(X_L/R) = \tan^{-1}(137.42/100) = 43.3°$

$$i_s = \frac{100\sin 377t}{137.42\angle 43.3°} = \frac{100}{137.42\angle 43.3°} = 0.72\angle -43.3° \text{ amperes}$$

$i_s = 0.72\sin(377t - 43.3°)$ amperes

1.34 RLC natural response

Determine the natural response current, i_n, for the circuit shown in Figure 1.34.

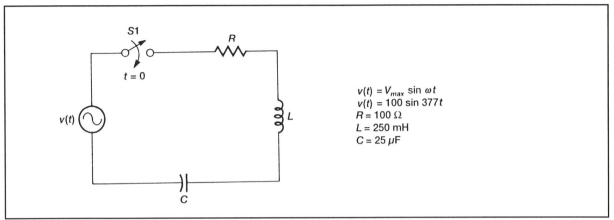

Figure 1.34   RLC natural response

Solution:

$i(t) = i_n + i_s$

$L\dfrac{d}{dt}i(t) + Ri(t) + \dfrac{1}{C}\int i(t)dt = V_{max}\sin(\omega t)$ The resulting differential equation is

$[\dfrac{d^2}{dt^2} + \dfrac{R}{L}\dfrac{d}{dt} + \dfrac{1}{LC}]i_n = 0$ let $s = \dfrac{d}{dt}$

$[s^2 + (\dfrac{R}{L})s + (\dfrac{1}{LC})]i_n = (As^2 + Bs + C)i_n = 0$

Case 1: $b^2 > 4ac$, $[R/(2L)]^2 > 1/(LC)$, $i_n = k_1 e^{s_1 t} + k_2 e^{s_1 t}$

Case 2: $b^2 = 4ac$, $[R/(2L)]^2 = 1/(LC)$, $i_n = k_1 e^{s_1 t} + k_2 e^{s_2 t}$

Case 3: $b^2 < 4ac$, $[R/(2L)]^2 < 1/(LC)$, $i_n = e^{bt}(k_3 \sin\omega_n t) = e^{-(R/2L)t}[V_{max}/(\omega_n L)]\sin\omega_n t$

$b = R/2L = 200$, $b^2 = 40{,}000$, $4ac = 1/LC = 160{,}000$

$\omega_n = (160\times 10^3 - 40\times 10^3)^{1/2} = 346.4$

$s_1, s_2 = \dfrac{-b \pm \sqrt{b^2 - 4ac}}{2a} = -\dfrac{R}{2L} \pm \sqrt{(\dfrac{R}{2L})^2 - (\dfrac{1}{LC})}$

$k_3 = V_{max}/(\omega_n L) = 100/(346.4 \times 250 \times 10^{-3}) = 1.15$

This results in case 3: $i_n = e^{-200t}(1.15\sin 346.4t)$

1.35 RLC forced response

Determine the value of the current, i_s, in the circuit shown in Figure 1.35. Assume that the natural response current, i_n, equals zero.

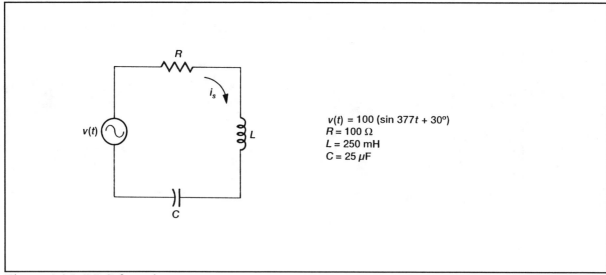

Figure 1.35 RLC forced response

Solution:

$$i_s = V/Z$$

$$|Z| = [(R)^2 + (X_L - X_C)^2]^{1/2}$$

$$|Z| = [(100)^2 + (94.25 - 106.1)^2]^{1/2} = 100.7 \text{ ohms}$$

$$\theta = \tan^{-1}[(X_L - X_C)/R] = \tan^{-1}(-11.85/100) = -6.75°$$

$$i_s = \frac{100\sin(377t + 30°)}{100.7 \angle -6.75°} = \frac{100 \angle 30°}{100.7 \angle -6.75°} = 0.99 \angle 36.75° \text{ amperes}$$

$$i_s = 0.99\sin(377t + 36.75°) \text{ amperes}$$

1.36 Power

Calculate (a) the apparent power, S, (b) the reactive power, Q, and (c) the real power, P, for the circuit shown in Figure 1.36.

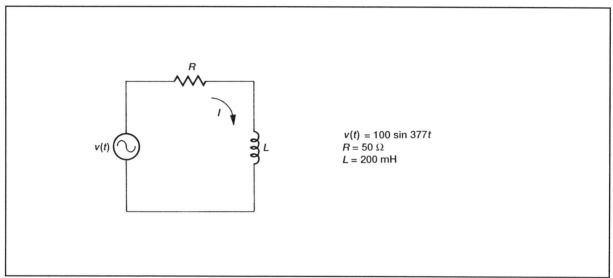

Figure 1.36 Power

Solution:

$$I = V/Z$$

$$|Z| = [(R)^2 + (X_L)^2]^{1/2} = [(50)^2 + (75.4)^2]^{1/2} = 90.47 \text{ ohms}$$

$$\theta = \tan^{-1}(X_L / R) = \tan^{-1}(1.51) = 56.45°$$

$$I = \frac{100 \sin(377t + \theta)}{90.47 \angle 56.45°} = 1.10 \sin(377t - 56.45°) \text{ amperes}$$

a) Apparent power S = VI   (volt-amperes)

   S = VI = 100×1.10 = 110 VA

b) Reactive power Q = VIsinθ = 100×1.1sin(56.45°) = 91.67 VARs

c) Real power P = VIcosθ = 100×1.1cos(56.45°) = 60.79 watts

$$S = [(P)^2 + (Q)^2]^{1/2} = [(12,100)]^{1/2} = 110 \text{ VA}$$

1.37 AC voltage divider

Determine the component relationships for the circuit shown in Figure 1.37 in order to achieve a frequency independent voltage divider.

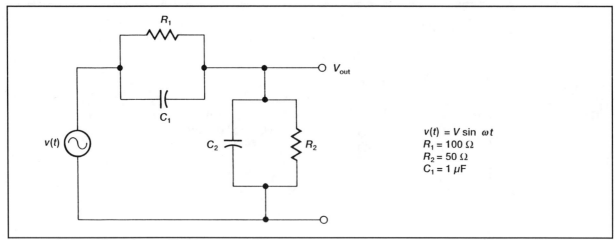

Figure 1.37 AC voltage divider

Solution:

$$V_{out} = V\left(\frac{R_2}{R_1+R_2}\right) = V\left(\frac{X_{c2}}{X_{c1}+X_{c2}}\right)$$

$$\left(\frac{R_2}{R_1+R_2}\right) = \left(\frac{1/\omega C_2}{\frac{1}{\omega C_1}+\frac{1}{\omega C_2}}\right) = \frac{\omega C_1}{\omega C_1 + \omega C_2}$$

$$\left(\frac{R_2}{R_1+R_2}\right) = \frac{C_1}{C_1+C_2}$$

$$R_2 C_1 + R_2 C_2 = R_1 C_1 + R_2 C_1$$

$$R_1 C_1 = R_2 C_2$$

$$C_2 = (R_1/R_2)C_1 = (100/50)(1 \times 10^{-6}) = 2 \times 10^{-6} \text{ farads}$$

$$V_{out} = V\left(\frac{50}{100+50}\right) = V\left(\frac{1 \times 10^{-6}}{1 \times 10^{-6} + 2 \times 10^{-6}}\right) = \frac{V}{3}$$

Given: $v(t) = V \sin \omega t$, $R_1 = 100\ \Omega$, $R_2 = 50\ \Omega$, $C_1 = 1\ \mu F$

1.38 Complex plane

Figure 1.38 shows two complex impedance vectors, $Z_1 = 3 + j4$ and $Z_2 = 2 - j3$, on the complex plane.

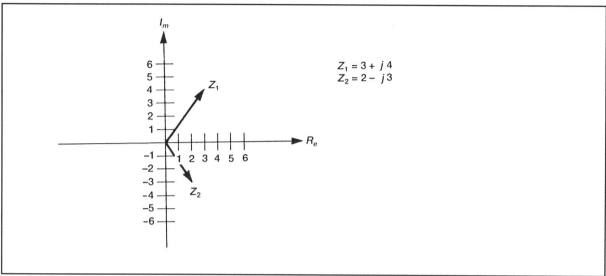

Figure 1.38 Complex plane

1. The sum of Z_1 and Z_2 equals:

   a) $5 + j5$ b) $1 + j5$ c) $5 + j1$ d) $1 + j1$

2. The difference of Z_1 minus Z_2 equals:

   a) $1 + j7$ b) $7 + j5$ c) $7 - j5$ d) $7 - j1$

3. The product of Z_1 and Z_2 equals:

   a) $12 - j6$ b) $18 - j1$ c) $6 - j8$ d) $8 - j9$

4. The quotient of Z_1 divided by Z_2 equals:

   a) $-6 + j17$ b) $1.5 - j1.33$ c) $2 - j1$ d) $-6/13 + j17/13$

1.38 Solution: Let $Z_1 = a + jb$ and $Z_2 = c + jd$

1. $Z_1 + Z_2 = (a+c) + j(b+d) = (3+2) + j(4-3) = 5 + j1$

   The correct answer is (c).

2. $Z_1 - Z_2 = (a-c) + j(b-d) = (3-2) + j(4-(-3)) = 1 + j7$

   The correct answer is (a).

3. $Z_1 \times Z_2 = (a+jb)(c+jd) = (ac - bd) + j(ad + bc)$

   $Z_1 \times Z_2 = (3 \times 2 - (4 \times -3)) + j((3 \times -3) + 4 \times 2) = 18 - j1$

   The correct answer is (b).

4. $\dfrac{Z_1}{Z_2} = \dfrac{(a+jb)(c-jd)}{(c+jd)(c-jd)} = \dfrac{(ac+bd)}{(c^2+d^2)} + j\dfrac{(bc-ad)}{(c^2+d^2)}$

   $\dfrac{Z_1}{Z_2} = \dfrac{(3 \times 2) + (4 \times -3)}{(4+9)} + j\dfrac{(4 \times 2) - (3 \times -3)}{(4+9)} = -\dfrac{6}{13} + j\dfrac{17}{13}$

   The correct answer is (d).

1.39 Laplace transform

For the circuit shown in Figure 1.39 determine (a) the admittance and (b) current expressions. Also, find (c) the initial and (d) final current values using the initial and final value theorems respectively.

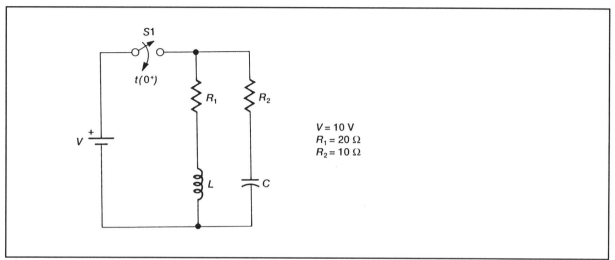

Figure 1.39 Laplace transform

Solution:

$\mathcal{L}(Z_L) = Ls, \qquad \mathcal{L}(Z_C) = 1/(Cs), \qquad V(s) = V/s$

a) $Y(s) = \dfrac{1}{R_1 + Ls} + \dfrac{1}{R_2 + \dfrac{1}{Cs}} = \dfrac{1}{R_1 + Ls} + \dfrac{Cs}{R_2 Cs + 1}$

$Y(s) = \dfrac{R_2 Cs + 1 + R_1 Cs + LCs^2}{R_2 LCs^2 + R_1 R_2 Cs + Ls + R_1}$

b) $I(s) = V(s)Y(s) = \left(\dfrac{V}{s}\right)\left(\dfrac{LCs^2 + R_1 Cs + R_2 Cs + 1}{R_2 LCs^2 + R_1 R_2 Cs + Ls + R_1}\right)$

c) Initial value $\lim\limits_{s \to \infty} sI(s) = (V)\left(\dfrac{LC + R_1 C/s + R_2 C/s + 1/s^2}{R_2 LC + R_1 R_2 C/s + L/s + R_1/s^2}\right) = \dfrac{V}{R_2} = \dfrac{10}{10} = 1$ ampere

d) Final value $\lim\limits_{s \to 0} sI(s) = \dfrac{V}{R_1} = \dfrac{10}{20} = 0.5$ ampere

1.40 Impulse function

Determine the current expression for the circuit shown in Figure 1.40 when subjected to an impulse function excitation.

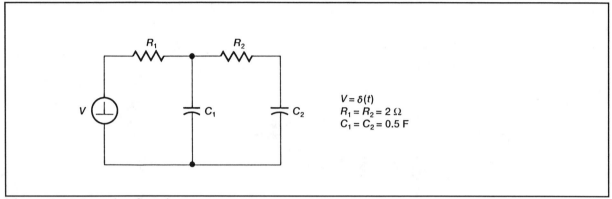

Figure 1.40 Impulse function

Solution:

$$Z(s) = R_1 + \cfrac{1}{C_1 s + \cfrac{1}{R_2 + \frac{1}{C_2 s}}} = R_1 + \frac{R_2 C_2 s + 1}{R_2 C_2 C_1 s^2 + C_1 s + C_2 s}$$

$$Z(s) = \frac{R_2 C_2 R_1 C_1 s^2 + (R_1 C_1 + R_1 C_2 + R_2 C_2)s + 1}{R_2 C_2 s + 1}$$

$I(s) = V(s)/Z(s) = V(s)Y(s)$ where $Y(s) = 1/Z(s)$ and $V(s) = 1$

Let $a = R_2 C_2 R_1 C_1$, $b = R_1 C_1 + R_1 C_2 + R_2 C_2$, $c = 1$, and $V(s) = 1$

$$I(s) = V(s)\left(\frac{s + \frac{1}{R_2 C_2}}{as^2 + bs + 1}\right), \quad s_1, s_2 = \frac{-b \pm \sqrt{b^2 - 4ac}}{2a}$$

$$I(s) = (1)\left(\frac{(s+1)}{s^2 + 3s + 1}\right) = \frac{(s+1)}{(s+2.62)(s+0.38)} = \frac{k_1}{(s+2.62)} + \frac{k_2}{(s+0.38)}$$

$$I(s) = \frac{0.72}{(s+2.62)} + \frac{0.28}{(s+0.38)}$$

$$i(t) = 0.72e^{-2.62t} + 0.28e^{-0.38t}$$

1.41 Step Function

The circuit shown in Figure 1.41 is subjected to a step function input. Determine the resulting current expression.

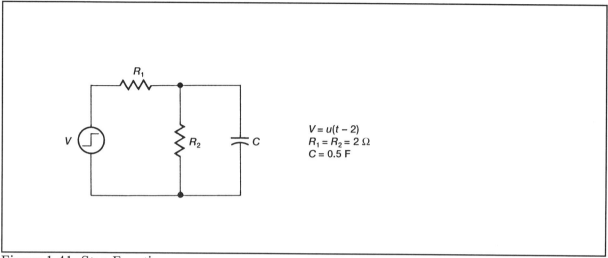

Figure 1.41 Step Function

Solution:

$$Z(s) = R_1 + \frac{1}{\frac{1}{R_2} + Cs} = \frac{R_1 R_2 Cs + R_1 + R_2}{R_2 Cs + 1}$$

$$I(s) = V(s)/Z(s) = V(s)Y(s) \quad \text{where} \quad Y(s) = 1/Z(s), \quad V(s) = e^{-2s}/s$$

$$I(s) = \left(\frac{e^{-2s}}{s}\right)\left(\frac{s + \frac{1}{R_2 C}}{s + \frac{R_1 + R_2}{R_1 R_2 C}}\right) = \left(\frac{e^{-2s}}{s}\right)\left(\frac{s + 1}{s + 2}\right)$$

$$I(s) = (e^{-2s})\left(\frac{s + 1}{s(s + 2)}\right) = (e^{-2s})\left(\frac{k_1}{s} + \frac{k_2}{s + 2}\right)$$

$$I(s) = (e^{-2s})\left(\frac{0.5}{s} + \frac{0.5}{s + 2}\right)$$

$$i(t) = 0.5u(t - 2) + 0.5e^{-2(t-2)}u(t - 2)$$

1.42 Ramp function

The circuit shown in Figure 1.42 is subjected to a ramp function input. Determine the resulting current expression.

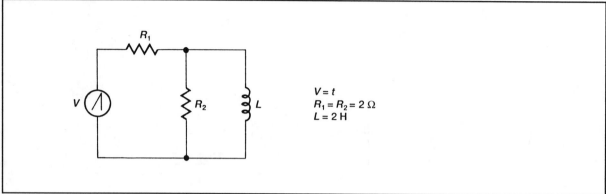

Figure 1.42 Ramp function

Solution:

$$Z(s) = R_1 + \frac{1}{\frac{1}{R_2} + \frac{1}{Ls}} = R_1 + \frac{R_2 Ls}{R_2 + Ls}$$

$$Z(s) = \frac{R_1 R_2 + R_1 Ls + R_2 Ls}{R_2 + Ls} = \frac{(R_1 L + R_2 L)s + R_1 R_2}{(S + \frac{R_2}{L})}$$

$$Z(s) = \frac{\left(s + \frac{R_1 R_2}{R_1 R_2 L}\right)}{\left(s + \frac{R_2}{L}\right)} = \frac{(s + 0.5)}{(s + 1)}$$

$$I(s) = V(s)/Z(s) = V(s)Y(s) \quad \text{where} \quad Y(s) = 1/Z(s) \,, \; V(s) = 1/s^2$$

$$I(s) = \left(\frac{1}{s^2}\right)\left(\frac{s+1}{s+0.5}\right) = \frac{k_1}{s^2} + \frac{k_2}{(s+0.5)}$$

$$I(s) = \frac{2}{s^2} + \frac{2}{(s+0.5)}$$

$$i(t) = 2t + 2e^{-0.5t}$$

1.43 Square wave

Figure 1.43 illustrates a square wave periodic signal.

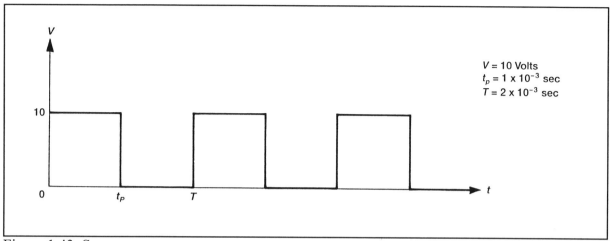

Figure 1.43 Square wave

1. The average value of the square wave is most nearly:

    a) 2.5   b) 7   c) 5   d) 10

2. The effective value of the square wave is most nearly:

    a) 5   b) 6.36   c) 3.18   d) 7

3. The duty cycle is most nearly:

    a) 0.5   b) 0.001   c) 0.002   d) 2

4. The pulse repetition rate most nearly:

    a) 2×10^3 b) 5×10^2 c) 2.5×10^2 d) 1×10^3

1.43 Solution:

1. $V_{avg} = \dfrac{1}{T}\int_0^{t_p} v(t)dt = \dfrac{1}{T}\left[\int_0^{1\times 10^{-3}} 10\,dt + \int_0^{1\times 10^{-3}} 0\,dt\right]$

$V_{avg} = \dfrac{1}{2\times 10^{-3}}[10t]_0^{1\times 10^{-3}} = \dfrac{10\times 10^{-3}}{2\times 10^{-3}} = 5$ volts

The correct answer is (c).

2. $V_{eff}^2 = \dfrac{1}{T}\int_0^{t_p}[v(t)]^2 dt = \dfrac{1}{T}\left[\int_0^{1\times 10^{-3}} (10)^2 dt\right]$

$V_{eff}^2 = \dfrac{1}{2\times 10^{-3}}[100t]_0^{1\times 10^{-3}} = \dfrac{100\times 10^{-3}}{2\times 10^{-3}} = 50$ volts

$V_{eff} = (50)^{1/2} = 7.07$ volts

The correct answer is (d).

3. $t_d = \dfrac{t_p}{T} = \dfrac{1\times 10^{-3}}{2\times 10^{-3}} \times 100 = 50\%$

The correct answer is (a).

4. $\text{PRR} = \dfrac{1}{T} = \dfrac{1}{2\times 10^{-3}} = 5\times 10^2$ pulses per second

The correct answer is (b).

1.44 Impulse response
Determine the impulse response of the circuit shown in Figure 1.44.

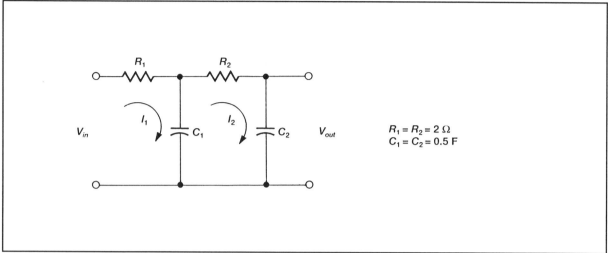

Figure 1.44 Impulse response

Solution:

$$V_{in}(s) = I_1(s)R_1 + I_1(s)/(C_1 s) - I_2(s)/(C_2 s) \quad (1)$$

$$0 = -I_1(s)/(C_1 s) + I_2(s)R_2 + I_2[1/(C_1 s) + 1/(C_2 s)] \quad (2)$$

$$I_2(s) = \frac{\begin{vmatrix} R_1 + 1/C_1 s & V_{in}(s) \\ -1/C_1 s & 0 \end{vmatrix}}{\begin{vmatrix} R_1 + 1/C_1 s & -1/C_1 s \\ -1/C_1 s & R_2 + 1/C_1 s + 1/C_2 s \end{vmatrix}} = \frac{V_{in}(s)C_2 s}{R_1 C_1 R_2 C_2 s^2 + (R_1 C_1 + R_1 C_2 + R_2 C_2)s + 1}$$

$$V_{out}(s) = I_2(s)/C_2 s$$

$$H(s) = \frac{V_{out}(s)}{V_{in}(s)} = \frac{1}{R_1 C_1 R_2 C_2 s^2 + (R_1 C_1 + R_1 C_2 + R_2 C_2)s + 1} = \frac{1}{s^2 + 3s + 1}$$

$$H(s) = \frac{1}{(s+2.62)(s+0.38)} = \frac{k_1}{(s+2.62)} = \frac{k_2}{(s+0.38)} = -\frac{0.45}{(s+2.62)} + \frac{0.45}{(s+0.38)}$$

$$h(t) = -0.45e^{-2.62t} + 0.45e^{-0.38t}$$

1.45 Phasor

Given the circuit in Figure 1.45a draw the effective voltage and current phasors and impedance vectors. Also, calculate (a) the resistance and (b) the inductance values.

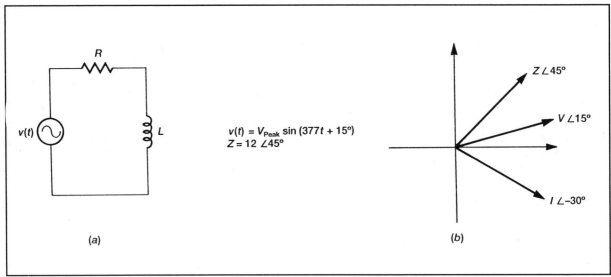

Figure 1.45 Phasor

Solution:

$$V_{eff} = (V_{pk}/\sqrt{2})\angle\theta° = (170/\sqrt{2})\angle 15° = 120\angle 15° \text{ volts}$$

$$I_{eff} = \frac{V_{eff}}{Z} = \frac{120\angle 15°}{12\angle 45°} = 10\angle -30° \text{ amperes}$$

$$Z = 12\angle 45° = 12\cos 45° + 12\sin 45°$$

$$Z = R + jX_L = R + j2\pi f L = 8.485 + j8.485 \text{ ohms}$$

a) $R = 8.485$ ohms

b) $X_L = 2\pi f L = 377L$

$L = X_L/377 = 8.485/377 = 22\times 10^{-3}$ henries

1.46 Frequency response

For the circuit shown in Figure 1.46 determine (a) the impedance expression. Also, calculate (b) the resonance frequency, f_0, and plot the magnitude and phase responses.

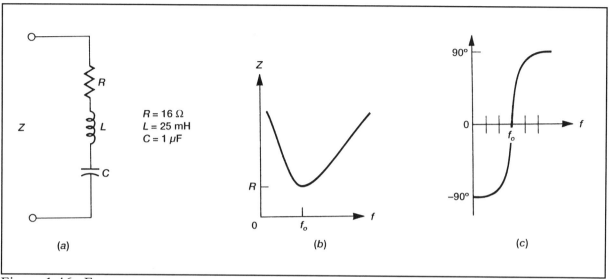

Figure 1.46 Frequency response

Solution:

a) $Z(s) = R + Ls + 1/(Cs)$

let $s = j\omega$

$Z(j\omega) = R + j\omega L + 1/(j\omega C) = R + j[\omega L - 1/(\omega C)]$

$|Z(j\omega)| = [R^2 + (\omega L - 1/(\omega C))^2]^{1/2}$

$\theta° = \tan^{-1}[(\omega L - 1/(\omega C)/R]$

b) $f_0 = \dfrac{1}{2\pi\sqrt{LC}} = \dfrac{1}{2\pi\sqrt{25 \times 10^{-3} \times 1 \times 10^{-6}}} = 1{,}000 \text{ Hz}$

Freq. f	100	500	900	1,000	1,110	2,000	5,000
Imp. Z	1,576	238	40	16	40	240	768
Angle $\theta°$	-89°	-86°	-64°	0°	64°	86°	89°

1.47 One port network

For the circuit shown in Figure 1.47 determine (a) the driving-point impedance, Z_{11}, using poles and zeroes. Also, calculate (b) $Z_{11}(0)$ and (c) $Z_{11}(\infty)$ respectively.

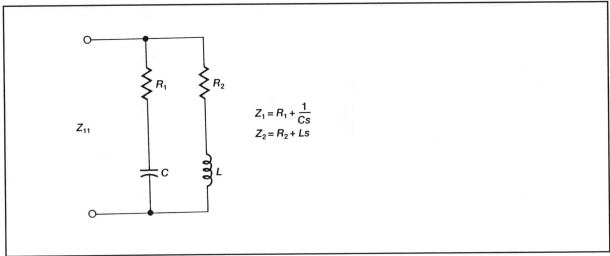

Figure 1.47 One port network

Solution:

a) $Y_{11} = \dfrac{1}{Z_1} + \dfrac{1}{Z_2} = \dfrac{1}{\dfrac{1}{Cs} + R_1} + \dfrac{1}{Ls + R_2}$

$Y_{11} = \dfrac{Cs}{R_1 Cs + 1} + \dfrac{1}{Ls + R_2} = \dfrac{LCs^2 + R_2 Cs + R_1 Cs + 1}{(R_1 Cs + 1)(Ls + R_2)}$

$Z_{11} = \dfrac{1}{Y_{11}} = \dfrac{(R_1 Cs + 1)(Ls + R_2)}{LCs^2 + R_2 Cs + R_1 Cs + 1}$

$Z_{11} = \dfrac{LCR_1 s^2 + Ls + R_1 R_2 Cs + R_2}{LCs^2 + R_2 Cs + R_1 Cs + 1}$

b) For $Z_{11}(0)$ let $s = 0$ and $Z_{11}(0) = R_2$

c) For $Z_{11}(\infty)$ divide by s^2 and let $s = \infty$

$Z_{11}(\infty) = R_1 [(LC)/(LC)] = R_1$

1.48 Two port network

Determine the transfer function, Z_{12}, of the voltage ratio, V_{out} / V_{in}, for the circuit shown in Figure 1.48.

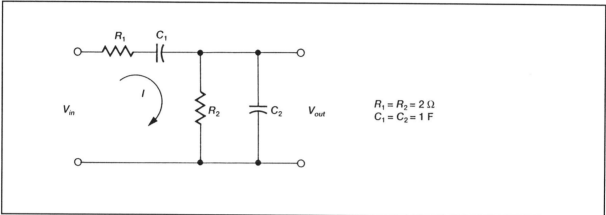

Figure 1.48 Two port network

Solution:

$$Z_{12} = \frac{V_{out}}{V_{in}} = \frac{IZ_2}{I(Z_1 + Z_2)} = \frac{Z_2}{Z_1 + Z_2}$$

$$Z_1 = R_1 + \frac{1}{C_1 s} = \frac{(R_1 C_1 s + 1)}{C_1 s}$$

$$Z_2 = \frac{1}{\frac{1}{R_2} + C_2 s} = \frac{R_2}{(R_2 C_2 s + 1)}$$

$$Z_1 + Z_2 = \frac{(R_1 C_1 s + 1)}{C_1 s} + \frac{R_2}{(R_2 C_2 s + 1)} = \frac{(R_1 C_1 s + 1)(R_2 C_2 s + 1) + R_2 C_1 s}{(C_1 s)(R_2 C_2 s + 1)}$$

$$Z_{12} = \frac{(R_2 C_2 s + 1)(C_1 s)(R_2)}{(R_2 C_2 s + 1)[(R_1 C_1 s + 1)(R_2 C_2 s + 1) + (R_2 C_1 s)]} = \frac{R_2 C_1 s}{R_1 C_1 R_2 C_2 s^2 + (R_1 C_1 + R_2 C_2 + R_2 C_1)s + 1} = \frac{2s}{4s^2 + 6s + 1}$$

1.49 Insertion loss

Determine the values of (a) X_L and (b) X_C respectively for the circuit shown in Figure 1.49 so that (c) the insertion loss equals zero.

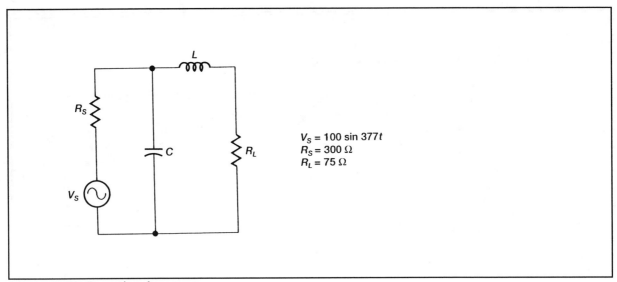

Figure 1.49 Insertion loss

Solution:

$$R_s = \frac{X_C(R_L + X_L)}{X_C + X_L + R_L} \quad \text{and} \quad R_L = X_L + \frac{X_C R_s}{X_C + R_s}$$

a) $X_L = [R_L(R_s - R_L)]^{1/2} = 130$ ohms

b) $X_C = (R_L R_s)/X_L = 173$ ohms

$Y_s = 1/R_s + 1/X_C = 1/300 + 1/173 = 0.0033 + j0.0057 = 6.66 \times 10^{-3} \angle 60°$ siemens

$Z_s = 1/Y_s = 150 \angle -60° = R'_s - jX_C = 75 - j130$ ohms

$Z_L = R_L + jX_L = 75 + j130$ ohms

$Z = R'_s - jX_C + R_L + jX_L = 75 - j130 + 75 + j130 = 150$ ohms

$V_s = IZ, \quad V_L = IR_L \quad \text{and}$

$V_s / V_L = Z/R_L = 150/75 = 2/1$

c) Insertion loss $= 20\log\left(\dfrac{R_L}{R'_s + R_L}\right)\left(\dfrac{V_s}{V_L}\right) = 20\log\left(\dfrac{75}{75+75}\right)\left(\dfrac{2}{1}\right) = \log(1) = 0$ dB

1.50 Switched capacitor network

A capacitor is switched periodically between a voltage source and a load every 1×10^{-3} seconds, as shown in Figure 1.50. Determine the effective capacitor resistance for a value of 1×10^{-6} farads.

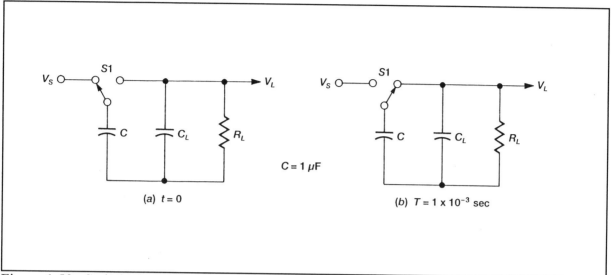

Figure 1.50 Switched capacitor network

Solution:

$$I = \frac{Q}{T}$$

$$I = \frac{Q(0) - Q(T)}{T} = \frac{CV_s - CV_L}{T}$$

$$I = \frac{C(V_s - V_L)}{T} \quad \text{since} \quad f = \frac{1}{T}$$

$$I = Cf(V_s - V_L) = \frac{V_s - V_L}{R}$$

$$R = \frac{V_s - V_L}{I} = \frac{1}{Cf} = \frac{1}{1 \times 10^{-6} \times 1 \times 10^3} = 1 \times 10^3 \quad \text{ohms}$$

1.51 Waveform analysis
A waveform contains the fundamental and second harmonic components as shown in Figure 1.51. The signal equation becomes $v(t) = 10\sin\omega t - 5\sin(\omega t + \pi/2)$.

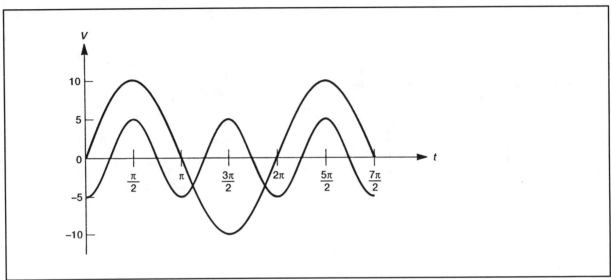

Figure 1.51 Waveform analysis

1. The maximum peak value of v(t) is:

   a) 10   b) 15   c) 5   d) 50

2. The average value of v(t) over half a cycle is:

   a) $10/\pi$ b) $15/\pi$ c) $5/\pi$ d) $20/\pi$

3. The rms value of v(t) equals:

   a) $100/\sqrt{2}$ b) $50/\sqrt{2}$ c) $(125/2)^{1/2}$ d) $125/\sqrt{2}$

4. The peak value of v(t) at $3\pi/2$ equals:

   a) 0   b) -15   c) -5   d) -10

5. The value of v(t) is most nearly zero at:

   a) $5\pi/6$ b) $6\pi/7$ b) $7\pi/9$ b) $15\pi/17$

1.51 Solution:

1. By inspection, $V_{max} = 10\sin\pi t - 5\sin(\pi t + \pi/2) = 10 + 5 = 15$ volts

   The correct answer is (b).

2. Using $\sin(2\omega t + \pi/2) = \cos 2\omega t$

   $$V_{avg} = \frac{10}{\pi}\int_0^\pi \sin\theta\, d\theta - \frac{5}{\pi}\int_0^\pi \cos 2\theta\, d\theta = \frac{20}{\pi} + 0 = \frac{20}{\pi} \text{ volts}$$

   The correct answer is (d).

3. $V_{rms} = [V^2_{rms1} + V^2_{rms2}]^{1/2} = [100/2 + 25/2]^{1/2} = (125/2)^{1/2}$

   The correct answer is (c).

4. $V_{pk}(3\pi/2) = 10\sin(3\pi/2) - 5\sin(3\pi + \pi/2) = -10 + 5 = -5$ volts

   The correct answer is (c).

5. $v(t) = 10\sin\omega t - 5\sin(\omega t + \pi/2)$, let $\omega t = 15\pi/17$

   $v(t) = 10\sin(15\pi/17) - 5\sin(30\pi/17 + \pi/2) = -0.08$ volts.

   The correct answer is (d).

1.52 Transmission constants

The symmetrical Pi network shown in Figure 1.52 can be defined in terms of four generalized constants A, B, C, and D respectively. As such $V_1 = AV_2 + BI_2$ and $I_1 = CV_2 + DI_2$.

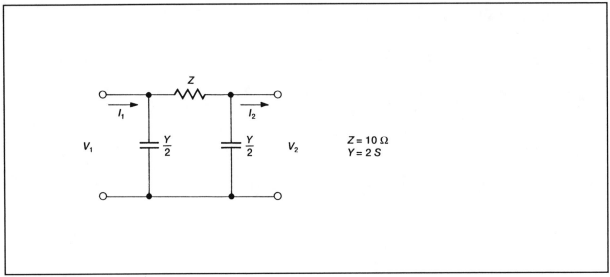

Figure 1.52 Transmission constants

1. The value of the A constant equals:

   a) 10   b) 2   c) 11   d) 12

2. The value of the B constant equals:

   a) 8   b) 12   c) 20   d) 10

3. The value of the C constant equals:

   a) 12   b) 5   c) 8   d) 2

4. The value of the D constant equals:

   a) 10   b) 16   c) 6   d) 11

5. The network characteristic impedance is:

   a) 5   b) $\sqrt{2}$ c) $(5/6)^{1/2}$ d) $2\sqrt{5}$

1.52 Solution:

1. $A = 1 + (ZY)/2 = 1 + (10 \times 2)/2 = 11$

    The correct answer is (c).

2. $B = Z = 10$ ohms

    The correct answer is (d).

3. $C = [1 + (ZY)/4] = 2[1 + (10 \times 2)/4] = 12$ siemens

    The correct answer is (a).

4. $D = 1 + (ZY)/2 = 1 + (10 \times 2)/2 = 11$

    The correct answer is (d).

5. $Z_0 = [Z_{oc} Z_{sc}]^{1/2} = [(A/C) \times (B/D)]^{1/2} = [B/C]^{1/2} = [10/2]^{1/2} = [5/6]^{1/2}$ ohms

    The correct answer is (c).

1.53 Hybrid Parameters

A BJT can be represented by a hybrid model network, with controlled sources, as shown in Figure 1.53.

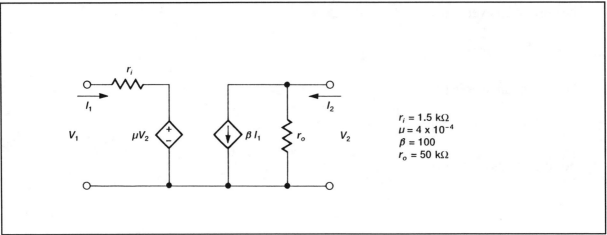

$r_i = 1.5\ k\Omega$
$\mu = 4 \times 10^{-4}$
$\beta = 100$
$r_o = 50\ k\Omega$

Figure 1.53 Hybrid Parameters

1. The equations for V_1 and I_2 are:

   a) $V_1 = h_{11} V_2 + h_{12} V_2$, $I_2 = h_{21} I_1 + h_{22} I_1$

   b) $V_1 = h_{11} I_1 + h_{12} I_2$, $I_2 = h_{21} I_1 + h_{22} V_2$

   c) $V_1 = h_{11} I_1 + h_{12} V_2$, $I_2 = h_{21} I_1 + h_{22} V_2$

   d) $V_1 = h_{11} I_1 + h_{12} V_2$, $I_2 = h_{21} I_1 + h_{22} I_2$

2. The value of h_{11} equals:

   a) 50×10^3 b) 1.5×10^3 c) 100   d) 4×10^{-4}

3. The value of h_{12} equals:

   a) 1.5×10^3 b) 100   c) 50×10^3 d) 4×10^{-4}

4. The value of h_{21} equals:

   a) 4×10^{-4} b) 100   c) 0.25×10^4 d) 0.01

5. The value of h_{22} equals:

   a) 50×10^3 b) 1.5×10^3 c) 0.2×10^{-4} d) 4×10^{-4}

1.53 Solution:

1. $V_1 = h_{11}I_1 + h_{12}V_2$, $\quad I_2 = h_{21}I_1 + h_{22}V_2$

   The correct answer is (c).

2. $h_{11} = V_1/I_1 \big|_{V_2=0} = r_i = 1.5 \times 10^3$ ohms

   The correct answer is (b).

3. $h_{12} = V_1/V_2 \big|_{I_1=0} = \mu(V_2/V_2) = \mu = 4 \times 10^{-4}$

   The correct answer is (d).

4. $h_{21} = I_2/I_1 \big|_{V_2=0} = \beta(I_1/I_1) = \beta = 100$

   The correct answer is (b).

5. $h_{22} = I_2/V_2 \big|_{I_1=0} = I_2/(I_2 r_0) = 1/r_0 = 0.2 \times 10^{-4}$ siemens

   The correct answer is (c).

1.54 Coupled circuit

The transformer coupled circuit shown in Figure 1.54 uses a short transmission line to provide power to a load. The primary-to-secondary impedance ratio of T_2 equals 4:1, and the turns ratio of T_1 equals 1:2 respectively.

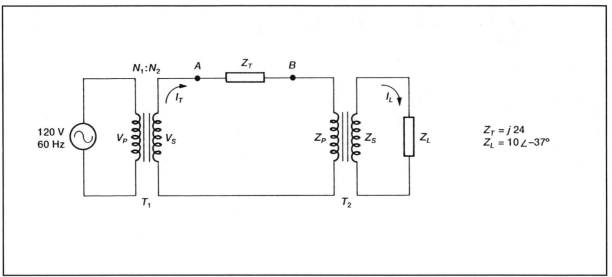

Figure 1.54 Coupled circuit

1. The load current, I_L, equals:

   a) $10\angle 37°$ b) $5\angle -37°$ c) $10\angle -37°$ d) $12\angle 37°$

2. The reflected impedance, Z_R, at point B equals:

   a) $20\angle 37°$ b) $40\angle -37°$ c) $20\angle 53°$ d) $40\angle -53°$

3. The transmission line current, I_T, equals:

   a) 15   b) $10\angle -53°$ c) 7.5   d) $10\angle 37°$

4. The inductance of Z_T is most nearly:

   a) 32×10^{-3} b) 24×10^{-3} c) 64×10^{-3} d) 12×10^{-3}

5. The capacitance of Z_L is most nearly:

   a) 8×10^{-6} b) 32×10^{-6} c) 10×10^{-6} d) 442×10^{-6}

1.54 Solution:

1. $\alpha_1 = N_1/N_2 = 1/2$

    $\alpha_2 = (Z_p/Z_s)^{1/2} = (4/2)^{1/2} = 2/1$

    $I_L = V_L/Z_L = V_P/(\alpha_1\alpha_2 Z_L) = V_P/Z_L = (120/10)\angle-37° = 12\angle 37°$ amps

    The correct answer is (d)

2. $Z_R(B) = (\alpha_2)^2 Z_L = 4Z_L = 40\angle-37°$ ohms

    The correct answer is (b).

3. $Z_s(A) = Z_R(B) + Z_T = 40\angle-37° + j24 = 32 - j24 + j24 = 32$ ohms

    $I_T = V_s/Z_s = V_P/(\alpha_1 Z_s) = 240/32 = 7.5$ amps

    The correct answer is (c).

4. $Z_T = 2\pi f L$

    $L = Z_T/(2\pi f) = 24/377 = 64 \times 10^{-3}$ henries

    The correct answer is (c).

5. $Z_L = 10\angle-37° = 8 - j6 = R - jX_C$

    $C = 1/(2\pi f X_C) = 1/(377 \times 6) = 442 \times 10^{-6}$ farads

    The correct answer is (d).

1.55 Attenuator

A symmetrical T network is used to provide a fixed insertion loss between the source and the load as shown in Figure 1.55.

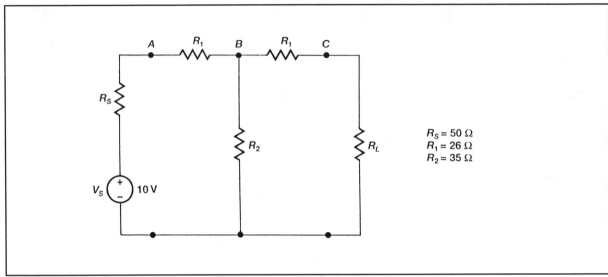

Figure 1.55 Attenuator

1. The input power, in dB, at point A is:

    a) 6　　　b) 3　　　c) 17　　　d) 20

2. The attenuator characteristic impedance is:

    a) 35　　　b) 52　　　c) 50　　　d) 15

3. The insertion loss, in dB, at point B is:

    a) -5　　　b) -10　　　c) -3.3　　　d) -6

4. The power loss at point, in dB, at point C with R_L = 50 ohms equals:

    a) -10　　　b) -12　　　c) -20　　　d) -16

5. The power loss at point, in dB, at point C with R_L = 75 ohms equals:

    a) -15　　　b) -12.7　　　c) -17　　　d) -20

1.55 Solution:

1. $P_{in}(dB) = 10\log(V_s^2/R_s) = 10\log(100/50) = 3$ dB

    The correct answer is (b).

2. $Z_{oc} = R_1 + R_2 = 61$ ohms

    $Z_{sc} = R_1 + R_2 \| R_1 = 40.9$ ohms

    $Z_0 = (Z_{oc}Z_{sc})^{1/2} = (61 \times 40.9)^{1/2} = 50$ ohms

    The correct answer is (c).

3. $V_A = V_s [R_2/(R_1 + R_2 + R_s)] = 3.15$ volts

    $\text{Loss}(dB) = 20\log(V_A/V_s) = 20\log(3.15/10) = -10$ dB

    The correct answer is (b).

4. $V_L = V_A[R_L/(Z_0 + R_L)] = 3.15 \times [50/100] = 1.575$ volts

    $P(dB) = 20\log(V_L/V_s) = 20\log(1.575/10) = -16$ dB

    The correct answer is (d).

5. $V_L = V_A[R_L/(Z_0 + R_L)] = 3.15 \times [75/125] = 1.89$ volts

    $P(dB) = 20\log[(V_A\sqrt{R_L})/(V_s\sqrt{R_s})] = 20\log[16.37/70.71] = -12.7$ dB

    The correct answer is (b).

Notes:

2.0 ELECTRONICS

2.1 Diode

Determine the expression for the voltage across R_L in Figure 2.1. Assume that the diode, D_1, has linear transfer characteristics.

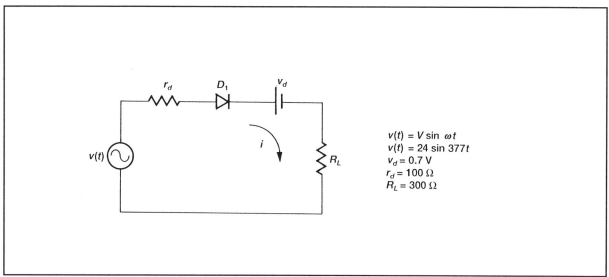

Figure 2.1 Diode

Solution:

$$V = ir_d + iR_L + v_d$$

$$V - v_d = i(r_d + R_L)$$

$$i = \frac{(V - v_d)}{(r_d + R_L)} = \frac{V}{(r_d + R_L)} - \frac{v_d}{(r_d + R_L)}$$

$$i = \frac{24 \sin 377t}{400} - \frac{0.7}{400} = 60 \times 10^{-3} \sin 377t - 1.75 \times 10^{-3} \quad (V > 0.7 \text{ volts})$$

$$i = 0 \quad (V < 0.7 \text{ volts})$$

$$v_L = iR_L = 18\sin 377t - 0.525 \quad \text{volts} \quad (V > 0.7 \text{ volts})$$

$$v_L = 0 \quad \text{volts} \quad (V < 0.7 \text{ volts})$$

2.2 Zener diode

Determine (a) the value and (b) power dissipation of R_S in the circuit shown in Figure 2.2. Assume that $I_z(\min)$ equals 10 ma, and it is an ideal zener diode.

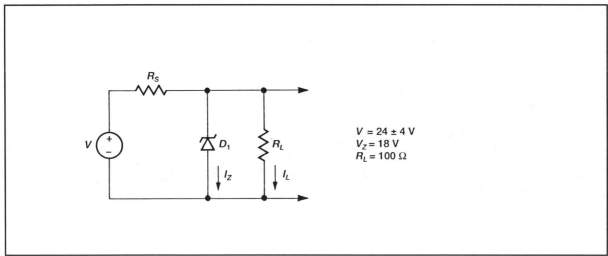

Figure 2.2 Zener diode

Solution:

a) $R_s = \dfrac{V_{min} - V_z}{I_s(\max)}$

$V_{min} = V - 4 = 24 - 4 = 20$ volts

$V_{max} = V + 4 = 24 + 4 = 28$ volts

$I_s(\max) = I_L(\max) + I_z = \dfrac{V_{max} - V_z}{R_L} + I_z$

$I_s(\max) = \dfrac{28 - 18}{100} + 10 \times 10^{-3} = 100 \times 10^{-3} + 10 \times 10^{-3} = 110 \times 10^{-3}$ amperes

$R_s = \dfrac{20 - 18}{110 \times 10^{-3}} = \dfrac{2}{110 \times 10^{-3}} = 18.8$ ohms

b) $P = I_S^2(\max) R_S = 0.22$ watts

2.3 Bipolar junction transistor

The NPN bipolar junction transistor (BJT) circuit shown in Figure 2.3 uses a voltage divider bias circuit. Find the voltage drop across the transistor using $\beta = 30$ and $V_{be} = 0.7$ volts. Assume that the transistor is not in saturation.

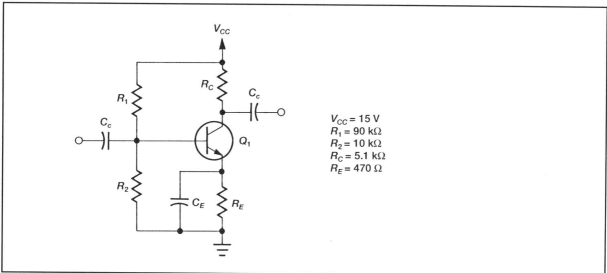

Figure 2.3 Bipolar junction transistor

Solution:

$$V_{th} = \frac{R_2}{(R_1 + R_2)}(V) = \frac{10 \times 10^3}{100 \times 10^3}(15) = 1.5 \text{ volts}$$

$$R_{th} = \frac{R_1 R_2}{(R_1 + R_2)} = \frac{900 \times 10^3}{100 \times 10^3} = 9 \times 10^3 \text{ ohms}$$

$$I_c = \frac{\beta(V_{th} - V_{be})}{R_{th} + (1 + \beta)R_E} = \frac{30(1.5 - 0.7)}{9 \times 10^3 + (31 \times 470)} = 1 \times 10^{-3} \text{ amperes}$$

$$V_{CE} = V_{cc} - V_L - V_E = 15 - I_C(R_C + R_E)$$

$$V_{CE} = 15 - 1 \times 10^{-3}(5.570 \times 10^3) = 15 - 5.57 = 9.43 \text{ volts}$$

2.4 Field effect transistor

The N-channel FET circuit shown in Figure 2.4 uses self-bias. Determine the voltage drop across the FET using a pinchoff voltage, V_P, of -5 volts and a saturation current, I_{DSS}, of 2 ma. Assume that the quiescent drain current, I_{DS}, equals 1 ma.

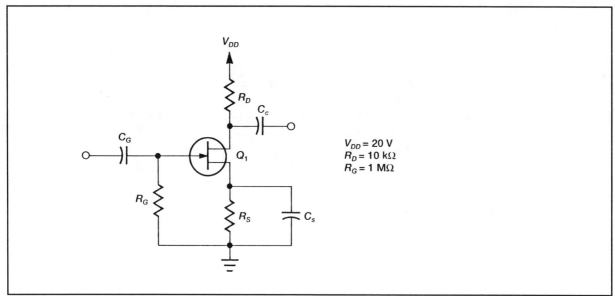

Figure 2.4 Field effect transistor

Solution:

$$I_{DS} = I_{DSS}(1 - V_{GS}/V_P)^2$$

$$V_{GS} = V_P[1 - (I_{DS}/I_{DSS})^{1/2}]$$

$$V_{GS} = -5[1 - (1/2)^{1/2}] = -1.46 \text{ volts}$$

$$R_S = \frac{V_{GS}}{I_{DS}} = \frac{1.46}{1 \times 10^{-3}} = 1.46 \times 10^3 \text{ ohms}$$

$$V_{DS} = V_{DD} - V_D - V_S = 20 - I_{DS}(R_D + R_S)$$

$$V_{DS} = 20 - 1 \times 10^{-3}(11.46 \times 10^3) = 20 - 11.46 = 8.54 \text{ volts}$$

2.5 MOSFET depletion mode

The N-channel depletion mode MOSFET circuit shown in Figure 2.5 uses a combination of self and fixed bias. Determine the values of (a) V_{GS} and (b) V_{DS} for a quiescent drain current, I_{DS}, of 1 ma.

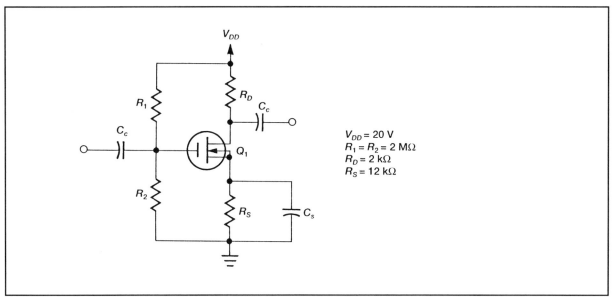

Figure 2.5 MOSFET depletion mode

Solution:

$$V_{th} = \left(\frac{R_2}{R_1 + R_2}\right)V = \left(\frac{2 \times 10^6}{4 \times 10^6}\right)20 = 10 \quad \text{volts}$$

$$V_S = I_{DS}R_S = 1 \times 10^{-3}(12 \times 10^3) = 12 \quad \text{volts}$$

a) $V_{GS} = V_{th} - V_S = 10 - 12 = -2$ volts

b) $V_{DS} = V_{DD} - V_D - V_S = V_{DD} - I_{DS}(R_D + R_S)$

$V_{DS} = 20 - 1 \times 10^{-3}(2 \times 10^3 + 12 \times 10^3) = 6$ volts

2.6 MOSFET enhancement mode

The N-channel enhancement mode MOSFET circuit shown in Figure 2.6 uses a drain-to-gate feedback bias. Determine the values of (a) V_{GS} and (b) V_{DS} for a quiescent drain current, I_{DS}, of 1 ma.

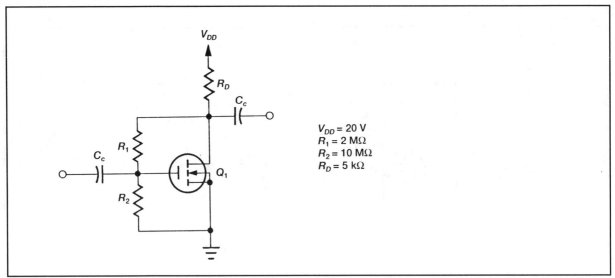

Figure 2.6 MOSFET enhancement mode

Solution:

a) $V_{DS} = V_{DD} - I_{DS}R_D = 20 - 1\times10^{-3}(5\times10^3) = 15$ volts

b) $V_{GS} = V_{DS}\left(\dfrac{R_2}{R_1 + R_2}\right) = 15\left(\dfrac{2 \times 10^6}{12 \times 10^6}\right) = 2.5$ volts

2.7 Common emitter amplifier

The common emitter BJT circuit shown in Figure 2.7 has the following hybrid parameters. Input impedance, h_{ie}, equals 2×10^3 ohms. Voltage feedback ratio, h_{re}, equals 4×10^{-4}. Current gain, h_{fe}, equals 50. Output admittance, h_{oe}, equals 20×10^{-6} siemens. Calculate the (a) Z_{in}, (b) Z_{out}, (c) β, and (d) A_v respectively.

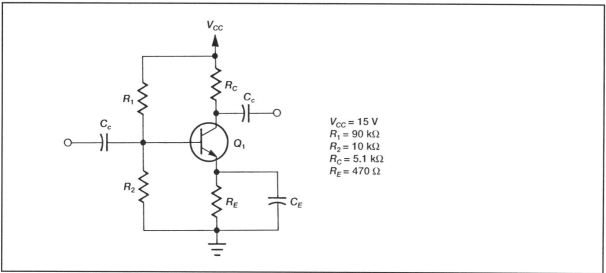

Figure 2.7 Common emitter amplifier

Solution:

Since $h_{oe}R_C \leq 0.1$

a) $Z_{in} = h_{ie} \parallel (R_1 \parallel R_2) = (2\times10^3) \parallel (90\times10^3 \parallel 10\times10^3)$ ohms

$Z_{in} = (2\times10^3) \parallel (9\times10^3) = 1{,}636$ ohms

b) $Z_{out} = (1/h_{oe}) \parallel R_C = (1/20\times10^{-6}) \parallel (5.1\times10^3) = 4{,}628$ ohms

c) $\beta = -h_{fe} = -50$

d) $A_v = -(h_{fe}R_C)/Z_{in} = -(50\times5.1\times10^3)/1{,}636 = -156$

2.8 Common base amplifier

The common base BJT circuit shown in Figure 2.8 has the following hybrid parameters. Input impedance, h_{ie}, equals 2×10^3 ohms. Voltage feedback ratio, h_{re}, equals 4×10^4. Current gain, h_{fe}, equals 50. Output admittance, h_{oe}, equals 20×10^{-6} siemens. Calculate the (a) Z_{in}, (b) Z_{out}, (c) α, and (d) A_v respectively.

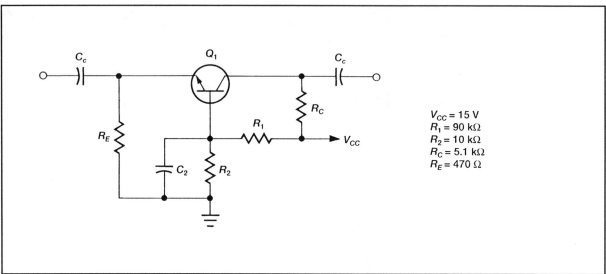

Figure 2.8 Common base amplifier

Solution:

Since $h_{oe} R_C \le 0.1$

a) $Z_{in} = h_{ie} / (1 + h_{fe}) \parallel R_E = (2 \times 10^3 / 51) \parallel 470 = 39 \parallel 470 = 36$ ohms

b) $Z_{out} = (1 + h_{fe}) / h_{oe} \parallel R_C = (51 / 20 \times 10^{-6}) \parallel (5.1 \times 10^3) = 5.1 \times 10^3$ ohms

c) $\alpha = h_{fe} / (1 + h_{fe}) = 50 / 51 = 0.98$

d) $A_v = (h_{fe} R_C) / h_{ie} = (50 \times 5.1 \times 10^3) / (2 \times 10^3) = 127$

2.9 Common collector amplifier

The common collector BJT circuit shown in Figure 2.9 has the following hybrid parameters. Input impedance, h_{ie}, equals 2×10^3 ohms. Voltage feedback ratio, h_{re}, equals 4×10^{-4}. Current gain, h_{fe}, equals 50. Output admittance, h_{oe}, equals 20×10^{-6} siemens. Calculate the (a) Z_{in}, (b) Z_{out}, (c) A_i, and (d) A_v respectively.

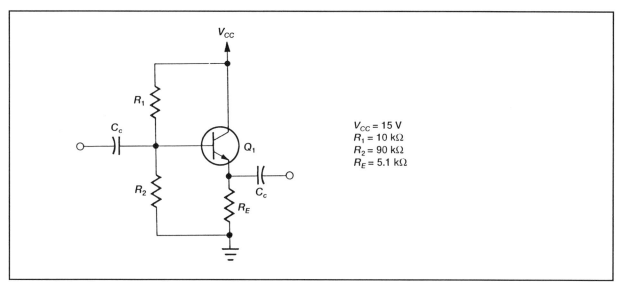

Figure 2.9 Common collector amplifier

Solution:

Since $h_{oe}R_E \leq 0.1$

a) $Z_{in} = [h_{ie} + (1 + h_{fe})R_E] \parallel R_1 \parallel R_2$

$Z_{in} = [2\times10^3 + (51)5.1\times10^3] \parallel (10\times10^3) \parallel (90\times10^3)$

$Z_{in} = (27.1\times10^4) \parallel (9\times10^3) = 8,700$ ohms

b) $Z_{out} = h_{ie}/(1 + h_{fe}) \parallel R_E = (2\times10^3/51) \parallel (5.1\times10^3)$

$Z_{out} = 39 \parallel (5.1\times10^3) = 39$ ohms

c) $A_i = (1 + h_{fe}) = 51$

d) $A_v = 1 - (h_{ie}/Z_{in}) = 1 - (2\times10^3/8,700) = 0.77$

2.10 Common source amplifier

The common source FET circuit shown in Figure 2.10 has the following parameters. Forward transadmittance, y_{fs}, equals 1.5×10^{-3} siemens. Output conductance, y_{os}, equals 10×10^{-6} siemens. Gate-to-source resistance, r_{gs}, equals 1×10^{8} ohms. Gate-to-drain resistance, r_{gd}, equals 1×10^{8} ohms. Calculate the (a) Z_{in}, (b) Z_{out}, and (c) A_v respectively.

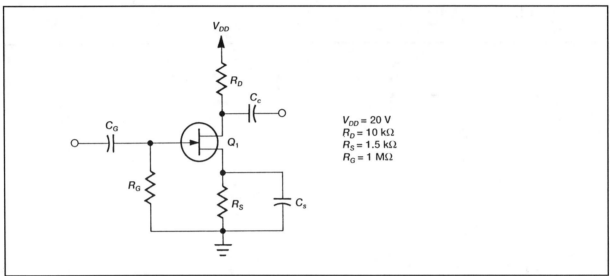

Figure 2.10 Common source amplifier

Solution:

$$r_d = 1/y_{os} = 1 \times 10^5 \text{ ohms}$$

$$g_m = y_{fs} = 1.5 \times 10^{-3}$$

$$\mu = y_{fs} r_d = g_m r_d = 1.5 \times 10^{-3} \times 1 \times 10^5 = 150$$

a) $Z_{in} = R_G \parallel r_{gs} = (1 \times 10^6) \parallel (1 \times 10^8) = 9.9 \times 10^5$ ohms

b) $Z_{out} = R_D \parallel r_d = R_D \parallel 1/y_{os} = (1 \times 10^4) \parallel (1 \times 10^5) = 9{,}090$ ohms

c) $A_v = -\mu [R_D/(r_d + R_D)] = -(g_m r_d)[R_D/(r_d + R_D)]$

$$A_v = -(150)\frac{10 \times 10^3}{1.1 \times 10^5} = -13.6$$

2.11 Common gate amplifier

The common gate FET circuit shown in Figure 2.11 has the following parameters. Forward transadmittance, y_{fs}, equals 1.5×10^{-3} siemens. Output conductance, y_{os}, equals 10×10^{-6} siemens. Gate-to-source resistance, r_{gs}, equals 1×10^8 ohms. Gate-to-drain resistance, r_{gd}, equals 1×10^8 ohms. Calculate the (a) Z_{in}, (b) Z_{out}, and (c) A_v respectively.

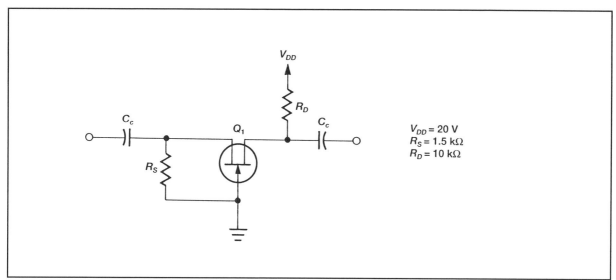

Figure 2.11 Common gate amplifier

Solution:

$r_d = 1/y_{os} = 1 \times 10^5$ ohms

$g_m = y_{fs} = 1.5 \times 10^{-3}$

$\mu = y_{fs} r_d = g_m r_d = 1.5 \times 10^{-3} \times 1 \times 10^5 = 150$

a) $Z_{in} = (r_d + R_D)/(\mu + 1) \parallel R_s$

$Z_{in} = (1.1 \times 10^5 / 151) \parallel (1.5 \times 10^3) = 728 \parallel (1.5 \times 10^3) = 490$ ohms

b) $Z_{out} = [\, r_d \parallel (\mu + 1)R_s \,] \parallel R_D = [1 \times 10^5 + (151)1.5 \times 10^3\,] \parallel (1 \times 10^4)$

$Z_{out} = (3.265 \times 10^5) \parallel (1 \times 10^4) = 9{,}969$ ohms

c) $A_v = (\mu + 1)R_D / [\, r_d + R_D + (\mu + 1)R_s \,]$

$$A_v = \frac{151 \times 10^4}{33.65 \times 10^4} = 4.5$$

2.12 Common drain amplifier

The common drain FET circuit shown in Figure 2.12 has the following parameters. Forward transadmittance, y_{fs}, equals 1.5×10^{-3} siemens. Output conductance, y_{os}, equals 10×10^{-6} siemens. Gate-to-source resistance, r_{gs}, equals 1×10^8 ohms. Gate-to-drain resistance, r_{gd}, equals 1×10^8 ohms. Calculate the (a) Z_{in}, (b) Z_{out}, and (c) A_v respectively.

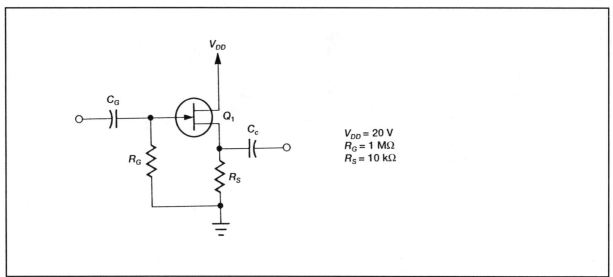

Figure 2.12 Common drain amplifier

Solution:

$$r_d = 1/y_{os} = 1 \times 10^5 \text{ ohms}$$

$$g_m = y_{fs} = 1.5 \times 10^{-3}$$

$$\mu = y_{fs} r_d = g_m r_d = 1.5 \times 10^{-3} \times 1 \times 10^5 = 150$$

a) $Z_{in} = R_G \parallel r_{gs} = (1 \times 10^6) \parallel (1 \times 10^8) = 9.9 \times 10^5$ ohms

b) $Z_{out} = r_d/(\mu+1) \parallel R_s = (1 \times 10^5 / 151) \parallel (1 \times 10^4) = 662 \parallel (1 \times 10^4) = 621$ ohms

c) $A_v = [(\mu/(\mu+1) R_s] / [r_d/(\mu+1) + R_s] = (\mu R_s) / [r_d + (\mu+1) R_s]$

$$A_v = (150) \frac{1 \times 10^4}{1 \times 10^5 + (151 \times 1 \times 10^4)} = \frac{1.5 \times 10^6}{1.61 \times 10^6} = 0.93$$

2.13 RC coupled amplifier

The RC coupled BJT circuit shown in Figure 2.13 has the following hybrid parameters. Input impedance, h_{ie}, equals 2×10^3 ohms. Current gain, h_{fe}, equals 25. Output admittance, h_{oe}, equals 10×10^{-6} siemens, and $C_{ce} = 32 \times 10^{-12}$ farads.

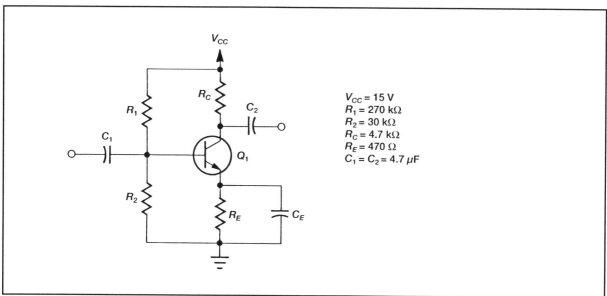

Figure 2.13 RC coupled amplifier

1. The midfrequency voltage gain, A_{vm}, is most nearly:

   a) -50   b) -25   c) -59   d) -63

2. The -3dB lower frequency, f_1, is most nearly:

   a) 1.25   b) 18   c) 12.5   d) 17

3. The -3dB upper frequency, f_2, is most nearly:

   a) 1×10^6 b) 5×10^4 c) 1.1×10^6 d) 1×10^5

4. The value of the bypass capacitor, C_E, is most nearly:

   a) 188×10^{-6} b) 94×10^{-6} c) 47×10^{-6} d) 18.8×10^{-6}

2.13 Solution:

1. $Z_{in} = h_{ie} \parallel (R_1 \parallel R_2) = (2\times10^3) \parallel (270\times10^3 \parallel 30\times10^3) = 1.86\times10^3$ ohms

   $A_{vm} = -(h_{fe}R_C)/Z_{in} = -(25\times4.7\times10^3)/(1.86\times10^3) = -63$

   The correct answer is (d).

2. At f_1 (-3dB) $X_C = Z_{in} = 1.86\times10^3$ ohms

   $f_1 = 1/(2\pi C_1 Z_{in}) = 1/(2\pi\times4.7\times10^{-6}\times1.86\times10^3) = 18$ Hz

   The correct answer is (b).

3. At f_2 (-3dB) $X_C = Z_{out} = 1/h_{oe} \parallel R_C = 4.5\times10^3$ ohms

   $f_2 = 1/(2\pi C_{ce} Z_{out}) = 1/(2\pi\times32\times10^{-12}\times4.5\times10^3) = 1.1\times10^6$ Hz

   The correct answer is (c).

4. At f_1 (-3dB) $X_{CE} = 0.1 R_E$

   $C_E = 10/(2\pi f_1 R_E) = 188\times10^{-6}$ farads

   The correct answer is (a).

2.14 Low frequency model

An ideal pulse is applied to the input of the amplifier shown in Figure 2.14. Determine (a) the filter cutoff frequency, f_L. Also, find (b) the output signal droop (%) if the input pulse duration, t_p, equals 1×10^{-3} seconds.

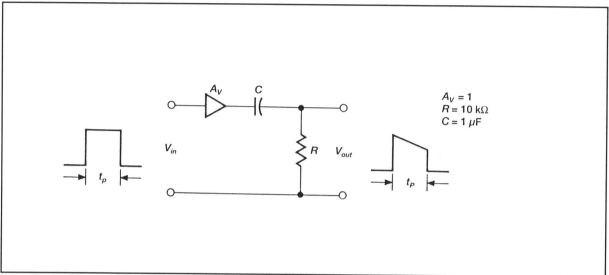

Figure 2.14 Low frequency model

Solution:

$$V_{out} = V_{in} e^{-t/RC}$$

$$V_{out} \approx V_{in}(1 - t/RC) \quad \text{for } t \ll RC$$

$$\text{Droop(\%)} = \frac{V_{in} - V_{out}}{V_{in}} \times 100 = \left[\frac{V_{in} - V_{in} + V_{in}(\frac{t}{RC})}{V_{in}}\right] \times 100$$

$$\text{Droop(\%)} = (t/RC) \times 100$$

Let $RC = 1/(2\pi f_L)$ and $t = t_p$

$$f = 1/t_p = 1 \times 10^3 \text{ Hz}$$

a) $f_L = 1/(2\pi RC) = 1/(2\pi \times 1 \times 10^{-2}) = 15.915$ Hz

b) $\text{Droop(\%)} = (t_p/RC) \times 100 = [(2\pi f_L)/f] \times 100 = 10\%$

2.15 High frequency model

An ideal pulse is applied to the input of the amplifier shown in Figure 2.15. Determine (a) the filter cutoff frequency, f_H. Also, find (b) the output signal rise time, t_r, if the input pulse duration, t_p, equals 1×10^{-5} seconds.

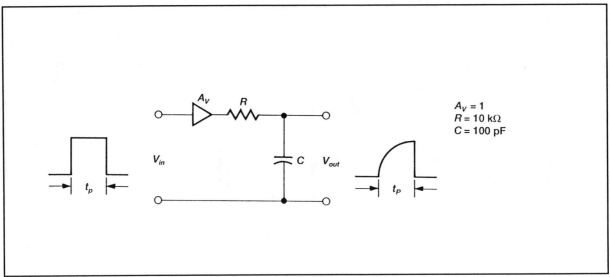

Figure 2.15   High frequency model

Solution:

$$V_{out} = V_{in}(1 - e^{-t/RC})$$

$$t_r = t(V_{out} = 0.9V_{in}) - t(V_{out} = 0.1V_{in})$$

$$t_r = (2.3RC) - (0.1RC) = 2.2RC$$

Let $RC = 1/(2\pi f_H)$

a) $f_H = 1/(2\pi RC) = 1/(2\pi \times 1 \times 10^{-6}) = 159{,}150$ Hz

b) $t_r = 2.2RC = (2.2)/(2\pi f_H) = 0.35/f_H = 2.2\times 10^{-6}$ seconds

2.16 Tuned amplifier

The tuned circuit FET amplifier shown in Figure 2.16 has the following parameters. Forward transadmittance, y_{fs}, equals 1.5×10^{-3} siemens. Output conductance, y_{os}, equals 10×10^{-6} siemens. And the drain-to-source capacitance, C_{DS}, equals 20×10^{-12} farads.

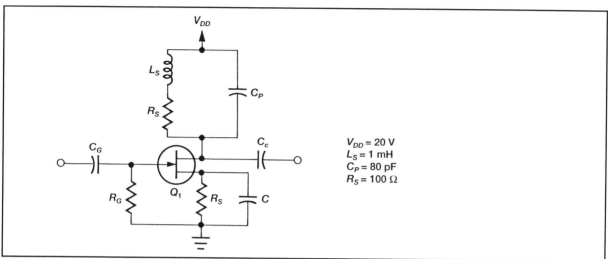

Figure 2.16 Tuned amplifier

1. The tank circuit impedance at resonance is most nearly:

   a) 1×10^7 b) 4×10^3 c) 1×10^5 d) 1.6×10^5

2. The circuit resonant frequency is most nearly:

   a) 50×10^4 b) 126×10^4 c) 57×10^4 d) 65×10^4

3. The circuit -3dB bandwidth is most nearly:

   a) 12.5×10^3 b) 16×10^3 c) 25×10^3 d) 32×10^3

4. The voltage gain at resonance is most nearly:

   a) -40 b) -75 c) -190 d) -240

2.16 Solution:

1. $C = C_p + C_{DS} = 80 \times 10^{-12} + 20 \times 10^{-12} = 100 \times 10^{-12}$

$$R_p = \frac{L_s}{R_s C} = \frac{1 \times 10^{-3}}{100 \times 100 \times 10^{-12}} = 1 \times 10^5 \text{ ohms}$$

The correct answer is (c).

2. $f_0 = \dfrac{1}{2\pi\sqrt{L_s C}} = \dfrac{1}{2\pi\sqrt{1 \times 10^{-3} \times 100 \times 10^{-12}}} = 50.33 \times 10^4$ Hz

The correct answer is (a).

3. $R = R_p \parallel 1/y_{os} = R_p \parallel r_d = (1 \times 10^5) \parallel (1 \times 10^5) = 5 \times 10^4$ ohms

$$BW(-3dB) = \frac{1}{2\pi RC} = \frac{1}{2\pi \times 5 \times 10^4 \times 100 \times 10^{-12}} = 31.83 \times 10^3 \text{ Hz}$$

The correct answer is (d).

4. $A_v(f_o) = -\mu = -y_{fs}R = -g_m R = -1.5 \times 10^{-3} \times 5 \times 10^4 = -75$

The correct answer is (b).

2.17 Wideband amplifier

The wideband FET amplifier circuit shown in Figure 2.17 has the following parameters. Forward transadmittance, y_{fs}, equals 1.5×10^{-3} siemens. Output conductance, y_{os}, equals 10×10^{-6} siemens. Drain-to-source capacitance, C_{DS}, equals 20×10^{-12} farads. The lower -3dB frequency, $f_1 = 10$ Hz, the upper -3dB frequency, $f_2 = 5 \times 10^6$ Hz, and $\zeta = 1$.

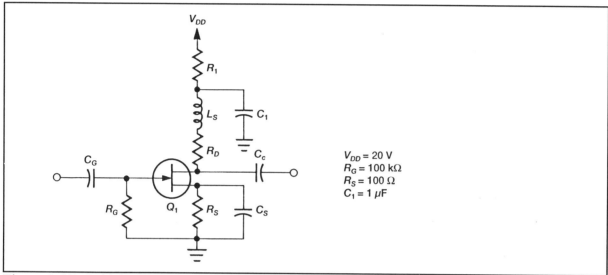

Figure 2.17 Wideband amplifier

1. The value of the bypass capacitor C_G is most nearly:

    a) 1.6×10^{-6} b) 0.8×10^{-6} c) 0.16×10^{-6} d) 0.32×10^{-6}

2. The value of the bypass capacitor C_s is most nearly:

    a) $1{,}592 \times 10^{-6}$ b) 320×10^{-6} c) 160×10^{-6} d) 800×10^{-6}

3. The value of resistor R_1 is most nearly:

    a) 1.6×10^3 b) 3.2×10^4 c) 3.2×10^3 d) 1.6×10^4

4. The value of resistor R_D is most nearly:

    a) 1×10^3 b) 1.6×10^3 c) 4.7×10^3 d) 5.1×10^3

5. The resonance frequency, f_0, is most nearly:

    a) 1.3×10^3 b) 1.14×10^5 c) 10×10^5 d) 10×10^6

2.17 Solution:

1. $C_G = \dfrac{1}{2\pi f_1 R_G} = \dfrac{1}{2\pi \times 10 \times 1 \times 10^5} = 0.16 \times 10^{-6}$ farads

   The correct answer is (c).

2. $X_{CS} = 0.1 R_S = 1/(2\pi f_1 C_S)$

   $C_S = 1/[2\pi f_1 (0.1 R_S)] = 1{,}592 \times 10^{-6}$ farads

   The correct answer is (a).

3. $R_1 C_1 = R_G C_G$

   $R_1 = R_G \left(\dfrac{C_G}{C_1}\right) = 1 \times 10^5 \left(\dfrac{0.16 \times 10^{-6}}{1 \times 10^{-6}}\right) = 1.6 \times 10^4$ ohms

   The correct answer is (d).

4. $R_D = 1/(2\pi f_2 C_{DS}) = 1{,}592$ ohms

   The correct answer is (b).

5. $\zeta = (R/2)(C/L)^{\frac{1}{2}} = 1$

   $L_s = \dfrac{R_D^2 C_{DS}}{(2\zeta)^2} = \dfrac{2.53 \times 10^6 \times 20 \times 10^{-12}}{4} = 12.65 \times 10^{-6}$ henries

   $f_0 = \dfrac{1}{2\pi \sqrt{L_s C_{DS}}} = 10 \times 10^6$ Hz

   The correct answer is (d)

2.18 Small signal amplifier

The small signal BJT amplifier circuit shown in Figure 2.18 has the following hybrid parameters. Input impedance, h_{ie}, equals 2×10^3 ohms. Voltage feedback ratio, h_{re}, equals 4×10^{-4}. Current gain, h_{fe}, equals 50. Output admittance, h_{oe}, equals 20×10^{-3} siemens. Calculate the (a) Z_{in}, (b) Z_{out}, (c) A_v, and (d) V_{out} respectively.

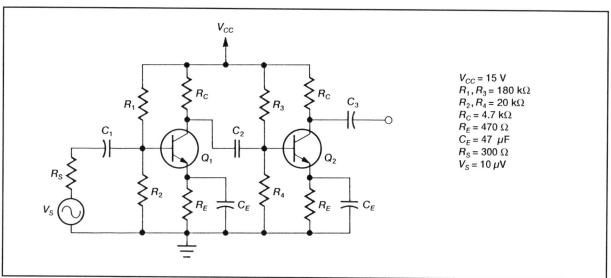

Figure 2.18 Small signal amplifier

$V_{CC} = 15$ V
$R_1, R_3 = 180$ kΩ
$R_2, R_4 = 20$ kΩ
$R_C = 4.7$ kΩ
$R_E = 470$ Ω
$C_E = 47$ μF
$R_S = 300$ Ω
$V_S = 10$ μV

Solution:

a) $Z_{in} = h_{ie} \| (R_1 \| R_2) = (2 \times 10^3) \| (180 \times 10^3 \| 20 \times 10^3)$

$Z_{in} = (2 \times 10^3) \| (18 \times 10^3) = 1.8 \times 10^3$ ohms

b) $Z_{out} = 1/h_{oe} \| R_C = (1/20 \times 10^{-6}) \| (4.7 \times 10^3) = 4,296$ ohms

c) $A_v(Q_1) = -h_{fe} \left(\dfrac{R_C}{Z_{in}} \right) \left(\dfrac{Z_{in}}{Z_{in} + R_s} \right) = -50 \left(\dfrac{4.7 \times 10^3}{1.8 \times 10^3} \right) \left(\dfrac{1.8 \times 10^3}{2.1 \times 10^3} \right) = -112$

$A_v(Q_2) = -h_{fe} \left(\dfrac{R_C}{Z_{in}} \right) \left(\dfrac{Z_{in}}{Z_{in} + Z_{out}} \right) = -50 \left(\dfrac{4.7 \times 10^3}{1.8 \times 10^3} \right) \left(\dfrac{1.8 \times 10^3}{6.1 \times 10^3} \right) = -38$

$A_v(Q_1 \times Q_2) = 4,256$

d) $V_{out} = V_s A_v(Q_1 \times Q_2) = 10 \times 10^{-6} \times 4,256 = 42.56 \times 10^{-3}$ volts

2.19 Large signal amplifier

The large signal BJT amplifier circuit shown in Figure 2.19 has the following hybrid parameters. Input impedance, h_{ie}, equals 100 ohms. Voltage feedback ratio, h_{re}, equals 80×10^{-6}. Current gain, h_{fe}, equals 50. Output admittance, h_{oe}, equals 1×10^{-3} siemens. The transistors are a matched pair with a quiescent collector current of 5×10^{-3} amperes and an output power of 10 watts.

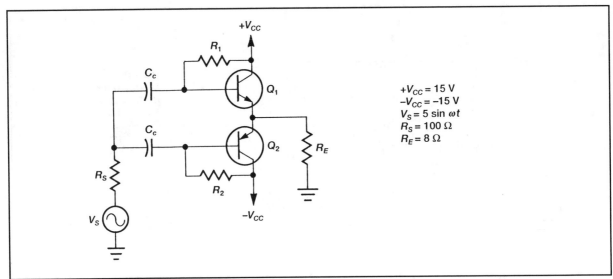

Figure 2.19 Large signal amplifier

1. The peak output current is most nearly:

   a) 2.5   b) 1.25   c) 1.13   d) 1.58

2. The minimum power supply voltage is most nearly:

   a) 12.65   b) 9   c) 10   d) 10.6

3. The value of R_1 and R_2 is most nearly:

   a) 126×10^3 b) 100×10^3 c) 153×10^3 d) 90×10^3

4. The maximum value of I_E is most nearly:

   a) 0.42   b) 2.5   c) 0.49   d) 1.0

2.19 Solution:

1. $I_{pk} = (2P/R_E)^{1/2} = 1.414(10/8)^{1/2} = 1.58$ amperes

   The correct answer is (d).

2. $V_{CC}(min) > I_{pk}R_E = 1.58 \times 8 = 12.65$ volts

   The correct answer is (a)

3. $I_B = I_C/(h_{fe}+1) = 5\times10^{-3}/51 = 98\times10^{-6}$ amperes

   $R_1 = R_2 = V_{CC}/I_B = 15/(98\times10^{-6}) = 153\times10^3$ ohms

   The correct answer is (c).

4. $Z_{in} = h_{ie} + (h_{fe}+1)R_E = 508$ ohms

   $I_B = V_S(max)/(R_S + Z_{in}) = 8.2\times10^{-3}$ amperes

   $I_E(max) = A_i \times I_B = (h_{fe}+1)I_B = 0.42$ amperes

   The correct answer is (a).

2.20 Voltage amplifier

The common source FET amplifier can be represented by the circuit model shown in Figure 2.20. The FET has the following parameters. Forward transadmittance, y_{fs}, equals 5×10^{-3} siemens. Output conductance, y_{os}, equals 10×10^{-6} siemens. Gate-to-source resistance, r_{gs}, equals 1×10^{6} ohms.

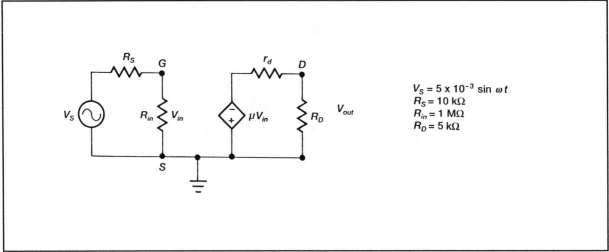

Figure 2.20 Voltage amplifier

1. The value of V_{in} is most nearly:

   a) 3.5×10^{-3} b) 4.95×10^{-3} c) 3.2×10^{-3} d) 2.5×10^{-3}

2. The value of Z_{out} is most nearly:

   a) 5×10^{3} b) 50×10^{4} c) 1×10^{5} d) 4.76×10^{3}

3. The value of A_v is most nearly:

   a) -24   b) -100   c) -250   d) -500

4. The value of V_{out} is most nearly:

   a) -495×10^{-3} b) -84×10^{-3} c) -118×10^{-3} d) -140×10^{-3}

2.20 Solution:

1. $V_{in} = V_s \left(\dfrac{R_{in}}{R_{in} + R_s} \right) = 5 \times 10^{-3} \left(\dfrac{1 \times 10^6}{1 \times 10^6 + 1 \times 10^4} \right) = 4.95 \times 10^{-3}$ volts

   The correct answer is (b).

2. $Z_{out} = r_d \parallel R_D = (1/y_{os}) \parallel R_D = (1/10 \times 10^{-6}) \parallel (5 \times 10^3) = 4.76 \times 10^3$ ohms

   The correct answer is (d).

3. $\mu = y_{fs}/y_{os} = y_{fs} r_d = g_m r_d = 500$

   $A_v = \dfrac{V_{out}}{V_{in}} = -\mu \left(\dfrac{R_D}{R_D + r_d} \right) = -23.8$

   The correct answer is (a).

4. $V_{out} = -V_{in} \mu \left(\dfrac{R_D}{R_D + r_d} \right) = -(4.95 \times 10^{-3} \times 500) \left(\dfrac{5 \times 10^3}{5 \times 10^3 + 1 \times 10^5} \right)$

   $V_{out} = -(4.95 \times 10^{-3} \times 500)(0.0476) = -117.86 \times 10^{-3}$ volts

   The correct answer is (c).

2.21 Current amplifier

The common emitter BJT amplifier can be represented by the circuit model shown in Figure 2.21. The transistor has the following hybrid parameters. Input impedance, h_{ie}, equals 1×10^3 ohms. Current gain, h_{fe}, equals 50. Output admittance, h_{oe}, equals 20×10^{-6} siemens.

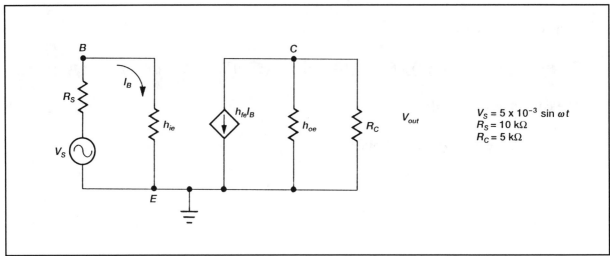

Figure 2.21 Current amplifier

1. The value of I_B is most nearly:

   a) 0.45×10^{-6} b) 5×10^{-6} c) 55×10^{-6} d) 0.5×10^{-6}

2. The value of Z_{out} is most nearly:

   a) 5×10^4 b) 5×10^3 c) 4.5×10^3 d) 10×10^3

3. The value of A_i is most nearly:

   a) -50   b) -45   c) -41   d) -25

4. The value of I_{out} is most nearly:

   a) -2×10^{-6} b) -22×10^{-6} c) -19×10^{-6} d) -10×10^{-6}

2.21 Solution:

Since $h_{oe}R_C \leq 0.1$

1. $I_B = \left(\dfrac{V_s}{R_s + h_{ie}}\right) = \left(\dfrac{5 \times 10^{-3}}{10 \times 10^3 + 1 \times 10^3}\right) = \dfrac{5 \times 10^{-3}}{11 \times 10^3} = 0.45 \times 10^{-6}$ amperes

   The correct answer is (a).

2. $Z_{out} = (1/h_{oe}) \parallel R_C = (5 \times 10^4) \parallel (5 \times 10^3) = 4.54 \times 10^3$ ohms

   The correct answer is (c).

3. $A_i = -I_C / I_B$

   $I_C = h_{fe}I_B - h_{oe}V_{out} = h_{fe}I_B - h_{oe}(R_C I_C)$

   $I_C = 1 + h_{oe}R_C = h_{fe}I_B$

   $A_i = -I_C / I_B = -h_{fe}/(1 + h_{oe}R_C) = -(50/1.01) = -45.45$

   The correct answer is (b).

4. $I_{out} = A_i I_B \left(\dfrac{1/h_{oe}}{1/h_{oe} + R_c}\right) = (-45.45)(0.45 \times 10^{-6})\left(\dfrac{5 \times 10^4}{5 \times 10^4 + 5 \times 10^3}\right)$

   $I_{out} = -18.6 \times 10^{-6}$ amperes

   The correct answer is (c).

2.22 Power amplifier

The common emitter BJT, transformer coupled load, amplifier can be represented by the circuit model shown in Figure 2.22. The transistor has the following hybrid parameters. Input impedance, h_{ie}, equals 100 ohms. Current gain, h_{fe}, equals 40. Output admittance, h_{oe}, equals 10×10^{-6} siemens. The transformer turns ratio, α, equals 4:1.

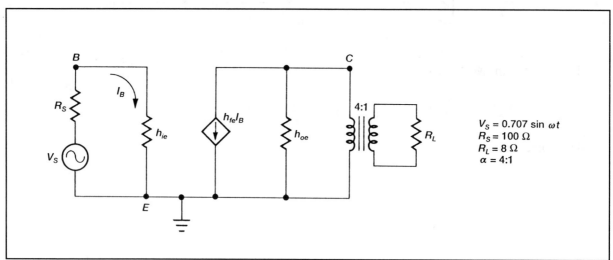

Figure 2.22   Power amplifier

1. The value of I_B is most nearly:

   a) 10×10^{-3} b) 3.5×10^{-3} c) 5×10^{-3} d) 7×10^{-3}

2. The value of A_v is most nearly:

   a) -40   b) -100   c) -51   d) -128

3. The value of A_p is most nearly:

   a) 1,600   b) 4,000   c) 5,120   d) 2,048

4. The value of P_{out} is most nearly:

   a) 2.55   b) 4.0   c) 3.6   d) 5.0

2.22 Solution:

Since $h_{oe}R_C \leq 0.1$

1. $I_B = V_S / (R_S + h_{ie}) = (0.707)/200 = 3.535 \times 10^{-3}$ amperes

   The correct answer is (b).

2. $R_C = \alpha^2 R_L = 16 \times 8 = 128$ ohms

   $A_v = -(h_{fe}/h_{ie})R_C = -(40/100)(128) = -51.2$

   The correct answer is (c).

3. $A_i = -I_C/I_B = -h_{fe}/(1 + h_{oe}R_C) \approx -h_{fe} = -40$

   $A_p = A_v A_i = (-51.2)(-40) = 2{,}048$

   The correct answer is (d).

4. $V_{CE} = A_V V_{in} = (51.2)(0.707)\left[\dfrac{h_{ie}}{h_{ie} + R_s}\right] = 18$ volts

   $I_C = A_i \times I_B = (40)(3.535 \times 10^{-3}) = 0.1414$ amperes

   $P_{out} = V_{CE}I_C = (18)(0.1414) = 2.55$ watts

   The correct answer is (a).

2.23 Feedback amplifier

The common emitter BJT feedback amplifier can be represented by the circuit model shown in Figure 2.23. The transistor has the following hybrid parameters. Input impedance, h_{ie}, equals 3×10^3 ohms. Current gain, h_{fe}, equals 50. Output admittance, h_{oe}, equals 20×10^{-6} siemens.

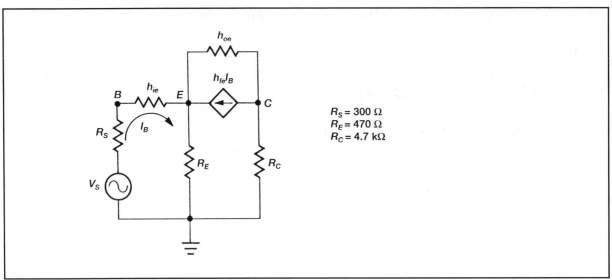

Figure 2.23   Feedback amplifier

1. The value of A_i is most nearly:

   a) -66   b) -50   c) -24   d) -33

2. The value of Z_{in} is most nearly:

   a) 3×10^3 b) 3.5×10^3 c) 25×10^3 d) 27×10^3

3. The value of Z_{out} is most nearly:

   a) 4.3×10^3 b) 5×10^4 c) 4.7×10^2 d) 4.7×10^3

4. The value of A_v is most nearly:

   a) -9.4   b) -7.8   c) -9   d) -10

2.23 Solution:

Since $h_{oe}R_C \leq 0.1$

1. $A_i = -I_C/I_B = -(h_{fe}I_B)/I_B \approx -h_{fe} = -50$

   The correct answer is (b).

2. $Z_{in} = \dfrac{V_{in}}{I_B} = \dfrac{h_{ie}I_B + R_E I_B + h_{fe}R_E I_B}{I_B}$

   $Z_{in} = h_{ie} + R_E + h_{fe}R_E = h_{ie} + R_E(1+h_{fe}) = 3\times 10^3 + (51)(470) = 26{,}970$ ohms

   The correct answer is (d).

3. $Z_{out} = (1/h_{oe} + R_E) \parallel R_C = 5.047\times 10^4 \parallel 4.7\times 10^3 = 4.3\times 10^3$ ohms

   The correct answer is (a).

4. $A_v = -\dfrac{A_i R_C}{Z_{in}} = -\dfrac{h_{fe}R_C}{h_{ie} + (1+h_{fe})R_E}$

   $A_v = -\dfrac{50 \times 4.7 \times 10^3}{3 \times 10^3 + (51 \times 470)} = -8.7$

   The correct answer is (c).

2.24 Differential amplifier

The BJT differential amplifier circuit shown in Figure 2.24 consists of a matched pair. The transistors have the following hybrid parameters. Input impedance, h_{ie}, equals 2×10^3 ohms. Current gain, h_{fe}, equals 50. Output admittance, h_{oe}, equals 5×10^{-6} siemens. Calculate the common mode rejection ratio (CMMR).

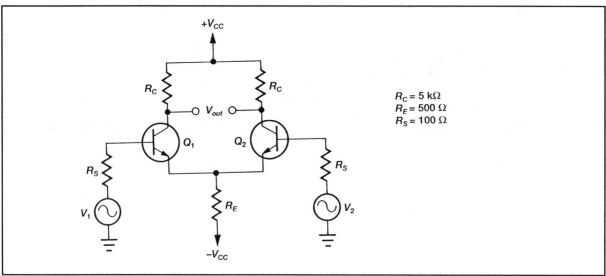

Figure 2.24 Differential amplifier

Solution:

$$\text{CMMR} = 20\log|A_d / A_c|$$

For a differential input $V_{in} = (V_1 - V_2)$

$$A_d = \frac{V_{out}}{V_{in}} = \frac{h_{fe} R_C}{R_s + h_{ie}} = \frac{50 \times 5 \times 10^3}{2.1 \times 10^3} = 119$$

For a common input $V_{in} = \tfrac{1}{2}(V_1 + V_2)$

$$A_c = \frac{V_{out}}{V_{in}} = \frac{h_{fe} R_C}{R_E(1 + h_{fe})} = \frac{50 \times 5 \times 10^3}{500 \times 51} = 9.8$$

$$\text{CMMR} = 20\log|A_d / A_c| = 20\log|119 / 9.8| = 21.7 \text{ dB}$$

2.25 Operational amplifier

The simplified operational amplifier (op amp) model shown in Figure 2.25 consists of a very high input impedance, high gain amplifier with negative feedback. Calculate the output-to-input voltage ratio for a gain, G, of 100, 1,000, and 10,000 respectively.

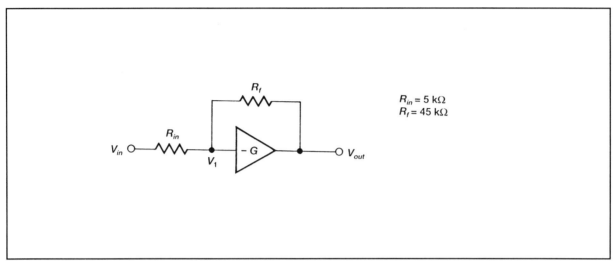

Figure 2.25 Operational amplifier

Solution:

$$\frac{V_{in} - V_1}{R_{in}} = \frac{V_1 - V_{out}}{R_f}$$

$$V_1 = -V_{out}/G$$

$$\frac{V_{out}}{V_{in}} = -\frac{GR_f}{R_f + R_{in} + GR_{in}} = -\left(\frac{R_f}{R_{in}}\right)\frac{GR_{in}}{GR_{in} + R_f + R_{in}}$$

$$\beta = R_{in}/(R_{in} + R_f) = (5 \times 10^3)/(50 \times 10^3) = 0.1$$

$$\frac{V_{out}}{V_{in}} = -\left(\frac{R_f}{R_{in}}\right)\frac{GR_{in}}{GR_{in} + R_f + R_{in}} = -\left(\frac{R_f}{R_{in}}\right)\left(\frac{1}{1 + \frac{1}{G\beta}}\right)$$

a) V_{out}/V_{in} (G = 100) = (-9)(1/1.1) = -8.18

b) V_{out}/V_{in} (G = 1,000) = (-9)(1/1.01) = -8.91

c) V_{out}/V_{in} (G = 10,000) = (-9)(1/1.001) = -8.99

2.26 Inverting op amp

The inverting operational amplifier (op amp) circuit model shown in Figure 2.26 consists of a very high input impedance, very high gain amplifier with resistive negative feedback. Calculate the actual voltage gain, and Z_{in}.

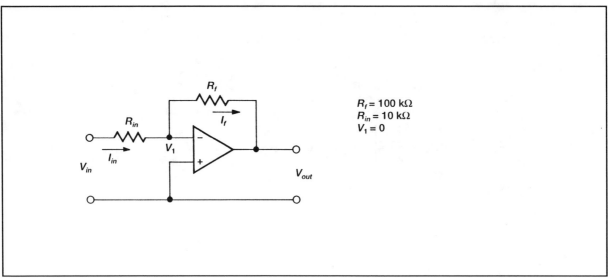

Figure 2.26 Inverting op amp

Solution:

$V_{in} = I_{in}R_{in}$

$I_f = -I_{in}$

$V_{out} = I_f R_f = -I_{in} R_f$

$V_{out}/V_{in} = (-I_{in}R_f)/(I_{in}R_{in}) = -R_f/R_{in} = -100/10 = -10$

$Z_{in} = V_{in}/I_{in} = R_{in} = 10 \times 10^3$ ohms

2.27 Noninverting op amp

The noninverting operational amplifier (op amp) circuit model shown in Figure 2.27 consists of a very high input impedance, very high gain amplifier with resistive negative feedback. Calculate the actual voltage gain.

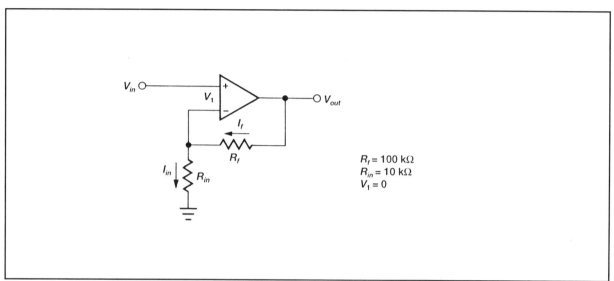

Figure 2.27 Noninverting op amp

Solution:

$V_{out} = I_{in} R_{in} + I_f R_f$

$I_f = I_{in}$

$V_{out} = I_{in}(R_{in} + R_f)$

$I_{in} = V_{in} / R_{in}$

$V_{out} = (V_{in} / R_{in})(R_{in} + R_f)$

$V_{out} / V_{in} = (R_{in} + R_f) / (R_{in}) = (100 + 10) / 10 = 11$

2.28 Summing op amp

The summing operational amplifier (op amp) circuit model shown in Figure 2.28 consists of a very high input impedance, very high gain amplifier with resistive negative feedback, and three different input signals. Calculate the output voltage value.

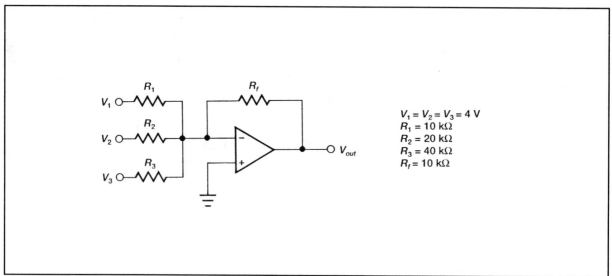

Figure 2.28   Summing op amp

Solution:

$$V_{out} = -R_f \sum I_i$$

$$V_{out} = -R_f (I_1 + I_2 + I_3)$$

$$V_{out} = -R_f (V_1/I_1 + V_2/I_2 + V_3/I_3)$$

$$V_{out} = -10(4/10 + 4/20 + 4/40)$$

$$V_{out} = -10(0.4 + 0.2 + 0.1) = -(4 + 2 + 1) = -7 \text{ volts}$$

2.29 Integrating op amp

The integrating operational amplifier (op amp) circuit model shown in Figure 2.29 consists of a very high input impedance, very high gain amplifier with capacitive negative feedback. Determine the output voltage value at t = 5 seconds, assuming the initial voltage across the capacitor is zero.

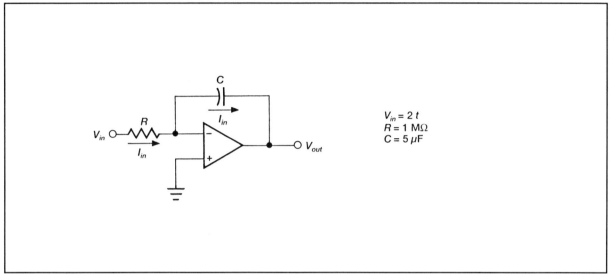

Figure 2.29 Integrating op amp

Solution:

$$CV = Q = It$$

$$C\Delta V = I\Delta t$$

$$(\Delta V_{out})/(\Delta t) = I/C = -I_{in}/C = -V_{in}/(RC)$$

$$V_{out} = -\frac{1}{RC}\int_0^t V_{in}\,dt = -\frac{1}{5}\int_0^5 2t\,dt$$

$$V_{out} = -0.2[t^2]_0^5 = -0.2[25-0] = -5 \text{ volts}$$

2.30 Differentiating op amp

The differentiating operational amplifier (op amp) circuit model shown in Figure 2.30 consists of a very high input impedance, very high gain amplifier, with a capacitive input and resistive negative feedback. Determine the output voltage value at t = 5 seconds, assuming the initial voltage across the capacitor is zero.

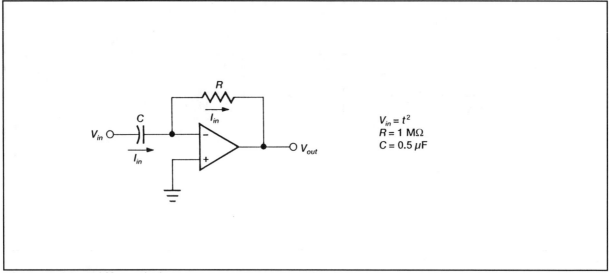

Figure 2.30   Differentiating op amp

Solution:

$Q = CV = It$

$I\Delta t = C\Delta V$

$I = C(\Delta V / \Delta t)$

$I_{in} = C(\Delta V_{in} / \Delta t)$

$V_{out} = -I_{in} R$

$$V_{out} = -(RC)\frac{\Delta V_{in}}{\Delta t} = -(RC)\frac{d}{dt}(V_{in})$$

$$V_{out} = -(0.5)\frac{d}{dt}(t^2) = -0.5(2t) = -t \quad \text{volts}$$

$V_{out}(t = 5) = -5$ volts

2.31 Transfer function op amp

The transfer function operational amplifier (op amp) circuit model shown in Figure 2.31 consists of a very high input impedance, very high gain amplifier, with a frequency sensitive negative feedback component. Determine (a) the actual voltage gain at $\omega = 0$ and at (b) $\omega = 1,000$ respectively.

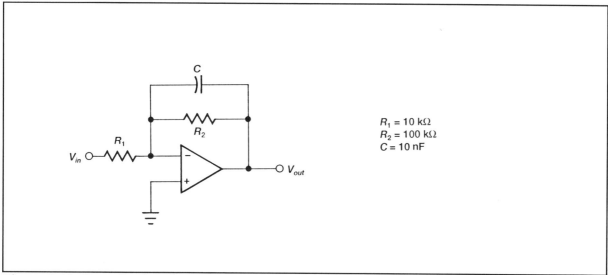

Figure 2.31 Transfer function op amp

Solution:

$$V_{out} / V_{in} = -Z_f / Z_{in}$$

$$Z_f = \frac{R_2 X_C}{R_2 + X_C} = \frac{R_2}{\frac{R_2}{X_C} + 1} = \frac{R_2}{1 - j\omega C R_2}$$

$$\frac{V_{out}}{V_{in}} = -\frac{Z_f}{Z_{in}} = -\frac{R_2}{R_1}\left(\frac{1}{1 - j\omega C R_2}\right)$$

a) V_{out} / V_{in} ($\omega = 0$) $= -(R_2 / R_1) = -100/10 = -10$

b) $\dfrac{V_{out}}{V_{in}}(\omega = 1,000) = -\dfrac{R_2}{R_1}\left(\dfrac{1}{\sqrt{1 + (\omega C R_2)^2}}\right) = -\dfrac{R_2}{R_1}(0.707)$

V_{out} / V_{in} ($\omega = 1,000$) $= -(10)(0.707) = -7.07$

2.32 Comparator

The comparator circuit model shown in Figure 2.32 consists of a very high input impedance, very high gain amplifier, with a fixed input threshold. Determine the output response for a sinusoidal input signal, $V_{in} = 5\sin 314t$.

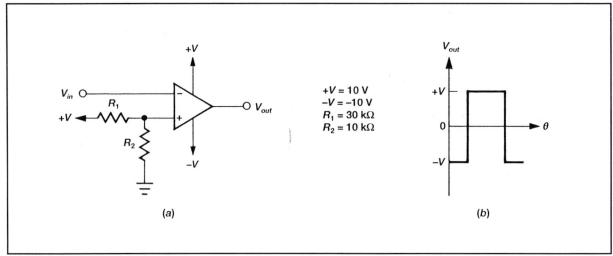

Figure 2.32 Comparator

Solution:

$$V_{th} = V\left(\frac{R_2}{R_1 + R_2}\right) = 10\left(\frac{10 \times 10^3}{30 \times 10^3 + 10 \times 10^3}\right) = 2.5 \quad \text{volts}$$

$V_{in} = 5\sin 314t = 5\sin 2\pi f t = 5\sin\omega t$

$f = \omega/2\pi = 314/2\pi = 50$ Hz and $T = 1/f = 20 \times 10^{-3}$ seconds

$2\pi/T = 360°/T = \theta/t$ and $t = (\theta/360°) \times T$

$V_{th} = 2.5 = 5\sin\theta$

$\theta = \sin^{-1}(2.5/5) = \sin^{-1}(0.5) = 30°, 150°$

$V_{out} = -10$ volts $\quad 0 < \theta < 30°$

$V_{out} = 10$ volts $\quad 30° < \theta < 150°$

$V_{out} = -10$ volts $\quad 150° < \theta < 360°$

2.33 Clipper

The voltage clipping circuit shown in Figure 2.33 consists of a resistive voltage divider and two zener diodes. Determine the output voltage, V_{out}, for an input voltage $V_{in} = 24\sin\omega t$. Assume that the forward diode voltage drop equals one volt.

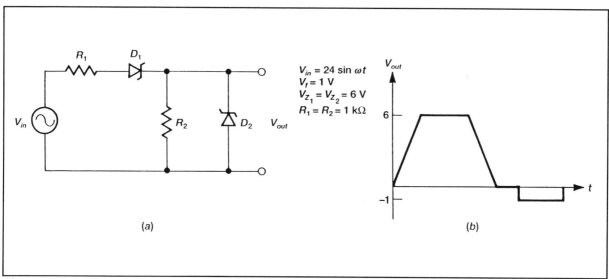

Figure 2.33 Clipper

Solution:

$$V_{in} = I_1(R_1 + R_2) + V_f$$

$$I_1 = \frac{(V_{in} - V_f)}{(R_1 + R_2)}$$

$$V_{out} = I_1 R_2 = \frac{(V_{in} - V_f)}{(R_1 + R_2)} R_2 \qquad V_{in} \leq 13 \text{ volts}$$

$$V_{out} = (V_{in} - 1)/2 \qquad V_{in} \leq 13 \text{ volts}$$

$$V_{out} = V_z = 6 \text{ volts} \qquad V_{in} > 13 \text{ volts}$$

$$V_{out} = 0 \text{ volts} \qquad -V_{in} \geq -6 \text{ volts}$$

$$V_{out} = -V_f = -1 \text{ volts} \qquad -V_{in} < -7 \text{ volts}$$

2.34 Clamper

The voltage clamping circuit shown in Figure 2.34 consists of a voltage divider, a diode and a six volt zener diode. Determine the output voltage, V_{out}, for an input voltage V_{in} = 12sin 377t. Assume that the forward diode voltage drop equals one volt.

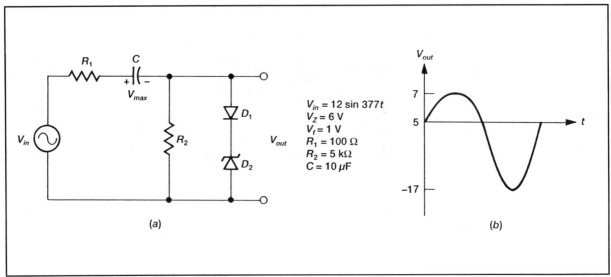

Figure 2.34 Clamper

Solution:

$X_c = 1/(2\pi fC) = 1/(377 \times 10 \times 10^{-6}) = 265$ ohms

$X_c \ll R_2$

$R_1 \ll R_2$

$V_{max} = 12$ volts

$V_{out} = V_{in} - V_{max} + (V_z + V_f) = V_{in} - 12 + (6 + 1) = V_{in} - 5$ volts

$V_{out} (V_{in} = 12\text{ v}) = 12 - 5 = 7$ volts

$V_{out} (V_{in} = 6\text{ v}) = 6 - 5 = 1$ volt

$V_{out} (V_{in} = -6\text{ v}) = -6 - 5 = -11$ volts

$V_{out} (V_{in} = -12\text{ v}) = -12 - 5 = -17$ volts

2.35 Limiter

The current limiting BJT circuit shown in Figure 2.35 uses negative feedback to limit the maximum collector current. Find (a) the current limit, I_L, value for $h_{fe} = 50$, (b) the value of the base resistor, R_B, and (c) find the transistor maximum power dissipation.

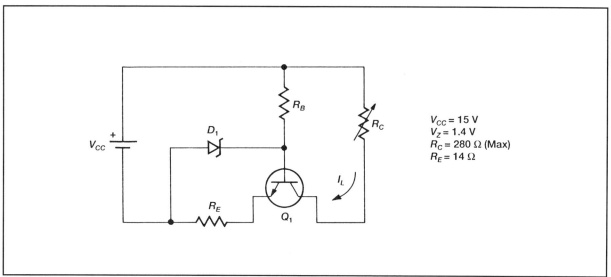

Figure 2.35 Limiter

Solution:

a) $V_z = I_L R_E + V_{be}$

$I_L = (V_z - V_{be}) / R_E = (1.4 - 0.7) / 14 = 50 \times 10^{-3}$ amperes

b) $I_B = I_L / h_{fe} = (50 \times 10^{-3}) / 50 = 1 \times 10^{-3}$ amperes

$R_B = (V_{cc} - V_z) / I_B = (15 - 1.4) / (1 \times 10^{-3}) = 13.6 \times 10^3$ ohms

c) $I_L R_C = 50 \times 10^{-3} \times 280 = 14$ volts

$V_{CE} = V_{cc} - I_L(R_C + R_E) = 15 - 14.7 = 0.3$ volts

For $(R_C = 0)$ $P_C = V_{cc} I_L = 15 \times 50 \times 10^{-3} = 0.75$ watts

2.36 Miller's theorem

The common source FET amplifier ac model shown in Figure 2.36 has the following parameters. Forward transadmittance, y_{fs}, equals 5×10^{-3} siemens. Output conductance, y_{os}, equals 10×10^{-6} siemens. Gate-to-source resistance, r_{gs}, equals 1×10^8 ohms. The C_{ds} equals 1×10^{-12} farads, and $C_{gd} = C_{gs}$ equal 10×10^{-12} farads. Calculate the equivalent input capacitance, C_{in}.

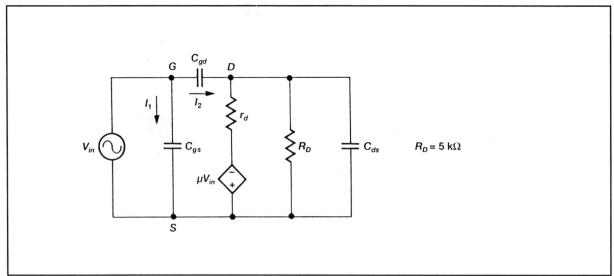

Figure 2.36 Miller's theorem

Solution:

$$I = I_1 + I_2$$

$$I_1 = V_{in} y_{gs}$$

$$I_2 = (V_{in} - A_v V_{in}) y_{gd}$$

$$I = V_{in} y_{gs} + (1 - A_v) V_{in} y_{gd}$$

$$Y_{in} = I / V_{in} = y_{gs} + (1 - A_v) y_{gd}$$

$$Y_{in} = j\omega [C_{gs} + (1 - A_v) C_{gd}]$$

$$\mu = y_{fs} / y_{os} = y_{fs} r_d = g_m r_d = 500$$

$$A_v = -\mu \left(\frac{R_D}{R_D + r_d} \right) = -\frac{y_{fs}}{y_{os}} \left(\frac{R_D}{R_D + r_d} \right) = -500 \left(\frac{5 \times 10^3}{5 \times 10^3 + 1 \times 10^5} \right) = -23.81$$

$$C_{in} = C_{gs} + (1 - A_v) C_{gd} = 10 \times 10^{-12} + (1 + 23.81) 10 \times 10^{-12} = 258.1 \times 10^{-12} \text{ farads.}$$

2.37 Voltage sweep generator

The BJT circuit shown in Figure 2.37 is used to generate a voltage sweep waveform. Determine (a) the output voltage amplitude, V_{out}, and (b) the sweep speed for an h_{fe} value of 50.

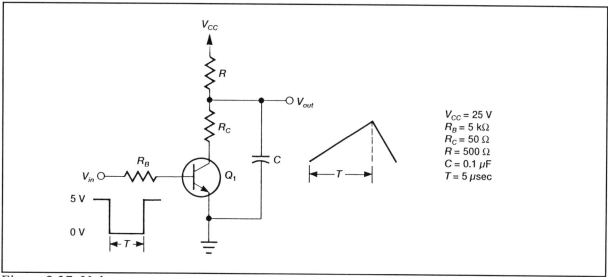

Figure 2.37 Voltage sweep generator

Solution:

$V_{in} = 0$ volts $\quad 0 < t < 5 \times 10^{-6}$ seconds

$V_{in} = 5$ volts $\quad 0 > t > 5 \times 10^{-6}$ seconds

$I_B = V_{in} / R_B = 5 / (5 \times 10^3) = 1 \times 10^{-3}$ amperes

$I_C(\max) = h_{fe} I_B = 50 \times 1 \times 10^{-3} = 50 \times 10^{-3}$ amperes

a) $V_{out}(t) = V(1 - e^{-t/RC})$

$t = T = 5 \times 10^{-6}$ seconds

$RC = 5 \times 10^2 \times 0.1 \times 10^{-6} = 50 \times 10^{-6}$ seconds

For $RC \gg T$

$V_{out}(T) \approx V_{cc}[T/(RC)] = 25(5 \times 10^{-6}) / (50 \times 10^{-6}) = 2.5$ volts

b) Sweep speed $= V_{out}(T) / T = V_{cc} / (RC) = 25 / (50 \times 10^{-6}) = 5 \times 10^5$ volts/second

2.38 Current sweep generator

The BJT circuit shown in Figure 2.38 is used to generate a current sweep waveform. Determine (a) the inductor current amplitude, i_L, and (b) the sweep speed for an h_{fe} value of 50.

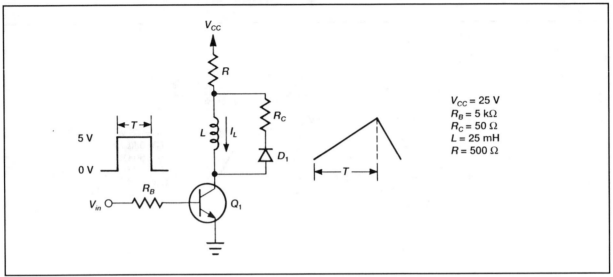

Figure 2.38 Current sweep generator

Solution:

$V_{in} = 0$ volts $\quad 0 > t > 5\times10^{-6}$ seconds

$V_{in} = 5$ volts $\quad 0 < t < 5\times10^{-6}$ seconds

$I_B = V_{in} / R_B = 5 / (5\times10^3) = 1\times10^{-3}$ amperes

$I_C \text{(max)} = h_{fe} I_B = 50\times1\times10^{-3} = 50\times10^{-3}$ amperes

a) $i_L(t) = I_C(1 - e^{-tR/L})$

$t = T = 5\times10^{-6}$ seconds

$L/R = (25\times10^{-3}) / (5\times10^2) = 50\times10^{-6}$ seconds

For $L/R \gg T$

$i_L(T) \approx I_C[(TR)/L] = 50\times10^{-3}[(5\times10^{-6}\times5\times10^2)/(25\times10^{-3})] = 5\times10^{-3}$ amperes

b) Sweep speed $= i_L(T)/T = (I_C R/L) = (50\times10^{-3}\times5\times10^2)/(25\times10^{-3}) = 1\times10^3$ amperes/second

2.39 Phase shift oscillator

The N-channel FET phase-shift oscillator circuit shown in Figure 2.39 produces a sinewave output. The FET parameters are $\mu = 150$ and $r_d = 1\times10^5$ ohms. Calculate (a) the frequency and (b) the condition for oscillation respectively.

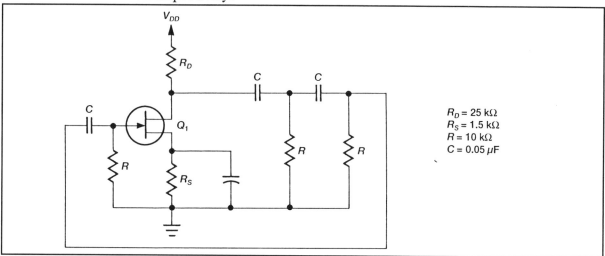

Figure 2.39 Phase shift oscillator

Solution:

$$\frac{V_D}{V_G} = A_f = \frac{R^3}{(R^3 - 5RX_C^2) + j(X_C^3 - 6R^2X_C)}$$

For $A_f \angle 180°$; $j(X_C^3 - 6R^2X_C) = 0$

$X_C^3 = 6R^2X_C$ or $X_C^2 = 6R^2$

a) $1/(2\pi fC)^2 = 6R^2$

$$f = \frac{1}{2\pi RC\sqrt{6}} = \frac{1}{2\pi \times 10 \times 10^3 \times 0.05 \times 10^{-6}\sqrt{6}} = 130 Hz$$

b) $A_f = -\dfrac{R^3}{R^3 - 5RX_C^2} = -\dfrac{1}{1 - 5(\frac{X_C}{R})^2} = -\dfrac{1}{1 - 5\times 6} = -\dfrac{1}{29}$

$$A_v = -\mu\left(\frac{R_D}{R_D + r_d}\right) = -150\left(\frac{25\times 10^3}{25\times 10^3 + 1\times 10^5}\right) = -30$$

For oscillation $|A_v A_f| \geq 1$

$A_v A_f = |(-30)(-1/29)| = 30/29 \geq 1$

2.40 Hartley oscillator

The N-channel FET Hartley oscillator circuit shown in Figure 2.40 produces a sinewave output. The FET parameters are $\mu = 150$ and $r_d = 1 \times 10^5$ ohms. Determine the expressions for (a) A_v, (b) A_f, and (c) calculate the frequency of oscillation.

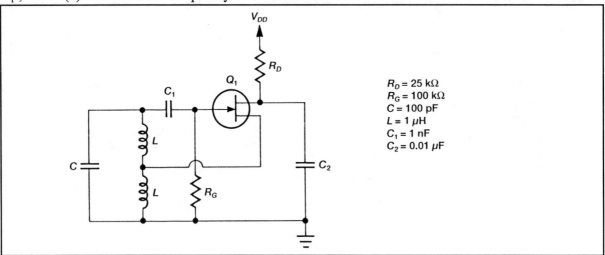

$R_D = 25\ k\Omega$
$R_G = 100\ k\Omega$
$C = 100\ pF$
$L = 1\ \mu H$
$C_1 = 1\ nF$
$C_2 = 0.01\ \mu F$

Figure 2.40 Hartley oscillator

Solution:

a) $A_v = -(\mu)\dfrac{X_L(X_L + X_C)}{r_d + (X_L + X_L + X_C)}$

b) $A_f = -\dfrac{X_L}{(X_L + X_C)}$

$A_v A_f = (\mu)\dfrac{X_L^2}{r_d + (X_L + X_L + X_C)}$

$A_v A_f = (\mu)\dfrac{(\omega L)^2}{r_D + j(\omega L + \omega L - 1/\omega C)}$

c) For oscillation $j[\omega L + \omega L - 1/(\omega C)] = 0$

$\omega L + \omega L = 1/(\omega C)$

$2\pi f(2L) = 1/(2\pi f C)$

$f = \dfrac{1}{2\pi\sqrt{2LC}} = \dfrac{1}{2\pi\sqrt{2 \times 10^{-6} \times 100 \times 10^{-12}}} = \dfrac{1}{2\pi\sqrt{2 \times 10^{-16}}} = 1.126 \times 10^7\ \text{Hz}$

2.41 Collpitts oscillator

The N-channel FET Collpitts oscillator circuit shown in Figure 2.41 produces a sinewave output. The FET parameters are $\mu = 150$ and $r_d = 1\times10^5$ ohms. Determine the expressions for (a) A_v, (b) A_f, and (c) calculate the frequency of oscillation.

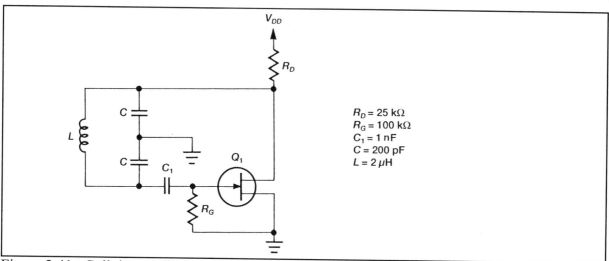

Figure 2.41 Collpitts oscillator

$R_D = 25\ k\Omega$
$R_G = 100\ k\Omega$
$C_1 = 1\ nF$
$C = 200\ pF$
$L = 2\ \mu H$

Solution:

a) $A_v = -(\mu)\dfrac{X_C(X_C + X_L)}{r_d + (X_C + X_C + X_L)}$

b) $A_f = -\dfrac{X_C}{(X_C + X_L)}$

$A_v A_f = (\mu)\dfrac{X_C^2}{r_d + j(X_L + X_C + X_C)}$

$A_v A_f = (\mu)\dfrac{(1/\omega C)^2}{r_d + j(\omega L - 1/(\omega C) - 1/(\omega C))}$

c) For oscillation $j[\omega L - 1/(\omega C) - 1/(\omega C)] = 0$

$\omega L = 1/[(\omega C) + 1/(\omega C)]$

$2\pi f L = 1/[(\omega C^2)/(2C)] = 1/[\omega C/2] = 1/[(2\pi f C)/2]$

$f = \dfrac{1}{2\pi\sqrt{0.5LC}} = \dfrac{1}{2\pi\sqrt{2\times 10^{-6}\times 100\times 10^{-12}}} = 1.126\times 10^7\ Hz$

2.42 Pierce oscillator

The N-channel FET Pierce oscillator circuit shown in Figure 2.42 produces a sinewave output. The FET parameters are $\mu = 150$, $r_d = 1\times 10^5$ ohms, $C_{gs} = 10\times 10^{-12}$ farads, and $C_{ds} = 1\times 10^{-12}$ farads.

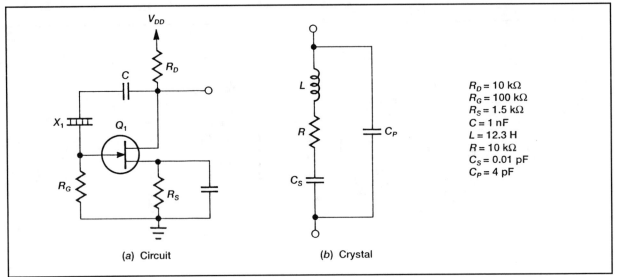

Figure 2.42 Pierce oscillator

1. The value of the crystal series resonance frequency is most nearly:

   a) 450×10^3 b) 300×10^3 c) 454×10^3 d) 350×10^3

2. The value of the crystal parallel resonance frequency is most nearly:

   a) 456×10^3 b) 230×10^3 c) 300×10^3 d) 450×10^3

3. The value of Q is most nearly:

   a) 2.6×10^3 b) 3.5×10^3 c) 0.9×10^3 d) 1.75×10^3

4. The value of A_v (min) for oscillation is most nearly:

   a) 14   b) 4   c) 2.5   d) 10

2.42 Solution:

1. $f_s = \dfrac{1}{2\pi\sqrt{LC_s}} = \dfrac{1}{2\pi\sqrt{12.3 \times 0.01 \times 10^{-12}}} = 454 \times 10^3$ Hz

   The correct answer is (c).

2. $C = (C_p C_s)/(C_p + C_s) = 0.0099 \times 10^{-12}$ farads

   $f_p = \dfrac{1}{2\pi\sqrt{LC}} = \dfrac{1}{2\pi\sqrt{12.3 \times 0.0099 \times 10^{-12}}} = 456 \times 10^3$ Hz

   The correct answer is (a).

3. $Q = X_L/R = (2\pi f_p L)/R = 3{,}524$

   The correct answer is (b).

4. $A_v = -(\mu)\dfrac{R_D}{R_D + r_d} = -150\left(\dfrac{1 \times 10^4}{1 \times 10^4 + 1 \times 10^5}\right) = -13.6$

   $A_v\text{(min)} = C_{gs}/C_{ds} = (10 \times 10^{-12})/(1 \times 10^{-12}) = 10$

   The correct answer is (d).

2.43 UJT oscillator

The unijunction transistor circuit shown in Figure 2.43 produces a sawtooth signal across the capacitor, C. Determine (a) the capacitor charging interval, T, for an intrinsic stand-off ratio, $\eta = 0.8$. Also, calculate (b) the sawtooth amplitude, V_s.

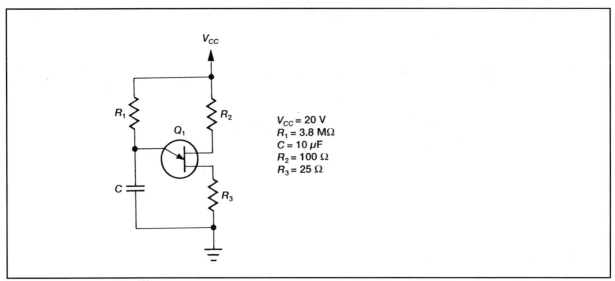

Figure 2.43 UJT oscillator

Solution:

$$\eta = (1 - e^{-T/R_1C})$$

$$1 - \eta = 1/(e^{T/R_1C})$$

$$e^{T/R_1C} = 1/(1 - \eta)$$

a) $T = R_1 C \ln[1/(1-\eta)] = 3.8 \times 10^6 \times 10 \times 10^{-6} \ln[1/(1-0.8)] = 38\ln(5) = 61.16$ seconds

b) $V_s = V_p - V_v$

$V_p = \eta V_{cc} = 0.8 \times 20 = 16$ volts

$$V_v = V\left(\frac{R_3}{R_3 + R_2}\right) = 20\left(\frac{25}{125}\right) = 4 \quad \text{volts}$$

$V_s = V_p - V_v = 16 - 4 = 12$ volts

2.44 Astable multivibrator

The collector coupled astable multivibrator circuit shown in Figure 2.44 produces a square wave output signal. The transistors are a matched pair and have an $h_{fe} = 50$. Calculate (a) the signal period, T, (b) the repetition rate. Also, determine (c) the collector resistor, R_C, value for transistor saturation.

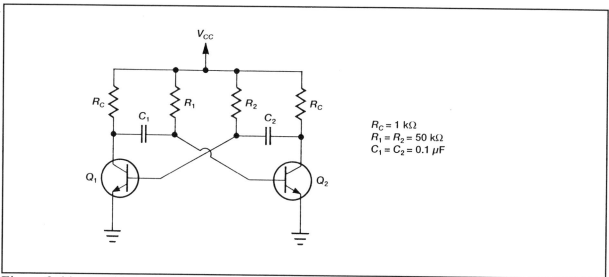

Figure 2.44 Astable multivibrator

Solution:

a) $T = \ln[(2V_{cc}/V_{cc})(R_1C_1)] + \ln[(2V_{cc}/V_{cc})(R_2C_2)]$

$T = t_1 + t_2 = \ln(2)(R_1C_1) + \ln(2)(R_2C_2)$

$T = 2\ln(2)(R_1C_1) = 1.386(50 \times 10^3 \times 0.1 \times 10^{-6})$

$T = 1.386(5 \times 10^{-3}) = 6.93 \times 10^{-3}$ seconds

b) $PRR = 1/T = 144.27$ pulses/second

c) For saturation $h_{fe}I_B \geq I_C$ or

$h_{fe}R_C \geq R_1, R_2$

$R_C \geq R_1/h_{fe} \geq 1 \times 10^3$ ohms

2.45 Monostable multivibrator

The monostable multivibrator circuit shown in Figure 2.45 produces an output pulse each time it is triggered from its stable state. The transistors are matched and have an $h_{fe} = 50$. Calculate (a) the pulse width, T_p. Also, determine (b) the collector resistor, R_C, value for transistor saturation.

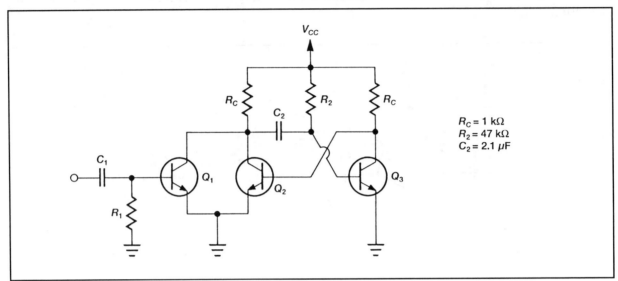

Figure 2.45 Monostable multivibrator

Solution:

a) $T_p = \ln(2)(R_2 C_2)$

$T_p = 0.693(47 \times 10^3 \times 2.1 \times 10^{-6})$

$T_p = 0.693(9.87 \times 10^{-2}) = 6.84 \times 10^{-2}$ seconds

b) For saturation $h_{fe} I_B \geq I_C$ or

$h_{fe} R_C \geq R_2$

$R_C \geq R_2 / h_{fe} \geq 940$ ohms

2.46 Bistable multivibrator

The bistable multivibrator circuit shown in Figure 2.46 has two stable states. The transistors are a matched pair and have an $h_{fe} = 50$. Calculate the stable state collector and base voltages, and the collector and base currents. Let V_{ce} (sat) = 0.3 v, V_{be} (ON) = 0.5 v, and V_{be} (sat) = 0.7 v.

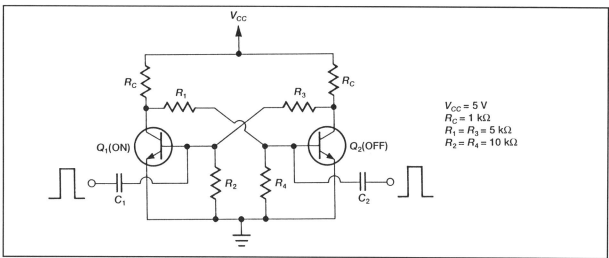

Figure 2.46 Bistable multivibrator

Solution: Assume Q_1 (ON), and Q_2 (OFF)

a) $V_{ce}(Q_1) = V_{ce}$ (sat) = 0.3 volts
b) $V_{be}(Q_1) = V_{be}$ (sat) = 0.7 volts
c) $I_C(Q_2) = 0$ amperes

$$I = \frac{V - V_{be}(Q_1)}{R_C + R_3} = \frac{5 - 0.7}{6 \times 10^3} = \frac{4.3}{6 \times 10^3} = 0.72 \times 10^{-3} \text{ amperes}$$

d) $V_{ce}(Q_2) = V_{cc} - IR_C = 5 - (0.72 \times 10^{-3})(1 \times 10^3) = 4.28$ volts

e) $V_{be}(Q_2) = V_{ce}(Q_1) \dfrac{R_4}{R_4 + R_1} = (0.3) \times \dfrac{10 \times 10^3}{15 \times 10^3} = 0.2$ volts

f) $I_B(Q_2) = 0$ amperes, Since $V_{be}(Q_2) < V_{be}$ (ON)

g) $I_{be}(Q_1) = I - V_{be}$ (sat)/ $R_2 = (0.72 \times 10^{-3}) - (7 \times 10^{-5}) = 0.65 \times 10^{-3}$ amperes

h) $I_C(Q_1) = \dfrac{V - V_{ce}(sat)}{R_C} - \dfrac{V_{ce}(sat)}{R_1 + R_4} = \dfrac{5 - 0.3}{1 \times 10^3} - \dfrac{0.3}{15 \times 10^3} = 4.68 \times 10^{-3}$ amperes

I_{be} (sat) = $I_C(Q_1) / h_{fe} = (4.68 \times 10^{-3}) / 50 = 0.094 \times 10^{-3}$ amperes

2.47 Schmitt trigger

The Schmitt trigger circuit shown in Figure 2.47 has two stable states as a function of the input signal level. The transistors are a matched pair and have an $h_{fe} = 50$. Calculate (a) the input threshold voltages V_1 and (b) V_2 respectively.

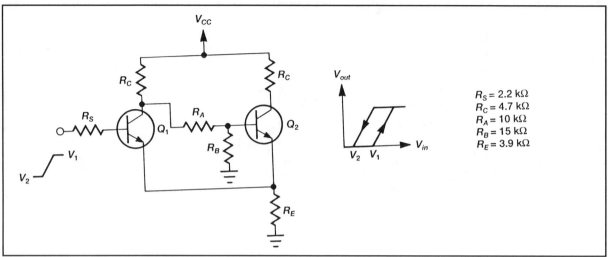

Figure 2.47 Schmitt trigger

Solution:

a) $V_1 = V\left(\dfrac{1}{1 + \dfrac{R_C + R_A}{R_B} + \dfrac{R_C + R_A}{h_{fe} R_E}}\right)$

$V_1 = 12\left(\dfrac{1}{1 + \dfrac{14.7 \times 10^3}{15 \times 10^3} + \dfrac{14.7 \times 10^3}{195 \times 10^3}}\right)$

$V_1 = 12\left(\dfrac{1}{1 + 0.98 + 0.075}\right) = 5.84 \quad volts$

b) $V_2 = V\left(\dfrac{1}{1 + \dfrac{R_C + R_A}{R_B} + \dfrac{R_C}{R_E}}\right)$

$V_2 = 12\left(\dfrac{1}{1 + 0.98 + 1.2}\right) = 3.77 \quad volts$

2.48 Log attenuator

The logarithmic attenuator circuit shown in Figure 2.48 uses a variable dc voltage to control the voltage gain, A_v. The three transistors are identical and have an $h_{fe} = 50$. Calculate the voltage gain for a voltage, V_{dc}, value of 0, 60, and 120 millivolts respectively.

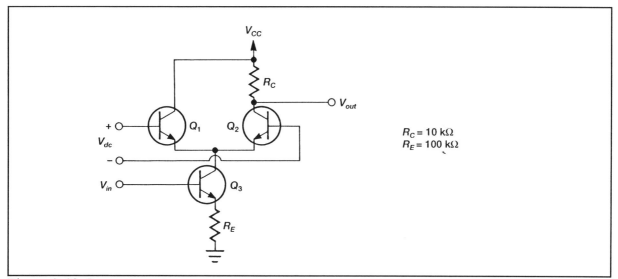

Figure 2.48 Log attenuator

Solution:

$$A_v = \frac{V_{out}}{V_{in}} \approx \left(\frac{R_C}{R_E}\right) \frac{1}{1 + e^{\left(\frac{V_{dc}}{26 \times 10^{-3}}\right)}}$$

a) $A_v(V_{dc} = 0) = \dfrac{1 \times 10^4}{1 \times 10^2} \times \dfrac{1}{1 + e^0} = 50$

b) $A_v(V_{dc} = 60 \times 10^{-3}) = 100 \times \dfrac{1}{1 + e^{2.3}}$

$A_v(V_{dc} = 60 \times 10^{-3}) = 100 \times \dfrac{1}{10.97} = 9.11$

c) $A_v(V_{dc} = 120 \times 10^{-3}) = 100 \times \dfrac{1}{e^{4.6}} = 0.995$

2.49 Multiplier

The linear multiplier circuit shown in Figure 2.49 consists of two cross coupled differential amplifiers with a constant current source. All the transistors are identical and have an $h_{fe} = 50$. Calculate the output, V_{out}, product of two sinewave input signals with different frequencies.

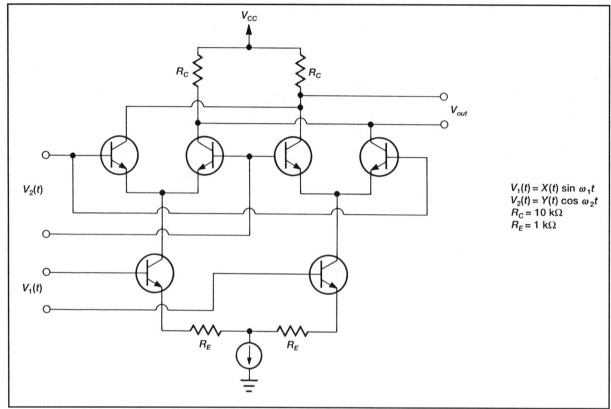

Figure 2.49 Multiplier

Solution:

$$V_{out} = [R_C/R_E][V_1(t)V_2(t)]$$

$$V_{out} = \left[\frac{1 \times 10^4}{1 \times 10^3}\right] x(t)y(t)\sin(\omega_1 t)\cos(\omega_2 t)$$

$$V_{out} = [10]x(t)y(t)\sin(\omega_1 t)\cos(\omega_2 t)$$

Using $\sin(A)\cos(B) = 1/2[\sin(A+B) + \sin(A-B)]$

$$V_{out} = [5]x(t)y(t)[\sin(\omega_1 + \omega_2)t + \sin(\omega_1 - \omega_2)t\,]$$

2.50 Hall effect

The Hall effect sensor circuit shown if Figure 2.50 consists of an n-type silicon semiconductor bar connected to a voltage follower. The bar dimensions are $3 \times 2 \times 12$ mm, and has a resistivity of $\rho = 1 \times 10^{-1}$ ohms/meter. A potential of 5 v is applied across its long ends, and a magnetic flux density of $B = 0.3$ webers/meter^2 is applied parallel to its 3 mm sides. Determine the resulting output voltage for a Hall effect coefficient of $R_H = 20 \times 10^{-3}$.

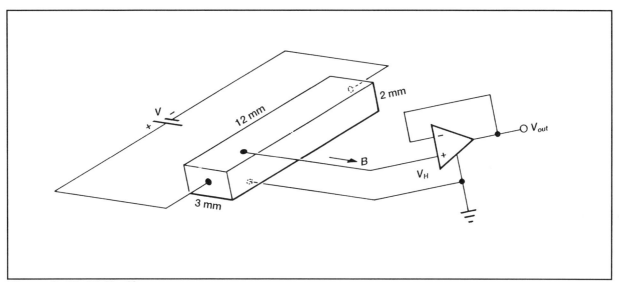

Figure 2.50 Hall effect

Solution:

$$R = \rho\left(\frac{l}{A}\right) = 1 \times 10^{-1}\left(\frac{12 \times 10^{-3}}{3 \times 10^{-3} \times 2 \times 10^{-3}}\right) = \frac{12 \times 10^{-4}}{6 \times 10^{-6}} = 200 \text{ ohms}$$

$I = V/R = 5/(200) = 25 \times 10^{-3}$ amperes

$$V_H = \frac{IBR_H}{W} = \frac{25 \times 10^{-3} \times 0.3 \times 20 \times 10^{-3}}{3 \times 10^{-3}}$$

$$V_H = \frac{150 \times 10^{-6}}{3 \times 10^{-3}} = 50 \times 10^{-3} \text{ volts}$$

Since $Z_{in} \to \infty$, $V_+ = V_-$, and $A_v = 1$

$V_{out} = V_H$

2.51 Darlington amplifier

The Darlington amplifier circuit shown in Figure 2.51 consists of two matched pair transistors. The transistors have the following hybrid parameters, input impedance, h_{ie}, equals 2×10^3 ohms, voltage feedback ratio, h_{re}, equals 4×10^{-4}, current gain, h_{fe}, equals 50, output admittance, h_{oe}, equals 20×10^{-6} siemens. The base-to emitter voltage, V_{be}, equals 0.7 volts.

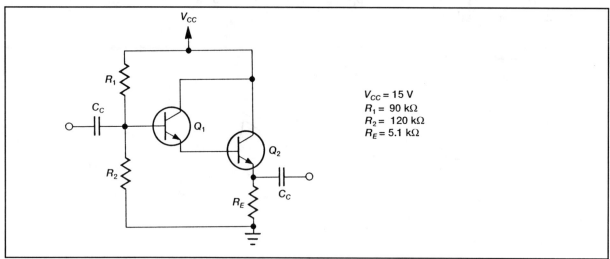

Figure 2.51 Darlington amplifier

1. The amplifier input impedance is most nearly:

   a) 120×10^3 b) 5.1×10^3 c) 50×10^3 d) 255×10^3

2. The base voltage of Q_2 is most nearly:

   a) 9.3   b) 7.0   c) 8.4   d) 7.9

3. The amplifier output impedance is most nearly:

   a) 5.1×10^3 b) 39   c) 2×10^3 d) 102

4. The overall voltage gain is most nearly:

   a) 50   b) 0.5   c) 1.0   d) 40

5. The overall current gain is most nearly:

   a) 426   b) 50   c) 510   d) 2.5×10^3

2.51 Solution:

Since $h_{oe} R_C \leq 0.1$

1. $R_{th} = (R_1 R_2)/(R_1 + R_2) = 51.4 \times 10^3$ ohms

   $Z_{in} = [(1 + h_{fe})^2 R_E / (1 + h_{oe} h_{fe} R_E)] || R_{th}$

   $Z_{in} = (2.175 \times 10^6) || (51.4 \times 10^3) = 50 \times 10^3$ ohms

   The correct answer is (c).

2. $V_{B1} = V_{CC}[R_2 / (R_1 + R_2)] = 8.6$ volts

   $V_{B2} = V_{B1} - V_{be} = 8.6 - 0.7 = 7.9$ volts

   The correct answer is (d).

3. $Z_{out} = [h_{ie} / (1 + h_{fe})] || R_E = 39.2 || (5.1 \times 10^3) = 39$ ohms

   The correct answer is (b).

4. $A_v = 1 - h_{ie} / [(1 + h_{fe}) R_E] = 1 - 7.7 \times 10^{-3} \approx 1$

   The correct answer is (c).

5. $A_i = [(1 + h_{fe})^2 / (1 + h_{oe} h_f R_E)] = 426$

   The correct answer is (a).

2.52 Cascode amplifier

A common emitter, common base configuration or cascode BJT amplifier is shown in Figure 3.52. The transistors are a matched pair and have the following parameters $h_{ie} = 2 \times 10^3$ ohms, $h_{re} = 4 \times 10^{-4}$, $h_{fe} = 50$, $h_{oe} = 20 \times 10^{-6}$ siemens. The base-to emitter voltage, V_{be}, equals 0.7 volts.

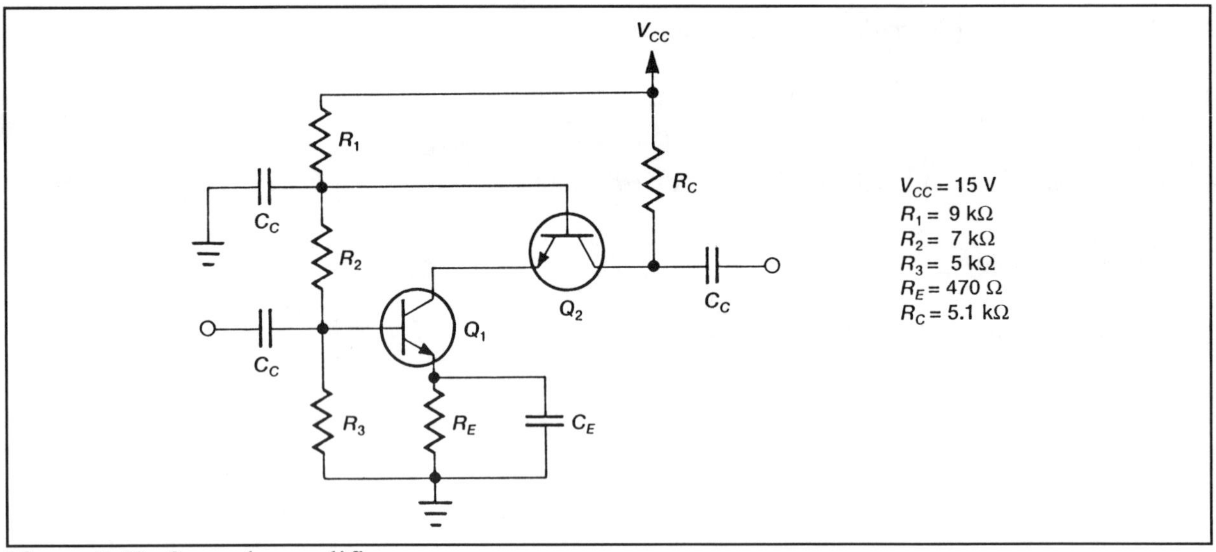

Figure 2.52 Cascode amplifier

$V_{CC} = 15$ V
$R_1 = 9$ kΩ
$R_2 = 7$ kΩ
$R_3 = 5$ kΩ
$R_E = 470$ Ω
$R_C = 5.1$ kΩ

1. The collector-to-emitter voltage of Q_2 is most nearly:

    a) 6   b) 4.5   c) 5.25   d) 8.3

2. The amplifier input impedance is most nearly:

    a) 2.9×10^3 b) 1.2×10^3 c) 2×10^3 d) 5×10^3

3. The amplifier output impedance is most nearly:

    a) 50×10^3 b) 255×10^3 c) 23.5×10^3 d) 5.1×10^3

4. The voltage gain of the first stage is most nearly:

    a) -50   b) -51   c) -1   d) -40

5. The voltage gain of the second stage is most nearly:

    a) 127.5   b) 2.5   c) 100   d) 150

2.52 Solution:

Since $h_{oe} R_C \leq 0.1$

1. $V_{B1} = V_{CC}[R_3 / (R_1 + R_2 + R_3)] = 3.75$ volts

   $V_{B2} = V_{CC}[(R_2 + R_3) / (R_1 + R_2 + R_3)] = 9.0$ volts

   $V_{CE2} = V_{B2} - V_{be} - (V_{B1} - V_{be}) = 5.25$ volts

   The correct answer is (c).

2. $R_{th} = R_2 \| R_3 = 2.9 \times 10^3$

   $Z_{in} = h_{ie} \| R_{th} = (2 \times 10^3) \| (2.9 \times 10^3) = 1.2 \times 10^3$ ohms

   The correct answer is (d).

3. $Z_{out} = [(1 + h_{fe}) / h_{oe}] \| R_C = (51/20 \times 10^{-6}) \| (5.1 \times 10^3) \approx 5.1 \times 10^3$ ohms

   The correct answer is (d).

4. $A_{v1} = -(h_{fe} / h_{ie}) \times (h_{ie} / (1 + h_{fe})) = 50/51 \approx 1$

   The correct answer is (c).

5. $A_{v2} = (h_{fe} / h_{ie}) R_C = (50/2 \times 10^{-3}) 5.1 \times 10^{-3} \approx 127.5$

   The correct answer is (a).

2.53 Hybrid amplifier

An FET/BJT amplifier configuration is shown in Figure 2.53. The FET parameters are $y_{fs} = 1.5 \times 10^{-3}$ siemens, $y_{os} = 10 \times 10^{-6}$ siemens, $V_p = -5$ volts, $I_{DS} = 1 \times 10^{-3}$ amps, and $I_{DSS} = 2 \times 10^{-3}$ amps. The BJT parameters are $h_{ie} = 2 \times 10^3$ ohms, $h_{re} = 4 \times 10^{-4}$, $h_{fe} = 50$, and $h_{oe} = 20 \times 10^{-6}$ siemens.

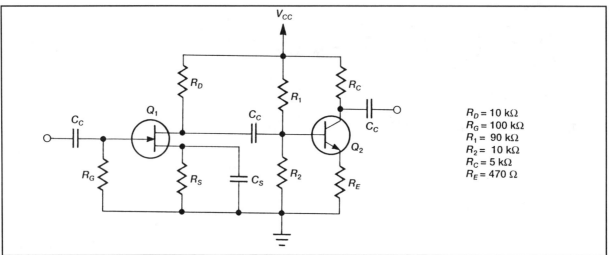

Figure 2.53 Hybrid amplifier

1. The value of R_s is most nearly:

   a) 1×10^3 b) 1.46×10^3 c) 470   d) 730

2. The second stage input impedance is most nearly:

   a) 10×10^3 b) 23.5×10^3 c) 6.7×10^3 d) 9×10^3

3. The first stage voltage gain is most nearly:

   a) -14.3   b) -6.85   c) -5.77   d) -150

4. The second stage voltage gain is most nearly:

   a) -38   b) -28.3   c) -10.75   d) -9.82

5. The amplifier output impedance is most nearly:

   a) 100   b) 5.57×10^3 c) 23.5×10^3 d) 5.1×10^3

2.53 Solution:

Since $h_{oe}R_C \le 0.1$

1. $V_{GS} = V_P[1 - (I_{DS}/I_{DSS})^{1/2}] = V_{GS} = -5[1 - (1/2)^{1/2}] = -1.46$ volts

   $R_S = V_{GS}/I_{DS} = 1.46 \times 10^3$ ohms

   The correct answer is (b).

2. $R_{th} = R_1 || R_2 = 9 \times 10^3$

   $Z_{in}(Q_2) = h_{ie} + R_E(1 + h_{fe}) || R_{th} = (25.97 \times 10^3) || (9 \times 10^3) = 6.7 \times 10^3$ ohms

   The correct answer is (c).

3. $R'_D = R_D || Z_{in}(Q_2) = (10 \times 10^3) || (6.7 \times 10^3) = 4 \times 10^3$ ohms

   $g_m = y_{fs} = 1.5 \times 10^{-3}$

   $r_d = 1/y_{os} = 1 \times 10^5$ ohms

   $\mu = y_{fs}/y_{os} = y_{fs}r_d = g_m r_d = 1.5 \times 10^{-3} \times 1 \times 10^5 = 150$

   $A_{v1} = -\mu[R'_D/(r_d + R'_D)] = -150[(4 \times 10^3)/(104 \times 10^3)] = -5.77$

   The correct answer is (c).

4. $A_{v2} = -h_{fe}[(R_C/Z_{in}(Q_2)] = -50[(5.1 \times 10^3)/(6.7 \times 10^3)] = -38$

   The correct answer is (a).

5. $Z_{out} = (1 + h_{fe})/h_{oe} || R_C = (2.55 \times 10^6) || (5.1 \times 10^3) \approx 5.1 \times 10^3$ ohms

   The correct answer is (d).

2.54 Buffer amplifier

The op amp circuit shown in Figure 2.54 is a non-inverting buffer amplifier. The op amp parameters are G = 100 dB, $r_i = 1 \times 10^8$ ohms, $r_o = 50$ ohms, gain bandwidth product, GBP, is 50×10^6 Hz, and the slew rate is 5 volts/μsec.

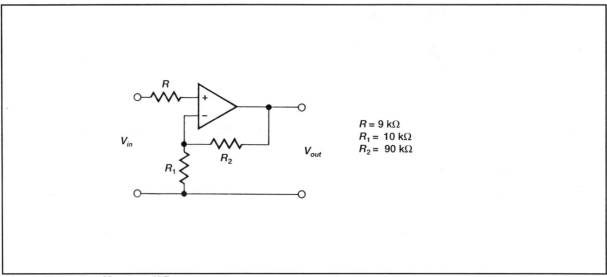

Figure 2.54 Buffer amplifier

1. The amplifier voltage gain equals:

    a) 9   b) 1   c) 4.7   d) 10

2. The amplifier input impedance equals:

    a) 1×10^8 b) 2×10^8 c) 9×10^3 d) 10×10^3

3. The amplifier output impedance equals:

    a) 25   b) 9×10^3 c) 50   d) 100

4. The amplifier small signal bandwidth equals:

    a) 50×10^6 b) 1×10^6 c) 5×10^6 d) 500×10^6

5. The amplifier power frequency, for v_{in} = 10sinωt, equals:

    a) 200×10^3 b) 5×10^5 c) 50×10^6 d) 79.6×10^3

2.54 Solution:

1. $A_v = (R_1 + R_2)/R_1 = (100 \times 10^3)/(10 \times 10^3) = 10$

   The correct answer is (d).

2. $Z_{in} = r_i [1 + (AR_1)/(R_1 + R_2)] = 2r_i = 2 \times 10^8$ ohms

   The correct answer is (b).

3. $Z_{out} = r_o / [1 + (AR_1)/(R_1 + R_2)] = r_o/2 = 25$ ohms

   The correct answer is (a).

4. $BW = (GBP)/A_v = (50 \times 10^6)/10 = 5 \times 10^6$ Hz

   The correct answer is (c).

5. $f_p = SR/(2\pi V_{peak}) = (5 \times 10^6)/(2\pi \times 10) = 79.6 \times 10^3$ Hz

   The correct answer is (d).

2.55 Op amp filter

The op amp filter amplifier shown in Figure 2.55 uses two RC networks to provide frequency dependent voltage gain.

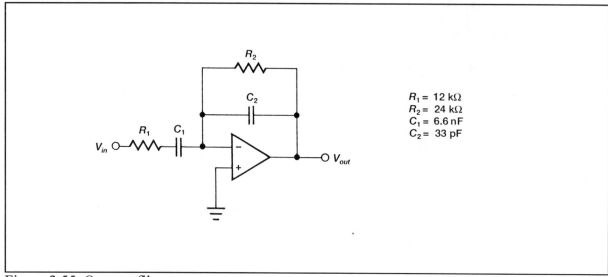

Figure 2.55 Op amp filter

1. The amplifier transfer function equation is:

   a) $-(R_2C_2s) / (R_1C_1s)$

   b) $-[R_2(1+ R_1C_1s)] / [(1+R_1C_1s)(1+ R_2C_2s)]$

   c) $-[R_2C_1s] / [(1+R_1C_1s)(1+R_2C_2)]$

   d) $-[(1+R_1C_1s)(1+R_2C_2s)] / [R_2C_1s]$

2. The half-power frequencies, f_1 and f_2, are most nearly:

   a) 1×10^3, 1×10^5 b) 2×10^3, 2×10^5 c) 1×10^3, 4×10^5 d) 320, 32×10^4

3. The approximate voltage gain at $f = 20\times10^3$ Hz is:

   a) -0.5   b) -20   c) -0.05   d) -2

4. The approximate voltage gain at $f = 200$ Hz is:

   a) -0.2   b) -0.8   c) -1.25   d) -5

5. The approximate voltage gain at $f = 1.8\times10^6$ Hz is:

   a) -4   b) -1.6   c) -0.25   d) -0.6

2.55 Solution:

1. $Z_1(s) = (1 + R_1C_1s)/C_1s$

   $Z_2(s) = R_2/(1 + R_2C_2s)$

   $G(s) = -Z_2(s)/Z_1(s) = -[R_2C_1s]/[(1+R_1C_1s)(1+R_2C_2s)]$

   The correct answer is (c).

2. $f_1 = 1/(2\pi R_1 C_1) \approx 2\times 10^3$ Hz

   $f_2 = 1/(2\pi R_2 C_2) \approx 2\times 10^5$ Hz

   The correct answer is (b).

3. Let $R_1C_1 = 1/\omega_1$, $R_2C_2 = 1/\omega_2$

   $G(j\omega) = -(j\omega R_2C_1)/[(1+j\omega/\omega_1)(1+j\omega/\omega_2)] = -(j\omega R_2C_1)/[(1+j\omega/\omega_1)(1+0.1)]$

   $G(j\omega) \approx -j\omega_1 R_2C_1 \approx -jR_2/R_1$

   $|G(j\omega)| \approx -R_2/R_1 \approx -2$

   The correct answer is (d)

4. $G(j\omega) = -(j\omega R_2C_1)/[(1+j\omega/\omega_1)(1+j\omega/\omega_2)] = -(j\omega R_2C_1)/[(1+0.1)(1+0.001)]$

   $G(j\omega) \approx -j\omega R_2C_1$

   $|G(j\omega)| \approx -\omega R_2C_1 \approx -2\pi\times 200 \times 24\times 10^3 \times 6.6\times 10^{-9} \approx -0.2$

   The correct answer is (a).

5. $G(j\omega) = -[j\omega R_2C_1]/[((\omega_1+j\omega)/\omega_1)((\omega_2+j\omega)/\omega_2)] \approx -(j\omega R_2C_1\omega_1\omega_2)/\omega^2$

   $G(j\omega) \approx -1/j\omega R_1C_2$

   $|G(j\omega)| \approx -1/(\omega R_1C_2) \approx -1/(2\pi\times 1.8\times 10^6 \times 12\times 10^3 \times 33\times 10^{-12}) \approx -0.25$

   The correct answer is (c).

NOTES:

3.0 CONTROL

3.1 Open-loop control

The open-loop controller shown in Figure 3.1 consists of an integrated circuit differential amplifier. Determine the open-loop gain.

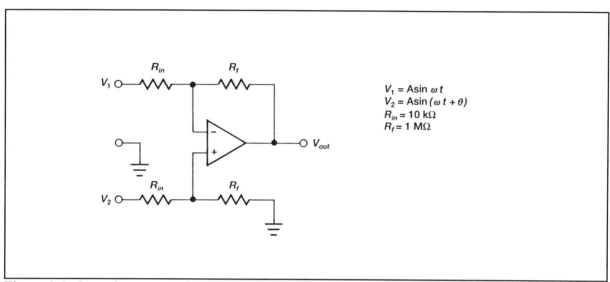

Figure 3.1 Open-loop control

Solution:

Let $V_{out} = V_a + V_b$

$V_a / V_1 = - R_f / R_{in}$

$V_b / V_2 = [R_f / (R_f + R_{in})] \times [(R_f + R_{in}) / R_{in}] = R_f / R_{in}$

$V_{out} = - V_1 / (R_f / R_{in}) + V_2 / (R_f / R_{in})$

$V_{out} = (R_f / R_{in}) \times (V_2 - V_1) = 100 \times A \sin(\omega t + \theta - \omega t) = 100 \times A \sin\theta$

$$\frac{V_{out}}{V_2 - V_1} = \frac{R_f}{R_{in}} = \frac{1 \times 10^6}{1 \times 10^4} = 100$$

3.2 Closed-loop control

The closed-loop controller shown in Figure 3.2 consists of an integrated circuit differential amplifier, and two cascaded inverting op amps. Determine the closed-loop gain.

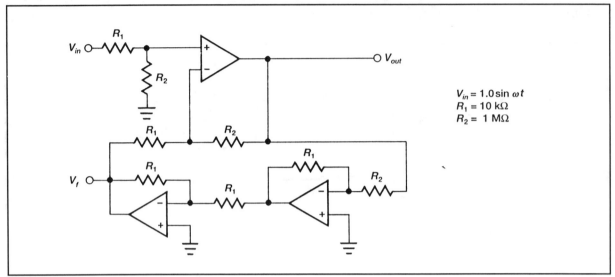

Figure 3.2 Closed-loop control

Solution:

$$A_v = R_2/R_1 = 100$$

$$A_f = (-R_1/R_2)(-R_1/R_1) = R_1/R_2 = 0.01$$

$$V_{out} = (V_{in} - V_f)A_v$$

$$V_f = V_{out}A_f$$

$$V_{out} = (V_{in} - V_{out}A_f)A_v$$

$$V_{out} = V_{in}A_v - V_{out}A_vA_f$$

$$V_{out} + (1 + A_vA_f) = V_{in}A_v$$

$$\frac{V_{out}}{V_{in}} = \frac{A_v}{1 + A_vA_f} = \frac{100}{1 + (100 \times 0.01)} = \frac{100}{2} = 50$$

3.3 Block diagrams

The feedback control block diagram shown in Figure 3.3 consists of two feedforward and one feedback elements. Determine (a) the resulting open-loop and (b) closed-loop gains.

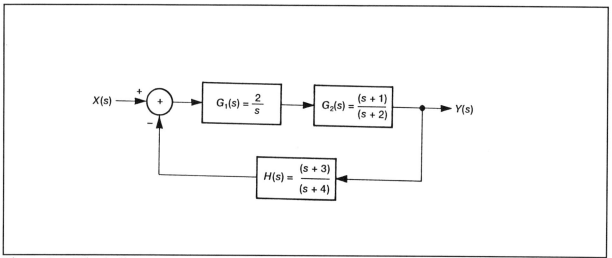

Figure 3.3 Block diagrams

Solution:

a) Open-loop gain = $G_1(s)G_2(s)H(s) = G(s)H(s)$

$$G(s)H(s) = \frac{2}{s} \times \frac{s+1}{s+2} \times \frac{s+3}{s+4}$$

$$G(s)H(s) = \frac{2s^2 + 8s + 6}{s(s^2 + 6s + 8)}$$

b) Closed-loop gain = $Y(s)/X(s)$

$$\frac{Y(s)}{X(s)} = \frac{G(s)}{1 + G(s)H(s)} = \frac{\left(\frac{2}{s}\right)\left(\frac{s+1}{s+2}\right)}{1 + G(s)H(s)}$$

$$\frac{Y(s)}{X(s)} = \frac{G(s)}{1 + G(s)H(s)} = \frac{2(s+1)(s+4)}{s(s+2)(s+4) + 2(s+1)(s+3)}$$

3.4 Signal flow graphs

The resistor network shown in Figure 3.4 represents a feedback control system. Determine the equivalent signal flow graph.

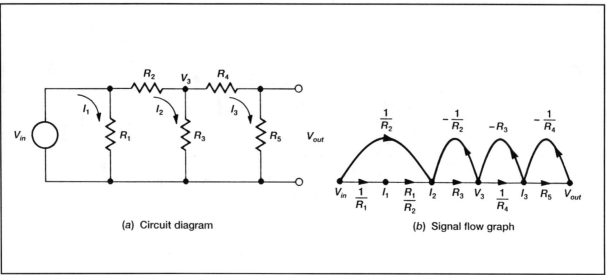

Figure 3.4 Signal flow graphs

Solution:

$I_1 = V_{in} / R_1$

$I_2 = (V_{in} - V_3) / R_2 = (I_1 R_1) / R_2 - V_3 / R_2$

$V_3 = I_2 R_3 - I_3 R_3$

$I_3 = (V_3 - V_{out}) / R_4$

$V_{out} = I_3 R_5$

3.5 Mason's rule

The block diagram shown in Figure 3.5a represents a feedback control system. Find the equivalent signal flow graph and determine the closed loop gain using Mason's rule.

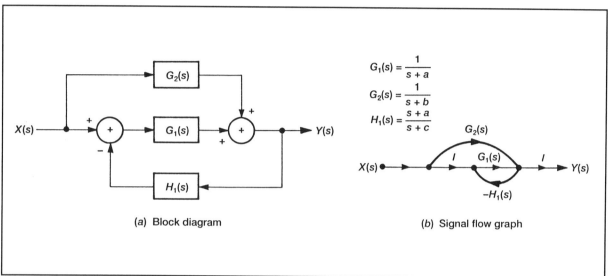

(a) Block diagram   (b) Signal flow graph

Figure 3.5 Mason's rule

Solution:

$$\frac{Y(s)}{X(s)} = \frac{\sum P_i \Delta_i}{\Delta} = \frac{P_1 \Delta_1 + P_2 \Delta_2}{1 - P_{11}}$$

$P_1 = G_1, \quad \Delta_1 = 1$

$P_2 = G_2, \quad \Delta_2 = 1$

$P_{11} = -G_1 H_1$

$$\frac{Y(s)}{X(s)} = \frac{G_1(s) + G_2(s)}{1 + G_1(s)H_1(s)} = \frac{\dfrac{1}{s+a} + \dfrac{1}{s+b}}{1 + \left(\dfrac{1}{s+a}\right)\left(\dfrac{s+a}{s+c}\right)}$$

$$\frac{Y(s)}{X(s)} = \frac{[(s+a) + (s+b)](s+c)}{[(s+a)(s+b)][(s+c) + 1]}$$

3.6 First order loop

The block diagram shown in Figure 3.6 represents a first order feedback control system. Determine (a) the open-loop gain, (b) the closed-loop gain, (c) the error ratio and (d) the feedback ratio.

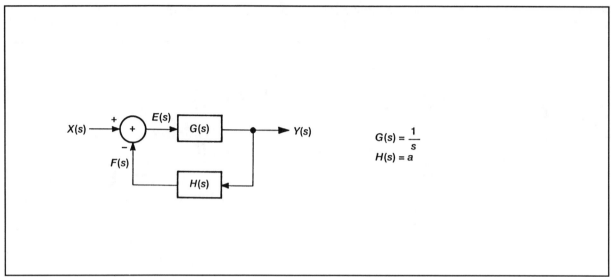

Figure 3.6 First order loop

Solution:

$$E(s) = X(s) - F(s) \qquad\qquad E(s) = X(s) - F(s)$$

$$X(s) = E(s) + F(s) \qquad\qquad X(s) = E(s) + F(s)$$

$$X(s) = F(s) / [(G(s)H(s)] + F(s) \qquad\qquad X(s) = E(s) + E(s)[G(s)H(s)]$$

$$X(s)[G(s)H(s)] = [1 + G(s)H(s)]F(s) \qquad\qquad X(s) = E(s)[1 + G(s)H(s)]$$

$$F(s) / X(s) = [G(s)H(s)] / [1 + G(s)H(s)] \qquad\qquad E(s)/X(s) = 1/[1 + G(s)H(s)]$$

a) Open-loop gain = $F(s) / E(s) = G(s)H(s) = (1/s)(a) = a/s$

b) Closed-loop gain = $Y(s) / X(s) = G(s) / [1 + G(s)H(s)] = 1/(s + a)$

c) Error ratio = $E(s) / X(s) = 1 / [1 + G(s)H(s)] = s/(s + a)$

d) Feedback ratio = $F(s) / X(s) = [G(s)H(s)] / [1 + G(s)H(s)] = a/(s + a)$

3.7 Second order loop

The block diagram shown in Figure 3.7 represents a second order feedback control system. Find (a) the open-loop gain, (b) the closed-loop gain, (c) the error ratio and (d) the feedback ratio.

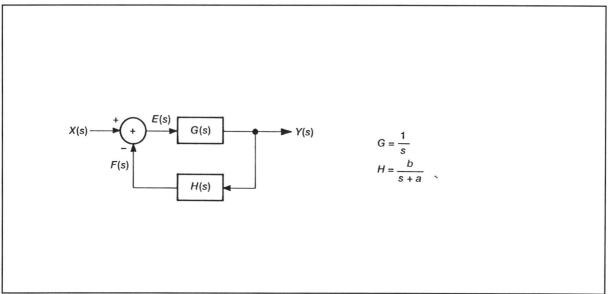

Figure 3.7 Second order loop

Solution:

a) Open – loop gain $= \dfrac{F(s)}{E(s)} = G(s)H(s) = \dfrac{b}{s(s+a)}$

b) Closed – loop gain $= \dfrac{Y(s)}{X(s)} = \dfrac{G(s)}{1+G(s)H(s)} = \dfrac{(s+a)}{s^2+as+b}$

c) Error ratio $= \dfrac{E(s)}{X(s)} = \dfrac{1}{1+G(s)H(s)} = \dfrac{s(s+a)}{s^2+as+b}$

d) Feedback ratio $= \dfrac{F(s)}{X(s)} = \dfrac{G(s)H(s)}{1+G(s)H(s)} = \dfrac{b}{s^2+as+b}$

3.8 Type 0 error

The block diagram shown in Figure 3.8 represents a type 0 feedback control system. Determine the steady state error, e_{ss}, for a unit step input.

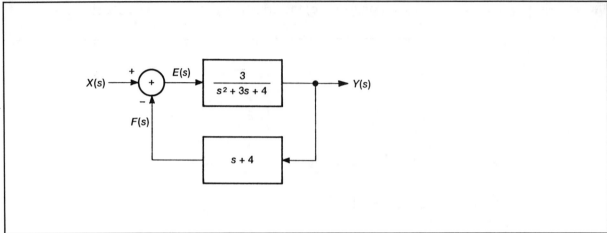

Figure 3.8 Type 0 error

Solution:

Error ratio = $E(s) / X(s) = 1 / [1 + G(s)H(s)]$

$E(s) = X(s) / [1 + G(s)H(s)]$

$$e_{ss} = \lim_{s \to 0} sE(s) = \frac{sX(s)}{1 + G(s)H(s)}$$

$X(s) = 1/s$ unit step

$$e_{ss} = \lim_{s \to 0} \frac{1}{1 + G(s)H(s)}$$

Let $k_p = \lim_{s \to 0} G(s)H(s)$

$$G(s)H(s) = \frac{3(s+4)}{s^2 + 3s + 4}$$

$$k_p = \lim_{s \to 0} \frac{3(s+4)}{s^2 + 3s + 4} = \frac{12}{4} = 3$$

$$e_{ss} = 1/(1+k_p) = 1/(1+3) = 1/4 = 0.25$$

3.9 Type 1 error

The block diagram shown in Figure 3.9 represents a type 1 feedback control system. Determine the steady state error, e_{ss}, for a unit ramp input.

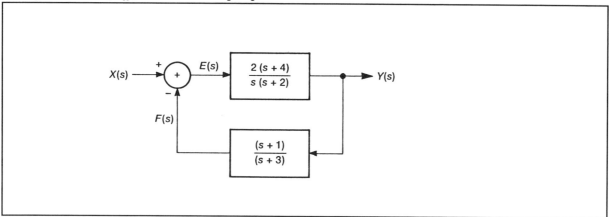

Figure 3.9 Type 1 error

Solution:

Error ratio = $E(s)/X(s) = 1/[1 + G(s)H(s)]$

$E(s) = X(s)/[1 + G(s)H(s)]$

$$e_{ss} = \lim_{s \to 0} sE(s) = \frac{sX(s)}{1 + G(s)H(s)}$$

$X(s) = 1/s^2$ unit ramp

$$e_{ss} = \lim_{s \to 0} \frac{1}{s[1 + G(s)H(s)]} = \lim_{s \to 0} \frac{1}{s[G(s)H(s)]}$$

Let $k_v = \lim_{s \to 0} sG(s)H(s)$

$$G(s)H(s) = \frac{2(s+4)(s+1)}{s(s+2)(s+3)}$$

$$k_v = \lim_{s \to 0} \frac{2(s+4)(s+1)}{(s+2)(s+3)} = \frac{8}{6}$$

$e_{ss} = 1/k_v = 6/8 = 0.75$

3.10 Type 2 error

The block diagram shown in Figure 3.10 represents a type 2 feedback control system. Determine the steady state error, e_{ss}, for a unit parabolic input.

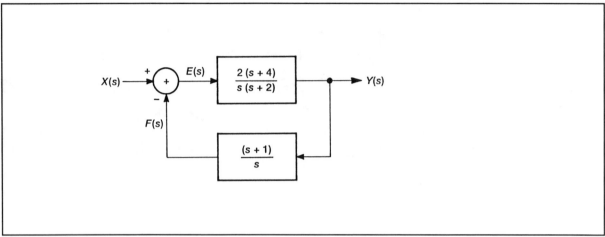

Figure 3.10 Type 2 error

Solution:

Error ratio = $E(s) / X(s) = 1 / [1 + G(s)H(s)]$

$E(s) = X(s) / [1 + G(s)H(s)]$

$$e_{ss} = \lim_{s \to 0} sE(s) = \frac{sX(s)}{1 + G(s)H(s)}$$

$X(s) = 1/s^3$ unit parabola

$$e_{ss} = \lim_{s \to 0} \frac{1}{s^2[1 + G(s)H(s)]} = \lim_{s \to 0} \frac{1}{s^2[G(s)H(s)]}$$

Let $k_a = \lim s^2 G(s)H(s)$

$$G(s)H(s) = \frac{2(s+4)(s+1)}{s^2(s+2)}$$

$$k_a = \lim_{s \to 0} \frac{2(s+4)(s+1)}{(s+2)} = \frac{8}{2} = 4$$

$e_{ss} = 1/k_a = 1/4 = 0.25$

3.11 Sensitivity

The block diagram shown in Figure 3.11 represents a feedback control system. Determine the system sensitivity with respect to G and H parameters. Assume a +10% variation in each case.

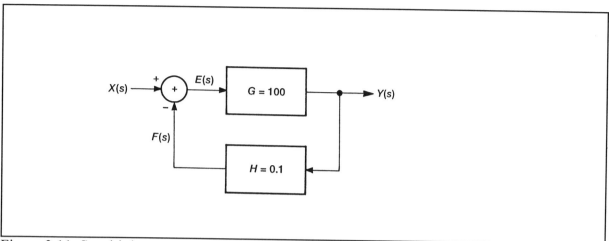

Figure 3.11 Sensitivity

Solution:

$$S_G = \frac{dT(G)}{dG}\left[\frac{G}{T(G)}\right]$$

$T(G) = G / [1 + GH]$

$S_G = [1/(1 + GH)^2][G/T(G)] = 1/[1 + GH] = 1/[1 + (100 \times 0.1)] = 1/[1+10] = 0.09$

$S_G (+10\%) = 1/[1 + (100 + 10) \times 0.1] = 1/[1 + 11] = 0.083$

Thus a 10% increase of G decreases S_G by 10%

$$S_H = \frac{dT(H)}{dH}\left[\frac{H}{T(H)}\right]$$

$T(H) = G / [1 + GH]$

$S_H = [-G/(1 + GH)][H/T(H)] = -H/[1 + GH] = -0.1/[1 + 10] = -0.009$

$S_H (+10\%) = -[0.1 + 0.01]/[1 + 11] = -0.01$

Thus a 10% increase of H increases S_H by 10%

3.12 Frequency domain

The block diagram shown in Figure 3.12 represents a frequency sensitive open-loop gain function. Find (a) the frequency domain magnitude and phase characteristics, and (b) the -3dB points.

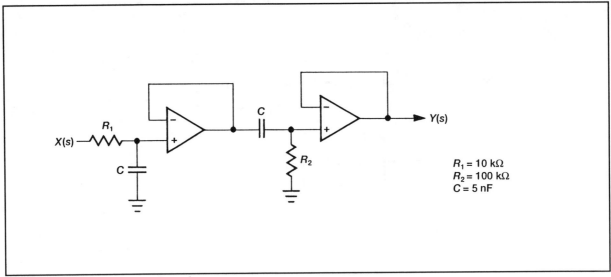

Figure 3.12 Frequency domain

Solution:

a) $\dfrac{Y(s)}{X(s)} = G_1(s)G_2(s) = \left(\dfrac{1}{R_1Cs + 1}\right)\left(\dfrac{R_2Cs}{R_2Cs + 1}\right)$ 　　let $s = j\omega$

$|G_1(j\omega)| = 1\angle 0°$, 　$\omega = 0$ 　　$|G_2(j\omega)| = 0\angle 90°$, 　$\omega = 0$

$|G_1(j\omega)| = (1/\sqrt{2})\angle -45°$, $\omega = 1/R_1C$ 　$|G_2(j\omega)| = (1/\sqrt{2})\angle 45°$, $\omega = 1/R_2C$

$|G_1(j\omega)| = 0\angle -90°$, 　$\omega = \infty$ 　　$|G_2(j\omega)| = 1\angle 0°$, 　$\omega = \infty$

b) $f_1(-3\text{dB}) = 1/(2\pi R_2 C) = 318.4$ Hz and

$f_2(-3\text{dB}) = 1/(2\pi R_1 C) = 3{,}184$ Hz

At $f_1 = 318.4$ Hz

$Y(s)/X(s) = |G_2(f_1)||G_1(f_1)| \angle (\theta_2° + \theta_1°) = [1/\sqrt{2} \times 0.995] \angle (45° - 5.7°)$

At $f_2 = 3{,}184$ Hz

$Y(s)/X(s) = |G_1(f_2)||G_2(f_2)| \angle (\theta_1° + \theta_2°) = [1/\sqrt{2} \times 0.995] \angle (-45° + 5.7°)$

3.13 Bode plot

Determine the Bode magnitude and phase values for the open-loop function, G(s)H(s), of Figure 3.13. Evaluate G(s)H(s) at ω = 0.1, 1, 5, and 10 respectively.

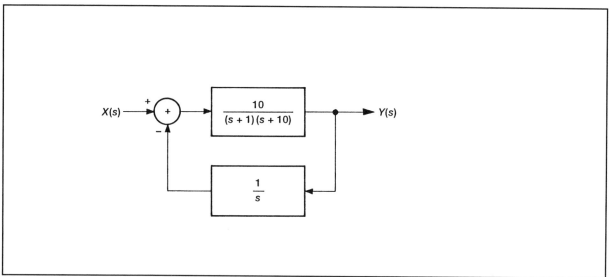

Figure 3.13 Bode plot

Solution:

The open-loop gain equals,

$$G(s)H(s) = \frac{10}{s(s+1)(s+10)} \qquad \text{let } s = j\omega$$

$$G(j\omega)H(j\omega) = \frac{10}{j\omega(j\omega+1)(j\omega+10)} = \frac{1}{j\omega(j\omega+1)(j\omega/10+1)}$$

$$\text{Mag(dB)} = 20\log|G(j\omega)H(j\omega)| = 20\log(1) - 20\log(\omega) - 20\log(\omega^2+1)^{1/2} - 20\log(\omega^2/100+1)^{1/2}$$

$$\theta° = \arg[G(j\omega)H(j\omega)] = 0° - 90° - \tan^{-1}(\omega) - \tan^{-1}(\omega/10)$$

ω	Mag(dB) = 20log |G(jω)H(jω)|	θ° = arg[G(jω)H(jω)]
0.1	0 + 20 + 0 + 0 = 20 dB	0° - 90° - 5.7° - 0.57° = -96.27°
1.0	0 + 0 - 3 + 0 = -3 dB	0° - 90° - 45° - 5.7° = -140.7°
5.0	0 - 14 - 14 - 1 = -29 dB	0° - 90° - 79° - 26° = -195°
10.0	0 - 20 - 20 - 3 = -43 dB	0° - 90° - 84° - 45° = -219°

3.14 Nichols chart

Determine the dB magnitude versus phase angle values of the closed-loop gain for Figure 3.14. The resulting plot is used with the Nichols chart to evaluate performance. Evaluate at ω = 0, 0.1, 0.5, 1, 2, 5, and 10 respectively.

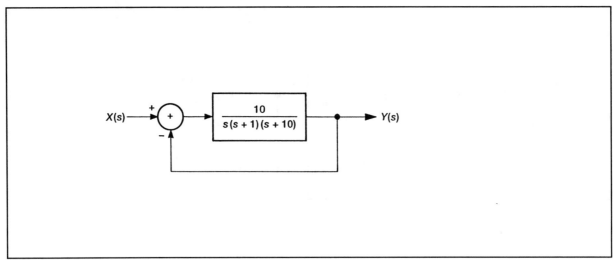

Figure 3.14 Nichols chart

Solution:

The closed-loop gain equals,

$$\frac{Y(s)}{X(s)} = \frac{G(s)}{G(s)H(s)+1} = \frac{10}{s(s+1)(s+10)+10} \qquad \text{let } s = j\omega$$

$$\frac{Y(j\omega)}{X(j\omega)} = \frac{10}{(j\omega)^3 + 11(j\omega)^2 + 10(j\omega) + 10} = \frac{10}{(10 - 11\omega^2) + j(10\omega - \omega^3)}$$

$$Mag(dB) = 20\log\left|\frac{Y(j\omega)}{X(j\omega)}\right| = 10\log\left|\frac{Y(j\omega)}{X(j\omega)}\right|^2 = 10\log\frac{100}{(10 - 11\omega^2)^2 + (10\omega - \omega^3)^2}$$

$$\theta° = \arg\left[\frac{Y(s)}{X(s)}\right] \tan^{-1}\left(\frac{10\omega - \omega^3}{10 - 11\omega^2}\right)$$

ω	0	0.1	0.5	1.0	2.0	5.0	10.0
Mag(dB)	0.0	0.05	0.125	0.861	-11.13	-28.79	-42.95
Phase (θ°)	0°	-5.76°	-33.71°	-96.34°	-160.6°	-195.8°	-219.5°

3.15 Nyquist plot

Determine the Nyquist imaginary versus real magnitude values of the open-loop gain for Figure 3.15. Evaluate at ω = 0.1, 1, 5, 10, and ∞ respectively.

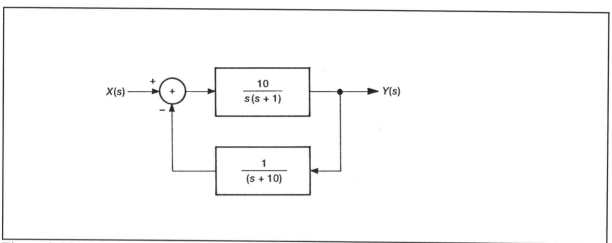

Figure 3.15 Nyquist plot

Solution:

The open-loop gain equals,

$$G(s)H(s) = \frac{10}{s(s+1)(s+10)} \qquad \text{let } s = j\omega$$

$$G(j\omega)H(j\omega) = \frac{10}{j\omega(j\omega+1)(j\omega+10)}$$

$$|G(j\omega)H(j\omega)| = \frac{10}{\omega[(\omega^2+1)(\omega^2+100)]^{1/2}}$$

$$\theta° = \arg[G(j\omega)H(j\omega)] = 0° - 90° - \tan^{-1}(\omega) - \tan^{-1}(\omega/10)$$

$$\text{Real}[G(j\omega)H(j\omega)] = |G(j\omega)H(j\omega)|\cos\theta$$

$$\text{Imag}[G(j\omega)H(j\omega)] = |(j\omega)(j\omega)|\sin\theta$$

ω	0.1	1.0	5.0	10.0	∞
|G(jω)H(jω)|	9.95	0.703	0.035	0.007	0
Phase (θ°)	-96.3°	-141°	-195°	-219°	-270°
Real	-1.07	-0.53	-0.033	-0.0054	0
Imag	-9.89	-0.45	0.009	0.0044	0

3.16 Stability

Determine the stability of the control system block diagram shown in Figure 3.16 by evaluating the roots of its characteristic equation.

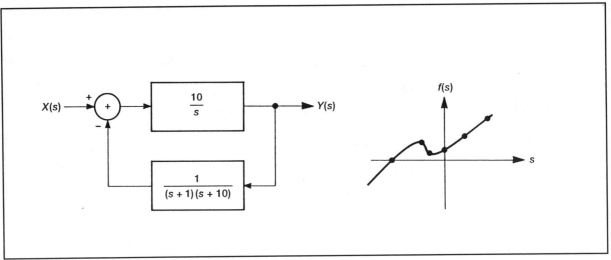

Figure 3.16 Stability

Solution:

$$\frac{Y(s)}{X(s)} = \frac{10(s+1)(s+10)}{s(s+1)(s+10)+10} = \frac{(s+1)(s+10)}{0.1s(s+1)(s+10)+1}$$

$f(s) = (0.1s)(s+1)(s+10) + 1 = 0.1s^3 + 1.1s^2 + s + 1 = 0$

$f'(s) = 0.3s^2 + 2.2s + 1 = 0$

$f''(s) = 0.6s + 2.2 = 0$

$$f'(s_1, s_2) = \frac{-b \pm \sqrt{b^2 - 4ac}}{2a}$$

$f'(s_1, s_2) = (-2.2 \pm 1.9) / 0.6 = -0.5, -6.83$

$f''(-0.5) = 1.9$ (mimimum)
$f''(-6.83) = -1.8$ (maximum)

s	-15	-10.11	-6.83	-0.5	0	1	10
f(s)	-104.5	0	13.62	0.7625	1	3.2	221

Since f(s) has only one root and it is negative, the control system is stable.

3.17 Gain and phase margins

Determine (a) the gain and (b) phase margins for the control system block diagram of Figure 3.17.

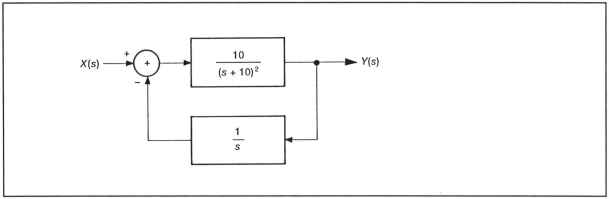

Figure 3.17 Gain and phase margins

Solution:

a) Gain margin = $1/(|G(j\omega)H(j\omega)|)$ when $\arg[G(j\omega)H(j\omega)] = -180°$

$$G(s)H(s) = \frac{10}{s(s+10)^2} \quad \text{let } s = j\omega$$

$$G(j\omega)H(j\omega) = \frac{10}{j\omega(j\omega+10)^2}$$

$$|G(j\omega)H(j\omega)| = \frac{10}{\omega(\omega^2+100)}$$

$\theta° = \arg[G(j\omega)H(j\omega)] = -180° = [0° - 90° - 2\tan^{-1}(\omega/10)]; \quad \tan^{-1}(\omega/10) = 45°$

$(\omega/10) = \tan 45°$ and $\omega = 10$

$$|G(j\omega)H(j\omega)| = \frac{10}{10[(10)^2+100]} = \frac{10}{2,000} = \frac{1}{200}$$

Gain margin = $1/(|G(j\omega)H(j\omega)|) = 200$

b) Phase margin = $180° + \arg[G(j\omega)H(j\omega)]$ when $|G(j\omega)H(j\omega)| = 1$

$$|G(j\omega)H(j\omega)| = \frac{10}{\omega(\omega^2+100)} = 1$$

$10 = \omega(\omega^2+100)$ or $\omega^3 + 100\omega - 10 = 0$ and $\omega = 0.01$

Phase margin = $180° - 90° - 2\tan^{-1}(0.01/10) = 180° - 90° - 0.11° = 89.89°$

3.18 Lead compensation

Determine the Bode plot for the open-loop function of Figure 3.18. Evaluate at $\omega = 0, 0.1, 1, 5, 10,$ and 100 respectively.

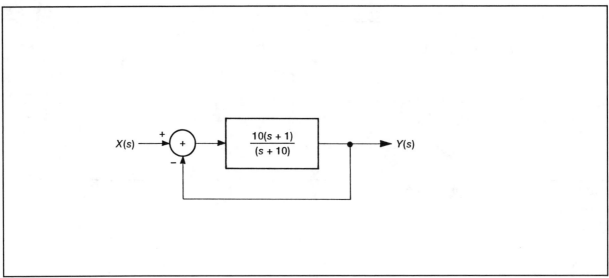

Figure 3.18 Lead compensation

Solution:

$$G(s)H(s) = \frac{10(s+1)}{(s+10)} \qquad \text{let } s = j\omega$$

$$G(j\omega)H(j\omega) = \frac{10(j\omega+1)}{(j\omega+10)} = \frac{(j\omega+1)}{(j\omega/10+1)}$$

$$\text{Mag(dB)} = 20\log(\omega^2+1)^{1/2} - 20\log(\omega^2/100+1)^{1/2}$$

$$\theta° = \arg[G(j\omega)H(j\omega)] = \tan^{-1}(\omega) - \tan^{-1}(\omega/10)$$

ω	0	0.1	1.0	5.0	10.0	100.0
Mag(dB)	0	0.042	0.29	13.18	17.04	20.0
$\theta°$	0°	0.513°	39.28°	52.19°	39.28°	5.13°

3.19 Lag compensation

Determine the Bode plot for the open-loop function of Figure 3.19. Evaluate at ω = 0, 0.1, 1, 5, 10, and 100 respectively.

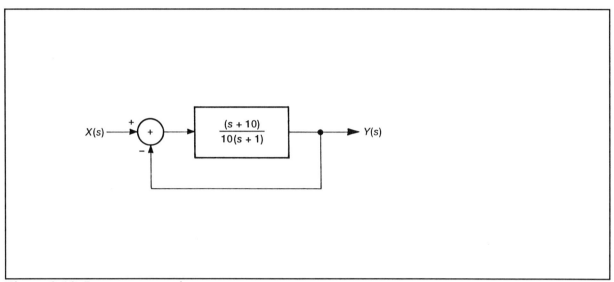

Figure 3.19 Lag compensation

Solution:

$$G(s)H(s) = \frac{(s+10)}{10(s+1)} \qquad \text{let } s = j\omega$$

$$G(j\omega)H(j\omega) = \frac{(j\omega+10)}{10(j\omega+1)} = \frac{(j\omega/10+1)}{(j\omega+1)}$$

$$\text{Mag(dB)} = 20\log(\omega^2/100+1)^{1/2} - 20\log(\omega^2+1)^{1/2}$$

$$\theta° = \arg[G(j\omega)H(j\omega)] = \tan^{-1}(\omega/10) - \tan^{-1}(\omega)$$

ω	0	0.1	1.0	5.0	10.0	100.0
Mag(dB)	0	-0.042	-0.29	-13.18	-17.04	-20.0
$\theta°$	0°	-0.513°	-39.28°	-52.19°	-39.28°	-5.13°

3.20 Lead-lag compensation

Determine the Bode plot for the open-loop function of Figure 3.20. Evaluate at $\omega = 0, 0.1, 1, 5, 10,$ and 100 respectively.

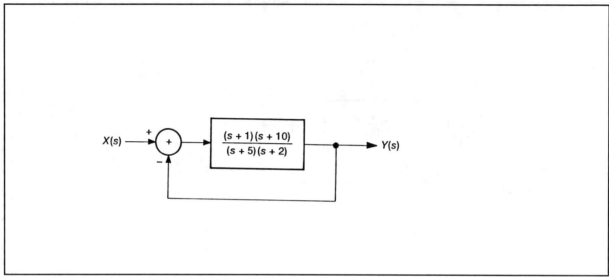

Figure 3.20 Lead-lag compensation

Solution:

$$G(s)H(s) = \frac{(s+1)(s+10)}{(s+5)(s+2)} \quad \text{let } s = j\omega$$

$$G(j\omega)H(j\omega) = \frac{(j\omega+1)(j\omega+10)}{(j\omega+5)(j\omega+2)} = \frac{(j\omega+1)(j\omega/10+1)}{(j\omega/5+1)(j\omega/2+1)}$$

$$\text{Mag(dB)} = 20\log(\omega^2+1)^{\frac{1}{2}} + 20\log(\omega^2/100+1)^{\frac{1}{2}} - 20\log(\omega^2/25+1)^{\frac{1}{2}} - 20\log(\omega^2/4+1)^{\frac{1}{2}}$$

$$\theta° = \arg[G(j\omega)H(j\omega)] = \tan^{-1}(\omega) + \tan^{-1}(\omega/10) - \tan^{-1}(\omega/5) - \tan^{-1}(\omega/2)$$

ω	0	0.1	1.0	2.0	5.0	10.0	100.0
Mag(dB)	0	0.03	1.91	3.51	3.51	1.91	0
$\theta°$	0°	2.28°	12.84°	7.92°	-7.92°	-12.84°	-2.28°

3.21 Time domain

The output response of a given control system to a unit step input function is expressed by, $y(t) = 1 - e^{-\zeta \omega_n t}[\cos\omega t + (\zeta/(1-\zeta^2)^{1/2})\sin\omega t]$. Determine the resulting (a) time delay, t_d, (b) rise time, t_r, (c) peak time, t_p, (d) maximum overshoot, m_p, and (e) settling time, t_s, for a 5% of steady state value. Let $\omega_n = 377$, and $\zeta = 1/2$.

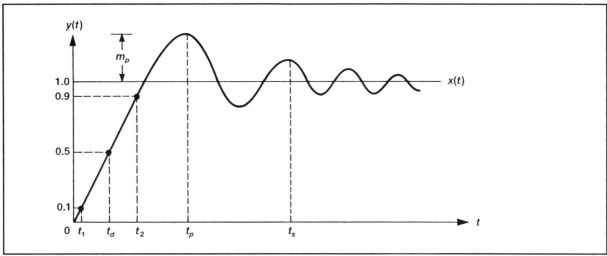

Figure 3.21 Time domain

Solution:

$$\omega = \omega_n(1-\zeta^2)^{1/2} = 377(0.75)^{1/2} = 326.5$$

$$\kappa = \zeta/(1-\zeta^2)^{1/2} = 0.5/(0.75)^{1/2} = 0.57$$

a) At $y(t) = 0.5 = 1 - e^{-\zeta \omega_n t_d}$

$$0.5 = \frac{1}{e^{\zeta \omega_n t_d}} \quad \text{or} \quad 2 = e^{\zeta \omega_n t_d}$$

$$t_d = \ln(2)/(\zeta \omega_n) = 3.67 \times 10^{-3} \quad \text{seconds}$$

b) $\theta = \tan^{-1}(1/\kappa) = \tan^{-1}(1.73) = 1.047$ radians

$$t_r = t_2 - t_1 = 1/\omega[\tan^{-1}(-1/\kappa)] = (\pi - \theta)/\omega = 6.41 \times 10^{-3} \quad \text{seconds}$$

c) $t_p = (\pi/\omega) = (3.1415/326.5) = 9.62 \times 10^{-3}$ seconds

d) $m_p = e^{-\pi(\kappa)} = e^{-3.4145(0.57)} = 0.163$

e) $t_s(5\%) = 3/(\zeta \omega_n) = 6/(377) = 15.9 \times 10^{-3}$ seconds

3.22 Complex plane

The closed-loop transfer function of a feedback control system is given as

$$\frac{Y(s)}{X(s)} = \frac{2s^2 + 8s + 6}{s(s^2 + 4s + 5)}$$

Draw the resulting pole-zero pattern on the complex plane.

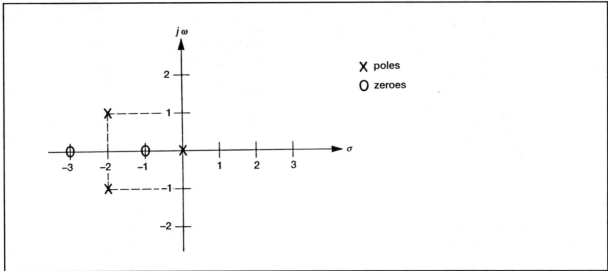

Figure 3.22 Complex plane

Solution:

$s = \sigma + j\omega$

a) Zeroes: $2s^2 + 8s + 6 = 0$

$$s_1, s_2 = \frac{-b \pm \sqrt{b^2 - 4ac}}{2a} = \frac{-8 \pm \sqrt{64 - 48}}{4}$$

$s_1, s_2 = -2 \pm 1 = -1, -3$

b) Poles: $s(s^2 + 4s + 5) = 0$

$$s_1, s_2 = \frac{-b \pm \sqrt{b^2 - 4ac}}{2a} = \frac{-4 \pm \sqrt{16 - 20}}{2}$$

$s_1, s_2 = -2 \pm j = -2 + j, -2 - j$

$s_3 = 0$

3.23 Partial fractions

Evaluate the unit impulse function response for the control system block diagram in Figure 3.23.

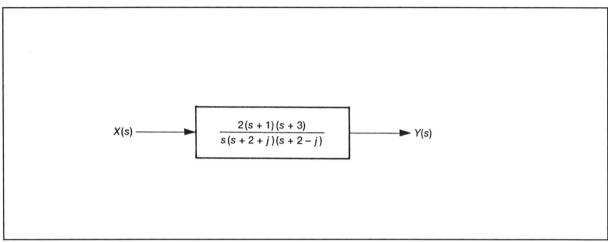

Figure 3.23 Partial fractions

Solution:

$X(s) = 1; \quad s = \sigma + j\omega$

$$Y(s) = \frac{2(s+1)(s+3)}{s(s+2+j)(s+2-j)} X(s) = \frac{k_1}{s} + \frac{k_2}{s+2+j} + \frac{k_3}{s+2-j}$$

$$k_1 = \frac{2(1)(3)}{(2+j)(2-j)} \bigg|_{s=0} = \frac{6}{5} = 1.2$$

$$k_2 = \frac{2(-2-j+1)(-2-j+3)}{(-2-j)(-2-j+2-j)} \bigg|_{s=-2-j} = \frac{-4}{-2+4j} = \frac{4}{10}(1+2j)$$

$$k_3 = \frac{2(-2+j+1)(-2+j+3)}{(-2+j)(-2+j+2+j)} \bigg|_{s=-2+j} = \frac{-4}{-2-4j} = \frac{4}{10}(1-2j)$$

$y(t) = 1.2 + 2|M|e^{-\sigma t}\cos(\omega t + \theta°)$

$|M| = (4/10)[1 + (2)^2]^{½} = (4/10)\sqrt{5} = 0.894$

$\theta° = \tan^{-1}(2/1) = 63.43°$

$s = \sigma + j\omega = -2 \pm j$

$y(t) = 1.2 + (1.788)e^{-2t}[\cos(t + 63.43°)]$

3.24 Transfer function

Determine the transfer function of the circuit in Figure 3.24 in terms of its passive elements.

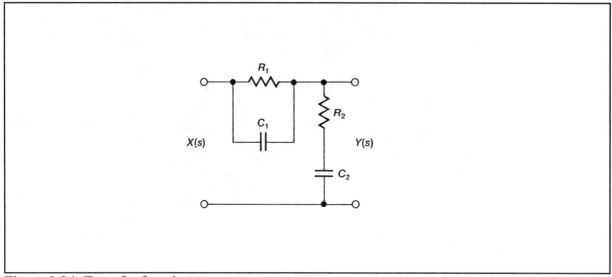

Figure 3.24 Transfer function

Solution:

$$Y(s) = X(s)\left(\frac{Z_2}{Z_1 + Z_2}\right)$$

$$\frac{Y(s)}{X(s)} = \frac{Z_2}{Z_1 + Z_2}$$

$$Z_1 = \frac{1}{\frac{1}{R_1} + C_1 s} = \frac{R_1}{R_1 C_1 s + 1}$$

$$Z_2 = R_2 + \frac{1}{C_2 s} = \frac{R_2 C_2 s + 1}{C_2 s}$$

$$Z_1 + Z_2 = \frac{(R_1 C_1 s + 1)(R_2 C_2 s + 1) + R_1 C_2 s}{(R_1 C_1 s + 1)(C_2 s)}$$

$$\frac{Z_2}{Z_1 + Z_2} = \frac{(R_2 C_2 s + 1)(R_1 C_1 s + 1)(C_2 s)}{[(R_1 C_1 s + 1)(R_2 C_2 s + 1) + R_1 C_2 s](C_2 s)}$$

$$\frac{Y(s)}{X(s)} = \frac{(R_2 C_2 s + 1)(R_1 C_1 s + 1)}{R_1 C_1 R_2 C_2 s^2 + (R_1 C_1 + R_2 C_2 + R_1 C_2)s + 1}$$

3.25 Ruth-Hurwitz

Determine the range of values for K in the control system block diagram of Figure 3.25 which results in a stable condition. And find the auxiliary equation.

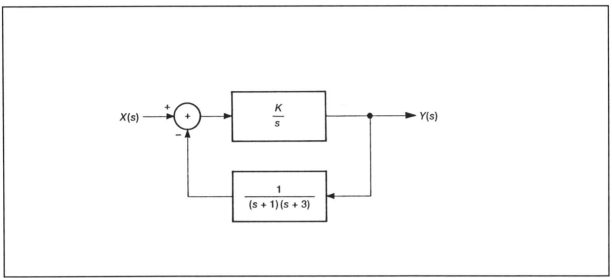

Figure 3.25 Ruth-Hurwitz

Solution:

$$\frac{Y(s)}{X(s)} = \frac{G(s)}{1 + G(s)H(s)} = \frac{K(s+1)(s+3)}{K + s(s+1)(s+3)}$$

$$s(s+1)(s+3) + K = s(s^2 + 4s + 3) + K = 0$$

$$s^3 + 4s^2 + 3s + K = a_3 s^3 + a_2 s^2 + a_1 s + a_0 = 0$$

s^3	a_3	a_1	1	3
s^2	a_2	a_0	4	K
s^1	$(a_2 a_1 - a_3 a_0)/a_2 = b_1$	b_2	$(12-K)/4$	0
s^0	$(b_1 a_0 - a_2 b_2)/b_1 = c_1$	c_2	K	0

Hence

$0 < K < 12$ will be stable

$K > 12$ will be unstable

$K = 12$ will oscillate

The auxiliary equation is $a_2 s^2 + a_0 = 4s^2 + 12 = 0$, $s = \pm j\sqrt{3}$

3.26 Root locus

Determine the root locus plot of a closed-loop control system whose transfer function is defined as,

$$\frac{Y(s)}{X(s)} = \frac{G(s)}{1 + G(s)H(s)} = \frac{K(s + 1)}{s(s^2 + 4s + 8)}$$

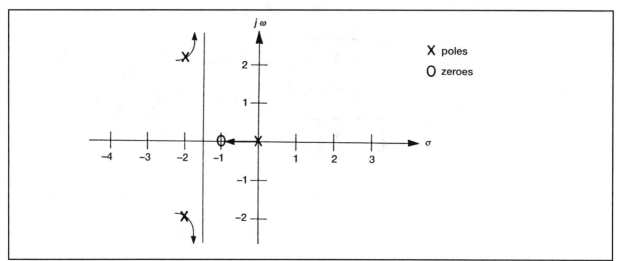

Figure 3.26 Root locus

Solution:

a) Zeroes; $(s + 1) = 0$; $s = -1$

b) Poles; $s(s^2 + 4s + 8) = 0$

$$s_1, s_2 = \frac{-b \pm \sqrt{b^2 - 4ac}}{2a} = \frac{-4 \pm \sqrt{16 - 32}}{2}$$

$s_1 = -2 + 2j$, $s_2 = -2 - 2j$

$s_3 = 0$

c) Number of branches = number of poles = 3

d) Real axis locus; $-1 < \sigma < 0$

e) Center of Asymptotes = $[\sum p_i - \sum z_i] / [n - m] = [(0-2-2) - (-1)] / [3-1] = -1.5$

f) Asymptotic angles = $\pm (q \times 180°)/(n - m) = \pm (q \times 180°)/2 = \pm 90°$ $q = 1, 3, 5, ...$

3.27 PD controller
Determine the output response of the control system shown in Figure 3.27 for a unit step input. Let $K = 1 \times 10^5$ and $T_D = 0.1$.

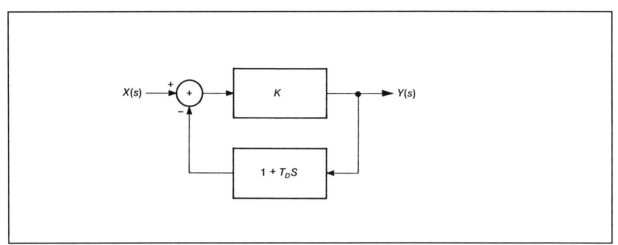

Figure 3.27 PD controller

Solution:

$$\frac{Y(s)}{X(s)} = \frac{G(s)}{1 + G(s)H(s)} = \frac{1}{\frac{1}{G(s)} + H(s)}$$

For $G(s) \gg 1$

$$\frac{Y(s)}{X(s)} = \frac{1}{H(s)} = \frac{1}{1 + T_D s} = \frac{1}{1 + 0.1s} = \frac{10}{s + 10}$$

$X(s) = 1/s$ unit step

$$Y(s) = \frac{X(s)}{H(s)} = \left(\frac{1}{s}\right)\frac{10}{s + 10} = \frac{k_1}{s} + \frac{k_2}{s + 10}$$

$$k_1 = \frac{10}{(s + 10)}\bigg|_{s=0} = 1$$

$$k_2 = \frac{10}{s}\bigg|_{s=-10} = -1$$

$$Y(s) = \frac{1}{s} + \frac{-1}{s + 10}$$

$y(t) = 1 - 1 \times e^{-10t}$

3.28 PI controller

Determine the output response of the control system shown in Figure 3.28 for a unit step input. Let $K = 1\times10^5$ and $T_I = 10$.

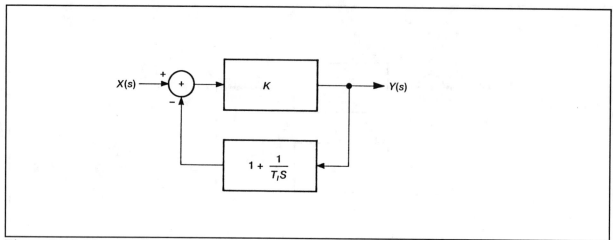

Figure 3.28 PI controller

Solution:

$$\frac{Y(s)}{X(s)} = \frac{G(s)}{1 + G(s)H(s)} = \frac{1}{\dfrac{1}{G(s)} + H(s)}$$

For $G(s) \gg 1$

$$\frac{Y(s)}{X(s)} = \frac{1}{H(s)} = \frac{T_I s}{T_I s + 1}$$

$X(s) = 1/s$ unit step

$$Y(s) = \frac{X(s)}{H(s)} = \left(\frac{1}{s}\right)\frac{T_I s}{T_I s + 1} = \frac{T_I}{T_I s + 1}$$

$$Y(s) = \frac{10}{10s + 1} = \frac{1}{s + 0.1}$$

$y(t) = 1 \times e^{-0.1t}$

3.29 PID controller

Determine the output response of the control system shown in Figure 3.29 for a unit step input. Let $K = 1\times 10^5$, $T_I = 10$, and $T_D = 0.1$.

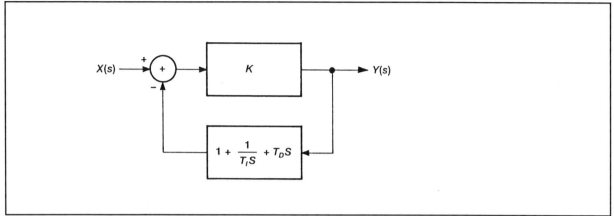

Figure 3.29 PID controller

Solution:

$$\frac{Y(s)}{X(s)} = \frac{G(s)}{1 + G(s)H(s)} = \frac{1}{\frac{1}{G(s)} + H(s)}$$

For $G(s) \gg 1$

$$\frac{Y(s)}{X(s)} = \frac{1}{H(s)} = \frac{T_I s}{1 + T_I s + T_I T_D s^2}$$

$X(s) = 1/s$ unit step

$$Y(s) = \frac{X(s)}{H(s)} = (\frac{1}{s})\frac{T_I s}{T_I T_D s^2 + T_I s + 1} = \frac{T_I}{T_I T_D s^2 + T_I s + 1} = \frac{10}{s^2 + 10s + 1}$$

$$s_1, s_2 = \frac{-b \pm \sqrt{b^2 - 4ac}}{2a} = \frac{-10 \pm \sqrt{100 - 4}}{2} = -0.1, -9.9$$

$$Y(s) = \frac{10}{s^2 + 10s + 1} = \frac{k_1}{s + 0.1} + \frac{k_2}{s + 9.9}$$

$$k_1 = \frac{10}{(s + 9.9)}\bigg|_{s=-0.1} = 1.02$$

$$k_2 = \frac{10}{(s + 0.1)}\bigg|_{s=-9.9} = -1.02$$

$y(t) = 1.02\times e^{-0.1t} - 1.02\times e^{-9.9t}$

3.30 State space model

Determine the state space representation of a linear system defined by

$$\frac{d^2y}{dt^2} + 5\frac{dy}{dt} + 6y = u(t)$$

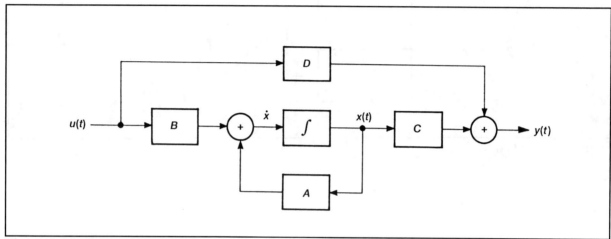

Figure 3.30 State space model

Solution:

$$\dot{x} = Ax(t) + Bu(t)$$
$$y(t) = Cx(t) + Du(t)$$

$$\ddot{y} + 5\dot{y} + 6y = u(t)$$

Let
$$x_1 = y, \quad x_2 = \dot{y}, \quad x_3 = \ddot{y}$$
$$\dot{x}_1 = x_2$$
$$\dot{x}_2 = x_3$$

Substituting,
$$x_3 + 5x_2 + 6x_1 = u(t)$$
$$\dot{x}_2 = x_3 = -6x_1 - 5x_2 + u(t)$$

$$\begin{bmatrix} \dot{x}_1 \\ \dot{x}_2 \end{bmatrix} = \begin{bmatrix} 0 & 1 \\ -6 & -5 \end{bmatrix} \begin{bmatrix} x_1 \\ x_2 \end{bmatrix} + \begin{bmatrix} 0 \\ 1 \end{bmatrix} u(t)$$

$$y(t) = \begin{bmatrix} 1 & 0 \end{bmatrix} \begin{bmatrix} x_1 \\ x_2 \end{bmatrix} + \begin{bmatrix} 0 \end{bmatrix} u(t)$$

3.31 State variable controller
Determine the signal flow graph and derive the corresponding state variable matrix equations for the closed-loop function

$$\frac{Y(s)}{U(s)} = \frac{(s+4)}{(s^2 + 5s + 6)}$$

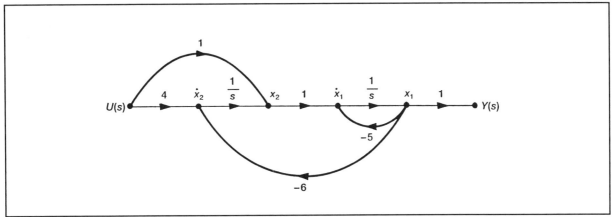

Figure 3.31 State variable controller

Solution:

$$\frac{Y(s)}{U(s)} = \frac{(s+4)}{(s^2 + 5s + 6)}$$

$$(\frac{d^2}{dt^2} + 5\frac{d}{dt} + 6)y(t) = (\frac{d}{dt} + 4)u(t)$$

$$\ddot{y} + 5\dot{y} + 6y = \dot{u} + 4u$$

Let
$x_1 = y, \quad x_2 = \dot{y}, \quad x_3 = \ddot{y}$
$\dot{x}_1 = x_2$
$\dot{x}_2 = x_3$
$\dot{x}_1 = -5x_1 + x_2 + u(t)$
$\dot{x}_2 = -6x_1 + 4u(t)$

$$\begin{vmatrix} \dot{x}_1 \\ \dot{x}_2 \end{vmatrix} = \begin{bmatrix} -5 & 1 \\ -6 & 0 \end{bmatrix} \begin{vmatrix} x_1 \\ x_2 \end{vmatrix} + \begin{vmatrix} 1 \\ 4 \end{vmatrix} u(t)$$

$$y(t) = \begin{bmatrix} 1 & 0 \end{bmatrix} \begin{vmatrix} x_1 \\ x_2 \end{vmatrix} + \begin{bmatrix} 0 \end{bmatrix} u(t)$$

3.32 Eigenvalues and eigenvectors

Determine the eigenvalues and eigenvectors, and plot the modal matrix for the state variable control system defined by

$$\begin{bmatrix} \dot{x}_1 \\ \dot{x}_2 \end{bmatrix} = \begin{bmatrix} 0 & -2 \\ -3 & -5 \end{bmatrix} \begin{bmatrix} x_1 \\ x_2 \end{bmatrix} + \begin{bmatrix} 1 \\ 1 \end{bmatrix} u(t)$$

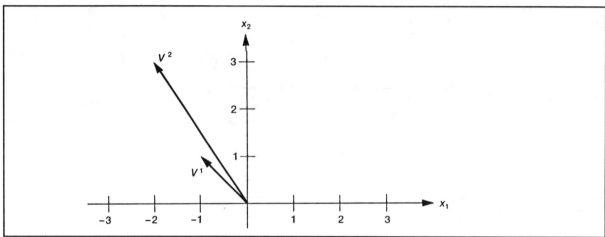

Figure 3.32 Eigenvalues and eigenvectors

Solution:

$$\begin{bmatrix} \dot{x}_1 \\ \dot{x}_2 \end{bmatrix} = \begin{bmatrix} 0 & -2 \\ -3 & -5 \end{bmatrix} \begin{bmatrix} x_1 \\ x_2 \end{bmatrix} + \begin{bmatrix} 1 \\ 1 \end{bmatrix} u(t)$$

$$|\lambda I - A| = \begin{vmatrix} \lambda & -2 \\ -3 & \lambda + 5 \end{vmatrix} = \lambda^2 + 5\lambda + 6 = 0$$

$(\lambda + 2)(\lambda + 3) = 0; \qquad \lambda_1, \lambda_2 = -2, -3$

$|\lambda_i I - A| V^i = 0$

$$\left(\begin{bmatrix} -2 & 0 \\ 0 & -2 \end{bmatrix} - \begin{bmatrix} 0 & 2 \\ -3 & -5 \end{bmatrix} \right) \begin{bmatrix} v_1 \\ v_2 \end{bmatrix} = \begin{bmatrix} -2 & -2 \\ 3 & 3 \end{bmatrix} \begin{bmatrix} v_1 \\ v_2 \end{bmatrix} = 0 \qquad \text{Let } v_2 = 1, \quad v_1 = -v_2 = -1$$

$$\left(\begin{bmatrix} -3 & 0 \\ 0 & -3 \end{bmatrix} - \begin{bmatrix} 0 & 2 \\ -3 & -5 \end{bmatrix} \right) \begin{bmatrix} v_1 \\ v_2 \end{bmatrix} = \begin{bmatrix} -3 & -2 \\ 3 & 2 \end{bmatrix} \begin{bmatrix} v_1 \\ v_2 \end{bmatrix} = 0 \qquad \text{Let } v_2 = 3, \quad v_1 = -(2/3) v_2 = -2$$

$$M = \begin{bmatrix} V^1 & V^2 \end{bmatrix} = \begin{bmatrix} -1 & -2 \\ 1 & 3 \end{bmatrix}$$

3.33 Diagonal matrix

Find the flowgraph of the modal domain equivalent for the state variable control system defined by

$$\begin{vmatrix} \dot{x}_1 \\ \dot{x}_2 \end{vmatrix} = \begin{bmatrix} 0 & 2 \\ -3 & -5 \end{bmatrix} \begin{vmatrix} x_1 \\ x_2 \end{vmatrix} + \begin{vmatrix} 1 \\ 1 \end{vmatrix} u(t) \quad \text{and} \quad y(t) = \begin{bmatrix} 1 & 2 \end{bmatrix} \begin{vmatrix} x_1 \\ x_2 \end{vmatrix}$$

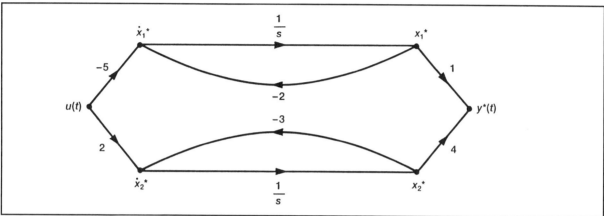

Figure 3.33 Diagonal matrix

Solution:

From 3.32 $M = \begin{bmatrix} V^1 & V^2 \end{bmatrix} = \begin{bmatrix} -1 & -2 \\ 1 & 3 \end{bmatrix}$ $\quad Adj\ M = |A_{ij}|^T = \begin{bmatrix} 3 & -1 \\ 2 & -1 \end{bmatrix}^T = \begin{bmatrix} 3 & 2 \\ -1 & -1 \end{bmatrix}$

$M^{-1} = \dfrac{Adj\ M}{\Delta} = \dfrac{\begin{bmatrix} 3 & 2 \\ -1 & -1 \end{bmatrix}}{(3)(-1) - (2)(-1)} = \begin{bmatrix} -3 & -2 \\ 1 & 1 \end{bmatrix}$ $\quad MM^{-1} = I = \begin{bmatrix} -1 & -2 \\ 1 & 3 \end{bmatrix}\begin{bmatrix} -3 & -2 \\ 1 & 1 \end{bmatrix} = \begin{bmatrix} 1 & 0 \\ 0 & 1 \end{bmatrix}$

$M^{-1}AM = \Lambda = \begin{bmatrix} -3 & -2 \\ 1 & 1 \end{bmatrix}\begin{bmatrix} 0 & 2 \\ -3 & -5 \end{bmatrix}\begin{bmatrix} -1 & -2 \\ 1 & 3 \end{bmatrix} = \begin{bmatrix} -2 & 0 \\ 0 & -3 \end{bmatrix}$

$M^{-1}B = \begin{bmatrix} -3 & -2 \\ 1 & 1 \end{bmatrix}\begin{bmatrix} 1 \\ 1 \end{bmatrix} = \begin{bmatrix} -5 \\ 2 \end{bmatrix}$ $\quad CM = \begin{bmatrix} 1 & 2 \end{bmatrix}\begin{bmatrix} -1 & -2 \\ 1 & 3 \end{bmatrix} = \begin{bmatrix} 1 & 4 \end{bmatrix}$

$\dot{x}^* = M^{-1}AMx^* + M^{-1}Bu(t) = \begin{bmatrix} -2 & 0 \\ 0 & -3 \end{bmatrix}\begin{vmatrix} x_1^* \\ x_2^* \end{vmatrix} + \begin{vmatrix} -5 \\ 2 \end{vmatrix} u(t)$

$y^*(t) = CMx^* = \begin{bmatrix} 1 & 4 \end{bmatrix}\begin{vmatrix} x_1^* \\ x_2^* \end{vmatrix} = x_1^* + 4x_2^*$

3.34 Transition matrix

Determine the natural response of a state variable control system, u(t) = 0, whose **A** matrix equals

$\begin{bmatrix} 0 & 2 \\ -3 & -5 \end{bmatrix}$ and an initial state of $x_0 = \begin{bmatrix} -2 \\ 3 \end{bmatrix}$

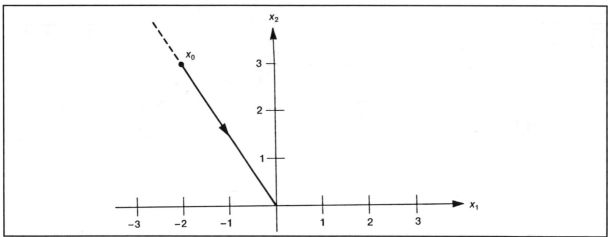

Figure 3.34 Transition matrix

Solution:

$\Phi(t) = e^{At}$

$A = \begin{bmatrix} 0 & 2 \\ -3 & -5 \end{bmatrix}$ and $\lambda_1, \lambda_2 = -2, -3$

$\Phi(t) = a_0 I + a_1 A = \begin{bmatrix} a_0 & 0 \\ 0 & a_0 \end{bmatrix} + \begin{bmatrix} 0 & 2a_1 \\ -3a_1 & -5a_1 \end{bmatrix} = \begin{bmatrix} a_0 & 2a_1 \\ -3a_1 & a_0 - 5a_1 \end{bmatrix}$

Using the Hamilton-Cayley remainder theorem,

$e^{\lambda t} = a_0 + a_1 \lambda;$ $\quad e^{-2t} = a_0 - 2a_1,$ $\quad e^{-3t} = a_0 - 3a_1$

Solving for a_0, a_1; $\quad a_0 = 3e^{-2t} - 2e^{-3t},$ $\quad a_1 = e^{-2t} - e^{-3t}$

$\Phi(t) = \begin{bmatrix} 3e^{-2t} - 2e^{-3t} & 2e^{-2t} - 2e^{-3t} \\ -3e^{-2t} + 3e^{-3t} & -2e^{-2t} + 3e^{-3t} \end{bmatrix}$

$x(t) = \Phi(t)x_0 = \Phi(t)\begin{vmatrix} -2 \\ 3 \end{vmatrix} = \begin{bmatrix} -2e^{-3t} \\ 3e^{-3t} \end{bmatrix}$

And $x_1 / x_2 = -2/3$ is an eigenvector

3.35 Controllability

A state variable control system is defined as

$$\begin{bmatrix} \dot{x}_1 \\ \dot{x}_2 \end{bmatrix} = \begin{bmatrix} 0 & 2 \\ -3 & -5 \end{bmatrix} \begin{bmatrix} x_1 \\ x_2 \end{bmatrix} + \begin{bmatrix} b_1 \\ b_2 \end{bmatrix} u(t) \quad \text{and} \quad y(t) = \begin{bmatrix} 1 & 1 \end{bmatrix} \begin{bmatrix} x_1 \\ x_2 \end{bmatrix}$$

Determine the conditions under which it is uncontrollable.

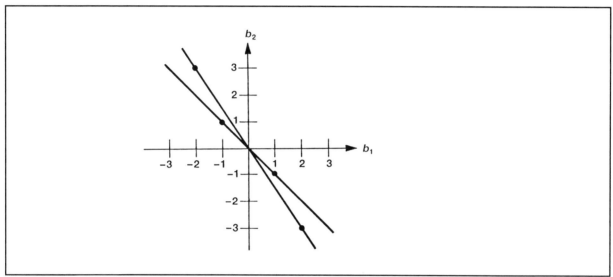

Figure 3.35 Controllability

Solution:

The criterion for controllability is $|\mathbf{B} : \mathbf{AB}| \neq 0$

$$AB = \begin{bmatrix} 0 & 2 \\ -3 & -5 \end{bmatrix} \begin{bmatrix} b_1 \\ b_2 \end{bmatrix} = \begin{bmatrix} 2b_2 \\ -3b_1 - 5b_2 \end{bmatrix}$$

$$[B : AB] = \begin{bmatrix} b_1 & 2b_2 \\ b_2 & -3b_1 - 5b_2 \end{bmatrix} = \Delta = -3b_1^2 - 5b_1 b_2 - 2b_2^2 \neq 0$$

$-(3b_1^2 + 5b_1 b_2 + 2b_2^2) = 0$
$-(3b_1 + 2b_2)(b_1 + b_2) = 0$

Uncontrollable when

$b_1 = -(2/3)b_2$ or

$b_1 = -b_2$

3.36 Observability
A state variable control system is defined as

$$\begin{vmatrix} \dot{x}_1 \\ \dot{x}_2 \end{vmatrix} = \begin{bmatrix} 0 & 2 \\ -3 & -5 \end{bmatrix} \begin{vmatrix} x_1 \\ x_2 \end{vmatrix} + \begin{vmatrix} 1 \\ 1 \end{vmatrix} u(t) \quad \text{and} \quad y(t) = \begin{bmatrix} c_1 & c_2 \end{bmatrix} \begin{vmatrix} x_1 \\ x_2 \end{vmatrix}$$

Determine the conditions under which it is unobservable.

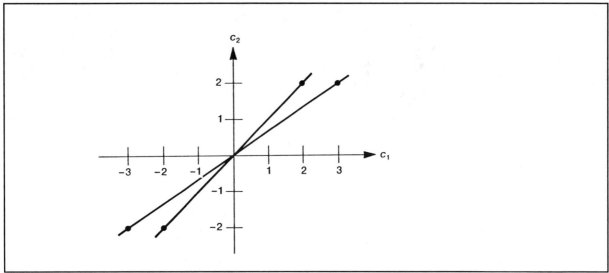

Figure 3.36 Observability

Solution:

The criterion for observability is $|\mathbf{C}^T : \mathbf{A}^T \mathbf{C}^T| \neq 0$

$$A^T C^T = \begin{bmatrix} 0 & -3 \\ 2 & -5 \end{bmatrix} \begin{vmatrix} c_1 \\ c_2 \end{vmatrix} = \begin{vmatrix} -3c_2 \\ 2c_1 - 5c_2 \end{vmatrix}$$

$$[C^T : A^T C^T] = \begin{bmatrix} c_1 & -3c_2 \\ c_2 & 2c_1 - 5c_2 \end{bmatrix} = \Delta = 2c_1^2 - 5c_1 c_2 + 3c_2^2 \neq 0$$

$(2c_1^2 - 5c_1 c_2 + 3c_2^2) = 0$
$(2c_1 - 3c_2)(c_1 - c_2) = 0$

Unobservable when

$c_1 = (3/2)c_2$ or

$c_1 = c_2$

3.37 Lyaponov criterion

A state variable control system is defined as

$$\begin{vmatrix} \dot{x}_1 \\ \dot{x}_2 \end{vmatrix} = \begin{bmatrix} 0 & 2 \\ -3 & -5 \end{bmatrix} \begin{vmatrix} x_1 \\ x_2 \end{vmatrix} + \begin{vmatrix} 1 \\ 1 \end{vmatrix} u(t)$$

Determine if the system is stable, and plot its Lyaponov function.

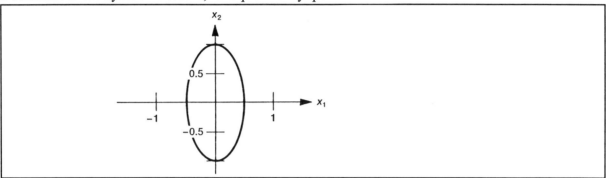

Figure 3.37 Lyaponov criterion

Solution: $A^T S + SA = -I$

$$\begin{bmatrix} 0 & -3 \\ 2 & -5 \end{bmatrix} \begin{bmatrix} a & b \\ b & c \end{bmatrix} + \begin{bmatrix} a & b \\ b & c \end{bmatrix} \begin{bmatrix} 0 & 2 \\ -3 & -5 \end{bmatrix} = \begin{bmatrix} -1 & 0 \\ 0 & -1 \end{bmatrix}$$

$$\begin{bmatrix} -6b & 2a-5b-3c \\ 2a-5b-3c & 4b-10c \end{bmatrix} = \begin{bmatrix} -1 & 0 \\ 0 & -1 \end{bmatrix}$$

$$S = \begin{bmatrix} \frac{4}{6} & \frac{1}{6} \\ \frac{1}{6} & \frac{1}{6} \end{bmatrix}, \quad s_{11} > 0, \quad \text{and} \quad \begin{bmatrix} s_{11} & s_{12} \\ s_{21} & s_{22} \end{bmatrix} > 0$$

Hence it is asymptotically stable at the origin

$$V(x) = X^T S X = \begin{vmatrix} x_1 & x_2 \end{vmatrix} \begin{bmatrix} \frac{4}{6} & \frac{1}{6} \\ \frac{1}{6} & \frac{1}{6} \end{bmatrix} \begin{vmatrix} x_1 \\ x_2 \end{vmatrix}$$

$V(x) = 1/6(4x_1^2 + 2x_1 x_2 + x_2^2) = k$, let $k = 6$
$V(x) = 4x_1^2 + 2x_1 x_2 + x_2^2$

x_1	0	0.1	0.2	0.3	0.4	0.5
x_2	1	0.88	0.74	0.55	0.32	0

3.38 Nonlinear control

Determine the effect of feedback on the nonlinear control system shown in Figure 3.38.

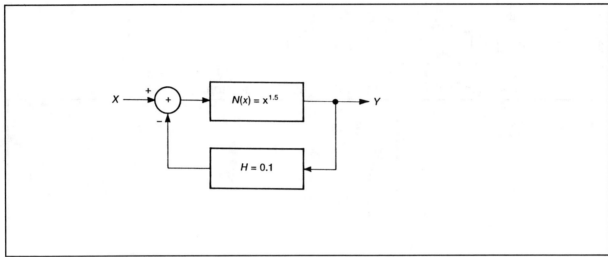

Figure 3.38 Nonlinear control

Solution:

$$\frac{Y}{X} = \frac{N(x)}{[1 + N(x)H]} = \frac{x^{1.5}}{1 + 0.1x^{1.5}}$$

x	0	0.5	1.0	1.5	2.0	2.5	3.0	3.5	4.0	4.5	5.0
N(x)	0	0.35	1.00	1.84	2.83	3.95	5.20	6.55	8.0	9.55	11.2
Y	0	0.34	0.9	1.55	2.2	2.83	3.42	3.95	4.44	4.88	5.28

The feedback tends to linearize the closed-loop gain

3.39 Phase plane

Determine the phase plane trajectory of the solution to the differential equation $\ddot{x} + 2\dot{x} + x = 0$.

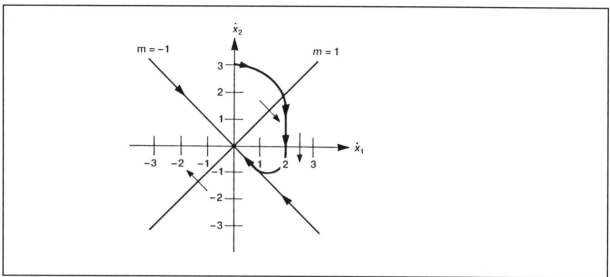

Figure 3.39 Phase plane

Solution:

$\ddot{x} + 2\dot{x} + x = 0.$

$x_1 = x$

$x_2 = \dot{x}_1$

$\dot{x}_2 = -2x_2 - x_1$

$$\begin{vmatrix} \dot{x}_1 \\ \dot{x}_2 \end{vmatrix} = \begin{bmatrix} 0 & 1 \\ -1 & -2 \end{bmatrix} \begin{vmatrix} x_1 \\ x_2 \end{vmatrix}$$

$$\frac{\dot{x}_2}{\dot{x}_1} = \frac{-2x_2 - x_1}{x_2} = \frac{-2m - 1}{m}, \quad \text{where } m = \frac{x_2}{x_1}$$

m	-2	-1	-0.5	0	0.5	1	2	5	∞
$\dot{x}_2/\dot{x}_1$	-1.5	-1	0	∞	-4	-3	-2.5	-2.2	-2

3.40 Limit cycles

Determine the limit cycle trajectory for the ON-OFF control system shown in Figure 3.40a. Also, find the condition for which the limit cycle is eliminated. Assume $M = 1$.

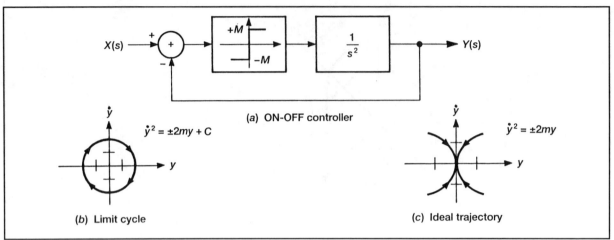

Figure 3.40 Limit cycles

Solution:

$$y = \pm M / s^2, \qquad s^2 y = \pm M$$

Let $\dot{y} = dy / dt$

$$\ddot{y} = \frac{d\dot{y}}{dt} = \frac{\frac{dy}{dt}(d\dot{y})}{dy} = \frac{\dot{y} d\dot{y}}{dy} = \pm M$$

$$\dot{y} d\dot{y} = \pm M dy$$

Solving

$$\dot{y}^2 = \pm 2My + C$$

$$\dot{y}^2 = +2My + C$$
$$\dot{y}^2 = -2My + C$$

$\dot{y}^2 = 2y + 4$ (C = 4) Limit cycle			$\dot{y}^2 = -2y + 4$ (C = 4) Limit cycle			$\dot{y}^2 = 2y$ (C = 0) Ideal trajectory			$\dot{y}^2 = -2y$ (C = 0) Ideal trajectory		
y	$\dot{y}^2$	$\dot{y}$	y	$\dot{y}^2$	$\dot{y}$	y	$\dot{y}^2$	$\dot{y}$	y	$\dot{y}^2$	$\dot{y}$
0	4	±2	0	4	±2	0	0	0	0	0	0
-1	2	±1.4	1	2	±1.4	1	2	±1.4	-1	2	±1.4
-2	0	0	2	0	0	2	4	±2	-2	4	±2

3.41 Describing function

Find the describing function of the ideal relay (ON-OFF) nonlinear function shown in Figure 3.41.

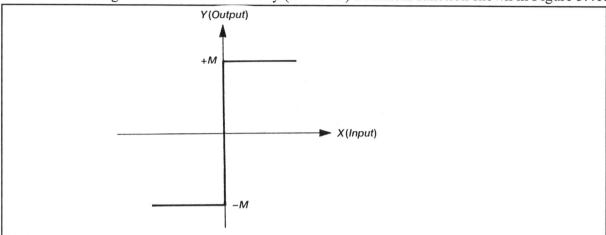

Figure 3.41 Describing function

Solution:

$N(X) = [Y_1 / X] \angle \theta_1$,

$Y_1 = (A_1^2 + B_1^2)^{1/2}$, $\qquad \theta_1 = \tan^{-1}(A_1 / B_1)$,

$x_{in}(t) = X\sin\omega t$

$n(t) = M\,\text{sgn}\, x_{in}(t)$

$A_1 = (1/\pi)\int_0^{2\pi} n(t)\cos\omega t\, d(\omega t)$, $\qquad$ let $\omega t = \theta$

$A_1 = (1/\pi)[\int_0^{\pi} M\cos\theta\, d\theta - \int_\pi^{2\pi} M\cos\theta\, d\theta] = M/\pi[\,[\sin\theta]_0^\pi - [\sin\theta]_\pi^{2\pi}\,]$

$A_1 = M/\pi[2\sin\pi - \sin 2\pi] = M/\pi[\,0 - 0\,] = 0$

$B_1 = (1/\pi)\int_0^{2\pi} n(t)\sin\omega t\, d(\omega t)$, $\qquad$ let $\omega t = \theta$

$B_1 = (1/\pi)[\int_0^{\pi} M\sin\theta\, d\theta - \int_\pi^{2\pi} M\sin\theta\, d\theta] = M/\pi[\,[-\cos\theta]_0^\pi + [\cos\theta]_\pi^{2\pi}\,]$

$B_1 = M/\pi[-\cos\pi + \cos 0 + \cos 2\pi - \cos\pi] = M/\pi[-2\cos\pi + 1 + \cos 2\pi]$

$B_1 = M/\pi[2 + 1 + 1] = (4M/\pi)$

$Y_1 = (A_1^2 + B_1^2)^{1/2} = B_1$

$\theta_1 = \tan^{-1}(A_1 / B_1) = 0°$

$N(X) = Y_1 / X = (4M / \pi X)$

3.42 Stability criterion

Find the stability of the nonlinear control system shown in Figure 3.42 using a Nyquist diagram.

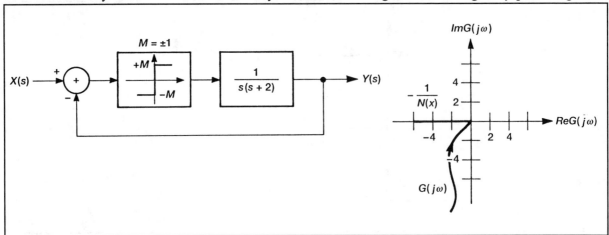

Figure 3.42 Stability criterion

Solution:

$$\frac{Y(s)}{X(s)} = \frac{N(x)G(s)}{1 + N(x)G(s)}$$

$$1 + N(X)G(s) = 0, \qquad N(X) = (4M)/(\pi X), \qquad G(s) = -1/N(X)$$

Let $s = j\omega$

$$G(j\omega) = \frac{1}{j\omega(j\omega + 2)}$$

$$|G(j\omega)| = \frac{1}{\omega(\omega^2 + 4)^{\frac{1}{2}}}$$

$$\theta° = [0° - 90° - \tan^{-1}(\omega/2)]$$

$$\text{Real}[G(j\omega)] = |G(j\omega)|\cos\theta, \qquad \text{Imag}[G(j\omega)] = |G(j\omega)|\sin\theta$$

ω	0.1	0.5	1.0	1.5	2.0	∞
|G(jω)|∠ θ°	5∠-93°	0.97∠-104°	0.45∠-116°	0.27∠-127°	0.17∠-135°	0∠-180°
Re G(jω)	-0.26	-0.23	-0.20	-0.16	-0.12	0
Im G(jω)	-4.99	-0.94	-0.40	-0.21	-0.12	0

X	0	0.5	1.0	1.5	2.0
-1/N(X) = -0.785X	0	-0.39	-0.785	-1.18	-1.57

Since G(jω) lies to the right of -1/N(X), the system is stable with the exception of one limit cycle where G(jω) intersects -1/N(X).

3.43 Stochastic control

Determine the resulting mean-square output of the linear control system of Figure 3.43 for a Gaussian white noise input.

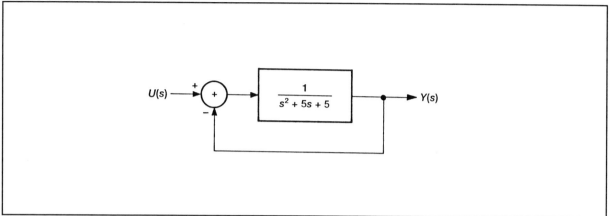

Figure 3.43 Stochastic control

Solution:

$$\begin{vmatrix} \dot{x}_1 \\ \dot{x}_2 \end{vmatrix} = \begin{bmatrix} -5 & 1 \\ -6 & 0 \end{bmatrix} \begin{vmatrix} x_1 \\ x_2 \end{vmatrix} + \begin{vmatrix} 0 \\ 1 \end{vmatrix} u(t)$$

$$y(t) = \begin{bmatrix} 1 & 0 \end{bmatrix} \begin{vmatrix} x_1 \\ x_2 \end{vmatrix}$$

$$E[u(t)u(t+\tau)] = R_{xx}(\tau) = k\delta(\tau)$$

$\mathbf{AX} + \mathbf{XA}^T + \mathbf{BUB}^T = 0;$ Due to symmetry $x_{12} = x_{21}$

$$\begin{bmatrix} -5 & 1 \\ -6 & 0 \end{bmatrix}\begin{bmatrix} x_{11} & x_{12} \\ x_{12} & x_{22} \end{bmatrix} + \begin{bmatrix} x_{11} & x_{12} \\ x_{12} & x_{22} \end{bmatrix}\begin{bmatrix} -5 & -6 \\ 1 & 0 \end{bmatrix} + \begin{bmatrix} 0 \\ 1 \end{bmatrix}\begin{bmatrix} 0 & 0 \\ 0 & k \end{bmatrix}\begin{bmatrix} 0 & 1 \end{bmatrix} = \begin{bmatrix} 0 & 0 \\ 0 & 0 \end{bmatrix}$$

$$\begin{bmatrix} -5x_{11}+x_{12} & -5x_{12}+x_{22} \\ -6x_{11} & -6x_{12} \end{bmatrix} + \begin{bmatrix} -5x_{11}+x_{12} & -6x_{11} \\ -5x_{12}+x_{22} & -6x_{12} \end{bmatrix} + \begin{bmatrix} 0 & 0 \\ 0 & k \end{bmatrix} = \begin{bmatrix} 0 & 0 \\ 0 & 0 \end{bmatrix}$$

Solving

$-5x_{11} + x_{12} = 0,$ $-6x_{11} - 5x_{12} + x_{22} = 0,$ $-12x_{12} + k = 0,$
$x_{12} = k/12,$ $-5x_{11} + k/12 = 0,$ $x_{11} = k/60$

$$E[y(t)^2] = CXC^T = \begin{bmatrix} 1 & 0 \end{bmatrix}\begin{bmatrix} x_{11} & x_{12} \\ x_{21} & x_{22} \end{bmatrix}\begin{vmatrix} 1 \\ 0 \end{vmatrix} = \begin{bmatrix} x_{11} & x_{12} \end{bmatrix}\begin{vmatrix} 1 \\ 0 \end{vmatrix} = x_{11} = \frac{k}{60}$$

3.44 Optimal control

For the feedback control system shown in Figure 3.44, find the value of K which results in a minimum integral square error performance index. Assume a unit step input function.

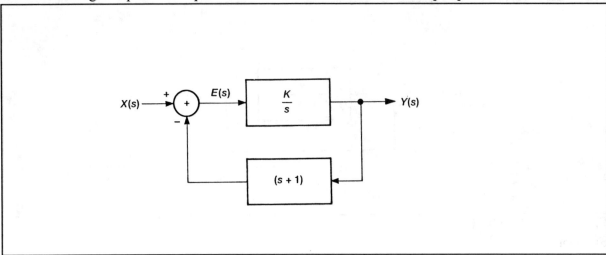

Figure 3.44 Optimal control

Solution:

$$\frac{E(s)}{X(s)} = \frac{1}{1 + G(s)H(s)} = \frac{1}{1 + \frac{K(s+1)}{s}} = \frac{s}{(1+K)s + 1}$$

$$X(s) = 1/s$$

$$E(s) = \frac{1}{s + \frac{1}{1+K}}$$

$$e(t) = e^{-\left(\frac{1}{1+K}\right)t}$$

$$e^2(t) = e^{-\left(\frac{2}{1+K}\right)t}$$

$$\int_0^\infty e^2(t)dt = \left(\frac{-2}{1+K}\right) e^{-\left(\frac{2}{1+K}\right)t} \Big|_0^\infty = \frac{2}{1+K}$$

$$\frac{d}{dK}\left[\frac{2}{1+K}\right] = \frac{-2}{(1+K)^2}$$

As $K \to \infty$ $\int_0^\infty e^2(t)dt = \frac{2}{1+K} = 0$

Thus for $K \gg 1$, the performance index is minimized and the error approaches zero.

3.45 Maximum principle

Determine the condition for the optimal control system shown in Figure 3.45. The performance index to be minimized is $J = \int_0^T (x_1 - u)^2 \, dt$. Assume $q_0 = -1$ to satisfy the maximum principle.

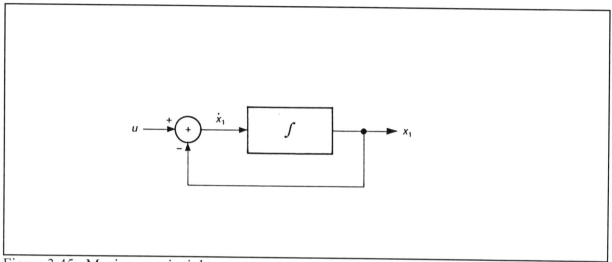

Figure 3.45 Maximum principle

Solution:

$\dot{x}_0 = f_0(x_1, u) = (x_1 - u)^2$

$J = \int_0^T f_0(x_1, u) \, dt = \int_0^T (x_1 - u)^2 \, dt$

$\dot{x}_1 = f_1(x_1, u) = -x_1 + u$

$q_0 = -1, \quad \dot{q}_0 = 0$

$H = q_0 f_0(x_1, u) + q_1 f_1(x_1, u) = -(x_1 - u)^2 + q_1(-x_1 + u)$

$\dot{q}_1 = -\dfrac{\partial H}{\partial x_1} = 2(x_1 - u) + q_1$

$\dot{x}_1 = -\dfrac{\partial H}{\partial u} = -2(x_1 - u) + q_1$

The optimal control condition becomes

$\dfrac{\partial H}{\partial u} = -2(x_1 - u) + q_1 = 0$

$-2(x_1 - u) + q_1 = 0$

$q_1 = 2x_1 - 2u$

3.46 Kalman filter

Estimate the value of the state variable, x, based on two output measurements, $y_1 = x_1 + v_1$ and $y_2 = x_2 + v_2$. Where x_i is the exact value, v_i is the random variable and $E(v_1, v_2) = 0$.

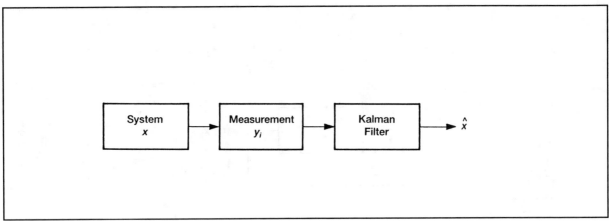

Figure 3.46 Kalman filter

Solution:

$$y_1 = x_1 + v_1, \qquad y_2 = x_2 + v_2$$

$$\hat{x} = k_1 y_1 + k_2 y_2$$

$$E(x) = \mu = x, \quad \text{deterministic}$$

$$E(v_1) = E(v_2) = \mu = 0, \quad \text{given}$$

$$E[\hat{x} - x] = E[k_1(x + v_1) + k_2(x + v_2) - x] = 0$$

$$E[\hat{x} - x] = E[k_1 x + k_2 x - x] = 0, \quad k_2 = 1 - k_1$$

$$E[(\hat{x} - x)^2] = \sigma^2 = E[(k_1(x + v_1) + k_2(x + v_2) - x)^2]$$

$$E[(\hat{x} - x)^2] = k_1^2 \sigma_1^2 + k_2^2 \sigma_2^2 = k_1^2 \sigma_1^2 + (1-k)^2 \sigma_2^2$$

$$\frac{dE}{dk_1}[(\hat{x} - x)^2] = 2k_1 \sigma_1^2 - 2\sigma_2^2 + 2k_1 \sigma_2^2 = 0$$

$$k_1 = \frac{\sigma_1^2}{\sigma_1^2 + \sigma_2^2} \quad \text{and} \quad k_2 = \frac{\sigma_2^2}{\sigma_1^2 + \sigma_2^2}$$

If the same measuring device was used, $\sigma_1^2 = \sigma_2^2$ and $k_1 = k_2$

$$\hat{x} = k_1 y_1 + k_2 y_2 \quad \text{or} \quad \hat{x} = 1/2(y_1 + y_2)$$

3.47 Adaptive control

Develop an adaptive control flowchart that sets a 12 hour timer to a new setpoint using a minimum distance rule (algorithm). The controller should address the following four cases.

a. $|T_n - T_0| \leq 6$ hrs, and $T_n > T_0$ 	b. $|T_n - T_0| \leq 6$ hrs, and $T_n < T_0$

c. $|T_n - T_0| > 6$ hrs, and $T_n > T_0$ 	d. $|T_n - T_0| > 6$ hrs, and $T_n < T_0$.

Solution:

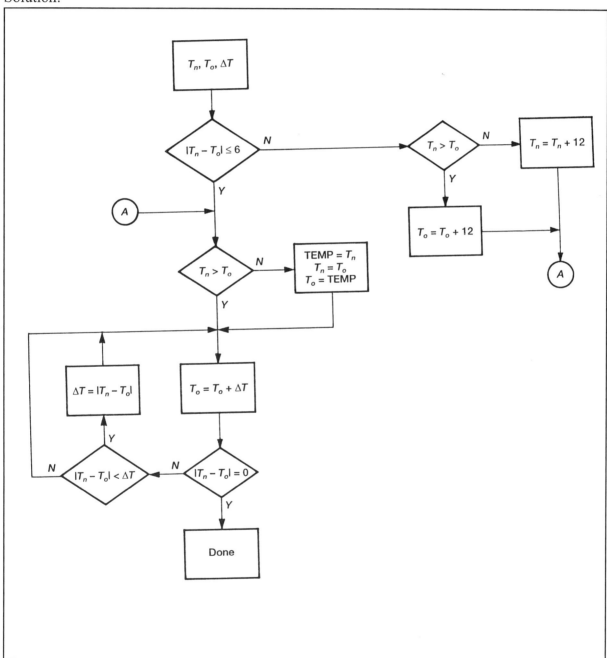

Figure 3.47 Adaptive control

3.48 z Transform

Find the unit impulse function response of the control system block diagram shown in Figure 3.48.

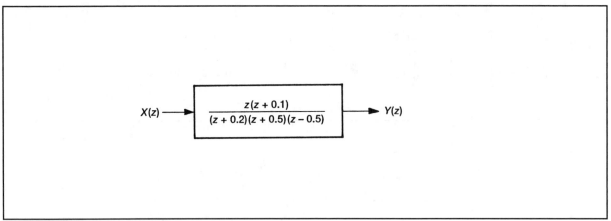

Figure 3.48 z Transform

Solution:

$$\frac{Y(z)}{X(z)} = H(z) = \frac{z(z+0.1)}{(z+0.2)(z+0.5)(z-0.5)}$$

$X(z) = 1$ and $z = e^{sT}$

$$\frac{H(z)}{z} = \frac{(z+0.1)}{(z+0.2)(z+0.5)(z-0.5)} = \frac{k_1}{(z+0.2)} + \frac{k_2}{(z+0.5)} + \frac{k_3}{(z-0.5)}$$

$$k_1 = \frac{(z+0.1)}{(z+0.5)(z-0.5)}\bigg|_{z=-0.2} = 0.476 = \frac{10}{21}$$

$$k_2 = \frac{(z+0.1)}{(z+0.2)(z-0.5)}\bigg|_{z=-0.5} = -1.33 = -\frac{4}{3}$$

$$k_3 = \frac{(z+0.1)}{(z+0.2)(z+0.5)}\bigg|_{z=0.5} = 0.857 = \frac{6}{7}$$

$$H(z) = \frac{(10/21)z}{(z+0.2)} - \frac{(4/3)z}{(z+0.5)} + \frac{(6/7)z}{(z-0.5)}$$

$h(k) = (10/21)(-0.2)^k - (4/3)(-0.5)^k + (6/7)(-0.5)^k$ \hspace{2em} $k = 0, 1, 2 ...$

3.49 Discrete-time control

A discrete-time control system is defined by the equation $y(k + 2) + 3y(k + 1) + 2y(k) = u(k)$. Determine the corresponding signal flow graph, derive the state space formulation, and find the solution matrix $\mathbf{S}(k)$.

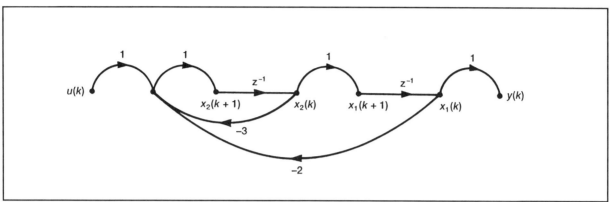

Figure 3.49 Discrete-time control

Solution: $y(k + 2) + 3y(k + 1) + 2y(k) = u(k)$

Let $x_1(k) = y(k)$, $x_2(k) = y(k + 1)$, $x_1(k + 1) = x_2(k)$
$x_2(k + 1) = -2x_1(k) - 3x_2(k) + u(k)$

$\mathbf{x}(k+1) = \mathbf{A}\mathbf{x}(k) + \mathbf{B}u(k)$
$\mathbf{y}(k) = \mathbf{C}\mathbf{x}(k) + \mathbf{D}u(k)$

$$\begin{vmatrix} x_1(k+1) \\ x_2(k+1) \end{vmatrix} = \begin{bmatrix} 0 & 1 \\ -2 & -3 \end{bmatrix} \begin{vmatrix} x_1(k) \\ x_2(k) \end{vmatrix} + \begin{vmatrix} 0 \\ 1 \end{vmatrix} u(k)$$

$$y(k) = \begin{bmatrix} 1 & 0 \end{bmatrix} \begin{vmatrix} x_1(k) \\ x_2(k) \end{vmatrix} + \begin{vmatrix} 0 \\ 0 \end{vmatrix} u(k$$

$\mathbf{S}(k) = \mathbf{A}^k = Z^{-1}([z\mathbf{I} - \mathbf{A}]^{-1} z)$

$$[z\mathbf{I} - \mathbf{A}]^{-1} = \frac{Adj \begin{bmatrix} z & -1 \\ 2 & z+3 \end{bmatrix}}{\Delta} = \frac{\begin{bmatrix} z+3 & -2 \\ 1 & z \end{bmatrix}^T}{z^2 + 3z + 2} = \frac{\begin{bmatrix} z+3 & 1 \\ -2 & z \end{bmatrix}}{(z+1)(z+2)}$$

$$\mathbf{S}(k) = Z^{-1} \frac{\begin{bmatrix} z+3 & 1 \\ -2 & z \end{bmatrix} z}{(z+1)(z+2)} = Z^{-1} \frac{\begin{bmatrix} z(z+3) & z \\ -2z & z^2 \end{bmatrix}}{z^2 + 3z + 2} = \begin{bmatrix} 2z_1^k - z_2^k & z_1^k - z_2^k \\ -2z_1^k + z_2^k & -z_1^k + z_2^k \end{bmatrix} \quad k = 0, 1, 2 \ldots$$

3.50 Direct digital control

A direct digital control (DDC) system is defined in state space by

$$\begin{vmatrix} x_1(k+1) \\ x_2(k+1) \end{vmatrix} = \begin{bmatrix} 0 & 1 \\ -2 & -3 \end{bmatrix} \begin{vmatrix} x_1(k) \\ x_2(k) \end{vmatrix} + \begin{vmatrix} 0 \\ 1 \end{vmatrix} u(k) \quad \text{and} \quad y(k) = \begin{bmatrix} 1 & 0 \end{bmatrix} \begin{vmatrix} x_1(k) \\ x_2(k) \end{vmatrix} + \begin{vmatrix} 0 \\ 0 \end{vmatrix} u(k)$$

Find the modal domain equivalent, draw the signal flow graph, and find the solution matrix $S(k)$.

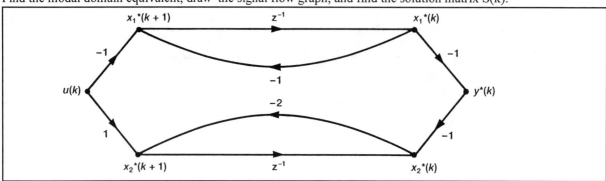

Figure 3.50 Direct digital control

Solution: $[zI-A] = z^2 + 3z + 2 = 0;$ $z_1 = -1,\ z_2 = -2$

$$[zI - A]V^1 = \begin{bmatrix} -1 & -1 \\ 2 & 2 \end{bmatrix} \begin{vmatrix} v_1 \\ v_2 \end{vmatrix} = 0, \quad v_1 = -v_2 \quad \text{let } v_2 = 1$$

$$[zI - A]V^2 = \begin{bmatrix} -2 & -1 \\ 2 & 1 \end{bmatrix} \begin{vmatrix} v_1 \\ v_2 \end{vmatrix} = 0, \quad v_1 = -v_2/2 \quad \text{let } v_2 = 2$$

$$M = [V^1 \ V^2] = \begin{bmatrix} -1 & -1 \\ 1 & 2 \end{bmatrix} \quad \text{and} \quad M^{-1} = \frac{AdjM}{\Delta} = \frac{\begin{bmatrix} 2 & -1 \\ 1 & -1 \end{bmatrix}^T}{\Delta} = \frac{\begin{bmatrix} 2 & 1 \\ -1 & -1 \end{bmatrix}}{-1} = \begin{bmatrix} -2 & -1 \\ 1 & 1 \end{bmatrix}$$

$$x^*(k+1) = M^{-1}AMx^*(k) + M^{-1}Bu(k) = \Lambda x^*(k) + M^{-1}Bu(k)$$

$$\begin{vmatrix} x_1^*(k+1) \\ x_2^*(k+1) \end{vmatrix} = \begin{bmatrix} -2 & -1 \\ 1 & 1 \end{bmatrix} \begin{bmatrix} 0 & 1 \\ -2 & -3 \end{bmatrix} \begin{bmatrix} -1 & -1 \\ 1 & 2 \end{bmatrix} \begin{vmatrix} x_1^*(k) \\ x_2^*(k) \end{vmatrix} + \begin{bmatrix} -2 & -1 \\ 1 & 1 \end{bmatrix} \begin{vmatrix} 0 \\ 1 \end{vmatrix} u(k) = \begin{bmatrix} -1 & 0 \\ 0 & -2 \end{bmatrix} \begin{vmatrix} x_1^*(k) \\ x_2^*(k) \end{vmatrix} + \begin{vmatrix} -1 \\ -1 \end{vmatrix} u(k)$$

$$y^*(k) = CM \begin{vmatrix} x_1^*(k) \\ x_2^*(k) \end{vmatrix} = \begin{bmatrix} 1 & 0 \end{bmatrix} \begin{bmatrix} -1 & -1 \\ 1 & 2 \end{bmatrix} \begin{vmatrix} x_1^*(k) \\ x_2^*(k) \end{vmatrix} = \begin{bmatrix} -1 & -1 \end{bmatrix} \begin{vmatrix} x_1^*(k) \\ x_2^*(k) \end{vmatrix}$$

$$S(k) = M\Lambda^k M^{-1} = \begin{vmatrix} -1 & -1 \\ 1 & 2 \end{vmatrix} \begin{bmatrix} z_1^k & 0 \\ 0 & z_2^k \end{bmatrix} \begin{bmatrix} -2 & -1 \\ 1 & 1 \end{bmatrix} = \begin{bmatrix} 2z_1^k - z_2^k & z_1^k - z_2^k \\ -2z_1^k + z_2^k & -z_1^k + z_2^k \end{bmatrix} \quad k = 0,1,2\ldots$$

3.51 Control loop ratios
The closed-loop control system shown in Figure 3.51 has a feedback and a feedforward path.

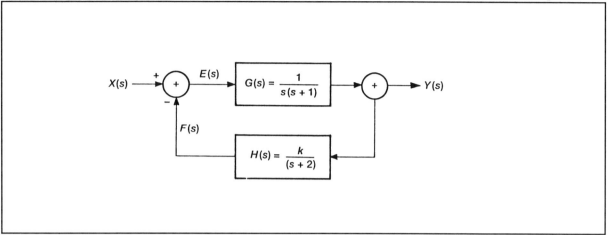

Figure 3.51 Control loop ratios

1. The generalized closed-loop gain equation is:

   a) G(s) / [1 + G(s)H(s)]   b) [G(s) + 1] / [1 + G(s)H(s)]

   c) 1 / [1 + G(s)H(s)]   d) [1 - G(s)] / [1 + G(s)H(s)]

2. The generalized error ratio equation is:

   a) H(s) / [1 + G(s)H(s)]   b) 1 / [1 + G(s)H(s)]

   c) H(s)G(s) / [1 + G(s)H(s)] d) [1 - H(s)] / [1 + G(s)H(s)]

3. The generalized feedback ratio equation is:

   a) [G(s)H(s) + H(s)] / [1 + G(s)H(s)]   b) [G(s)H(s)] / [1 + G(s)H(s)]

   c) H(s) / [1 + G(s)H(s)]   d) [G(s) + H(s)] / [1 + G(s)H(s)]

4. The open-loop gain equation is:

   a) $1 / [s(s+1)]$ b) $(s+2) / [Ks(s+1)]$

   c) $K / [s(s+1)(s+2)]$ d) $K / (s+2)$

5. The system characteristic equation is:

   a) $s^3 + 3s^2 + 2s + K$ b) $s^3 + 3s^2 + 2s$

   c) $s^2 + 3s + 2$ d) $K / (s^3 + 3s^2 + 2s)$

3.51 Solution:

1. $Y(s) = E(s)G(s) + X(s) = [X(s) - Y(s)H(s)]G(s) + X(s)$

   $Y(s)[1 + G(s)H(s)] = X(s)[G(s) + 1]$

   $Y(s) / X(s) = [G(s) + 1] / [1 + G(s)H(s)]$

   The correct answer is (b).

2. $E(s) = X(s) - Y(s)H(s) = X(s) - [E(s)G(s) + X(s)]H(s)$

   $E(s)[1 + G(s)H(s)] = X(s)[1 - H(s)]$

   $E(s) / X(s) = [1 - H(s)] / [1 + G(s)H(s)]$

   The correct answer is (d).

3. $F(s) = Y(s)H(s) = [E(s)G(s) + X(s)]H(s)$

   $F(s) = [X(s) - F(s)]G(s)H(s) + X(s)H(s)$

   $F(s)[1 + G(s)H(s)] = X(s)[G(s)H(s) + H(s)]$

   $F(s) / X(s) = [G(s)H(s) + H(s)] / [1 + G(s)H(s)]$

   The correct answer is (a).

4. $G_{ol}(s) = G(s)H(s) = K / [s(s+1)(s+2)]$

   The correct answer is (c).

5. $G(s)H(s) + 1 = 0$

   $s(s+1)(s+2) + K = s^3 + 3s^2 + 2s + K = 0$

   And $0 < K < 6$ for a stable system

   The correct answer is (a).

3.52 Time analysis

The linear system shown in Figure 3.52 is excited with a unit steo function. The resulting output response is $y(t) = 1 - e^{-500t}[\cos 866t + 0.577\sin 866t]$.

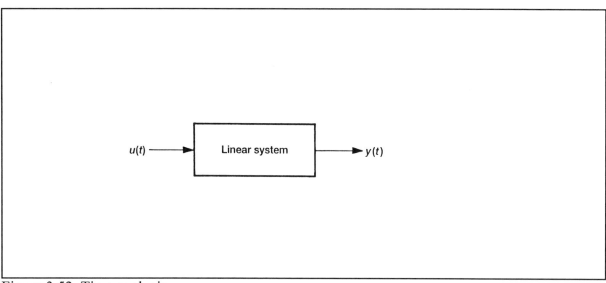

Figure 3.52 Time analysis

1. The value of the damping constant, ζ, is most nearly:

   a) 0.2 b) 0.57 c) 0.5 d) 1.73

2. The system time delay, t_d, is most nearly:

   a) 1.38×10^{-3} b) 0.7×10^{-3} c) 1×10^{-3} d) 2×10^{-3}

3. The output response rise time, t_r, is most nearly:

   a) 3.14×10^{-3} b) 2×10^{-3} c) 1.16×10^{-3} d) 2.42×10^{-3}

4. The output response maximum overshoot, m_p, (%) is most nearly:

   a) 4.3 b) 16.3 c) 5.6 d) 17.8

5. The output response settling time, t_s, for a 2% of steady state value is most nearly:

   a) 2×10^{-3} b) 6×10^{-3} c) 4×10^{-3} d) 8×10^{-3}

3.52 Solution:

1. $\zeta\omega_n = 500$

   $\omega = \omega_n(1-\zeta^2)^{1/2} = 866$

   $\zeta / (1-\zeta^2)^{1/2} = 500/866 = 0.577$

   $\zeta = 0.5$

   The correct answer is (c).

2. $y(t) = 0.5 = 1 - e^{-\zeta\omega_n \times t_d}$

   $1/2 = 1/e^{\zeta\omega_n \times t_d}$ or $2 = e^{\zeta\omega_n \times t_d}$

   $t_d = \ln(2) / (\zeta\omega_n) = 1.38 \times 10^{-3}$ seconds

   The correct answer is (a).

3. $\kappa = \zeta / (1-\zeta^2)^{1/2} = 0.5 / (0.75)^{1/2} = 0.57$

   $\theta = \tan^{-1}(1/\kappa) = \tan^{-1}(1.73) = 1.047$ radians

   $t_r = t_2 - t_1 = 1/\omega[\tan^{-1}(-1/\kappa)] = (\pi - \theta)/\omega = (3.14 - 1.047)/866 = 2.42 \times 10^{-3}$ seconds

   The correct answer is (d).

4. $m_p = 100 \times e^{-\pi\kappa} = 100 \times e^{-3.4145(0.577)} = 16.3\%$

   at $t_p = (\pi/\omega) = 3.63 \times 10^{-3}$ seconds

   The correct answer is (b).

5. $t_s(2\%) = 4/(\zeta\omega_n) = 8/(1,000) = 8 \times 10^{-3}$ seconds

   $t_s(5\%) = 3/(\zeta\omega_n) = 6/(1,000) = 6 \times 10^{-3}$ seconds

   The correct answer is (d).

3.53 S-plane analysis

A typical closed-loop feedback control system is shown in Figure 3.53. The closed-loop gain is used for s-plane analysis.

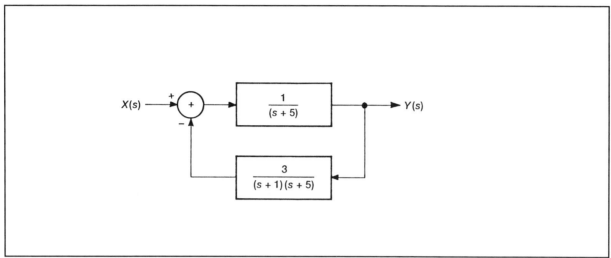

Figure 3.53 S-plane analysis

1. The closed-loop gain equation is:

    a) $3 / [(s+1)(s+5)^2]$ b) $(s+1)/(s+8)$

    c) $(s+1)/[(s+2)(s+4)]$ d) $3/[(s+1)(s+5)]$

2. The poles and zeroes of the closed-loop gain equation are:

    a) 1, 5 and 5   b) -2, -4 and -1   c) 2, 4 and 1   d) -1, -5 and -5

3. The center of asymptotes with the real axis on the s-plane is:

    a) -5   b) -7/3   c) -7   d) -5/3

4) The angle of asymptotes with the real axis on the s-plane is:

    a) ± 60°   b) ± 270°   c) ± 90°   d) ± 180°

5) The breakaway point on the real axis of the s-plane is:

    a) -1   b) -2   c) -3   d) -4

3.53 Solution:

1. $Y(s)/X(s) = G(s) / [1 + G(s)H(s)] = [1 / (s+5)] / [1 + 3 / (s+2)(s+5)]$

   $Y(s)/X(s) = (s+1) / [(s+2)(s+4)]$

   The correct answer is (c).

2. Poles: $(s+2)(s+4) = 0 \quad s_1 = -2, \quad s_2 = -4$

   Zeroes: $(s+1) = 0, \quad s = -1$

   The correct answer is (b).

3. $\sigma_c = [\sum p_i - \sum z_i] / [n - m] = [(-2-4) - (-1)] / [2-1] = -5$

   The correct answer is (a).

4. $\theta = \pm (q \times 180°) / (n - m) = \pm (q \times 180°) / 1 = \pm 180° \qquad q = 1, 3, 5...$

   The correct answer is (d).

5. $\sigma_b = 1/(\sigma_b + s_1) + 1/(\sigma_b + s_2) = 1/(\sigma_b + 2) + 1/(\sigma_b + 4) = 0$

   $2\sigma_b = -6$

   $\sigma_b = -3$

   The correct answer is (c).

3.54 Frequency analysis

A feedback control system block diagram is shown in Figure 3.54. The parameters T_1 and T_2 equal 1 and 0.1 respectively. The open-loop gain is used for frequency analysis.

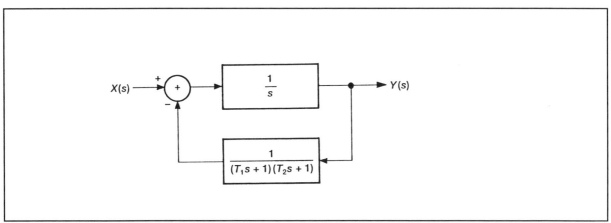

Figure 3.54 Frequency analysis

1. The values of ω_1 and ω_2 are most nearly:

    a) 0.628 and 6.28   b) 0.1 and 1   c) 1 and 10   d) 1.6 and 0.16

2. The dB magnitude equation of the open-loop gain is:

    a) $20\log(1) - 20\log(\omega) - 20\log(\omega^2 + 1)^{1/2} - 20\log(10\omega + 1)^{1/2}$

    b) $20\log(1) - 20\log(\omega) - 20\log(\omega + 1) - 20\log(10\omega + 1)$

    c) $20\log(1) - 20\log(\omega) - 20\log(\omega^2 + 1)^{1/2} - 20\log(0.01\omega^2 + 1)^{1/2}$

    d) $20\log(1) - 20\log(\omega) - 20\log(\omega^2 + 1)^{1/2} - 20\log(10\omega^2 + 1)^{1/2}$

3. The phase angle function of the open-loop gain is:

    a) $0° - 90° - \tan^{-1}(\omega) - \tan^{-1}(10\omega)$ b) $0° - 90° - \tan^{-1}(\omega) - \tan^{-1}(0.1\omega)$

    c) $0° - 90° - \tan^{-1}(\omega)^2 - \tan^{-1}(\omega/10)^2$ d) $0° - 90° - \tan^{-1}(\omega + 1) - \tan^{-1}(10\omega + 1)$

4. The open-loop function gain margin is most nearly:

    a) 10   b) 0.1   c) 0.9   d) 10.7

5. The open-loop gain phase margin is most nearly:

    a) 46.8   b) 43   c) -133   d) -313

3.54 Solution:

1. $\omega_1 = 1/T_1 = 1$

   $\omega_2 = 1/T_2 = 10$

   The correct answer is (c).

2. Let $s = j\omega$

   $G(j\omega)H(j\omega) = 10/[j\omega(j\omega+1)(j\omega+10)] = 1/[j\omega(j\omega+1)(0.1j\omega+1)]$

   $Mag(dB) = 20\log|G(j\omega)H(j\omega)| = 20\log(1) - 20\log(\omega) - 20\log(\omega^2+1)^{1/2} - 20\log(0.1\omega^2+1)^{1/2}$

   The correct answer is (c).

3. $\theta° = \arg[G(j\omega)H(j\omega)] = 0° - 90° - \tan^{-1}(\omega) - \tan^{-1}(0.1\omega)$

   The correct answer is (b).

4. Gain margin $= 1/(|G(j\omega)H(j\omega)|)$ when $\arg[G(j\omega)H(j\omega)] = -180°$

   $\theta = \arg[G(j\omega)H(j\omega)] = 0° - 90° - \tan^{-1}(\omega) - \tan^{-1}(0.1\omega) = -180°$

   at $\omega \approx 3.2$

   $|G(j\omega)H(j\omega)| = 10/[\omega(\omega^2+1)^{1/2}(\omega^2+100)^{1/2}] = = 10/[3.2(3.35)(10)] = 1/10.7$

   Gain margin $= 1/(|G(j\omega)H(j\omega)|) = 10.7$

   The correct answer is (d).

5. Phase margin $= 180° + \arg[G(j\omega)H(j\omega)]$ when $|G(j\omega)H(j\omega)| = 1$

   $|G(j\omega)H(j\omega)| = 10/[\omega(\omega+1)^{1/2}(\omega+100)^{1/2}] = 1$

   at $\omega \approx 0.8$

   Phase margin $= 180° + \theta = 180° + 0° - 90° - \tan^{-1}(0.8) - \tan^{-1}(0.08) = 180° - 133.2° = 46.8°$

   The correct answer is (a).

3.55 Transfer function analysis

The op amp transfer function shown in Figure 3.55 uses two RC networks to provide frequency dependent magnitude and phase parameters.

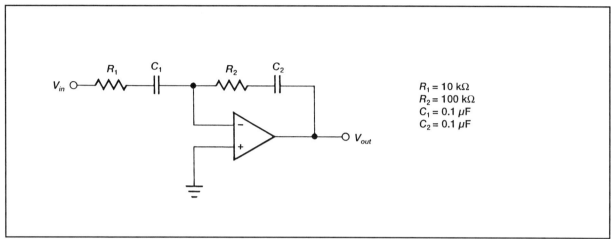

Figure 3.55 Transfer function analysis

1. The resulting transfer function equation is :

   a) $-(R_2/R_1)[(s + 1/R_2 C_2)/(s + 1/R_1 C_1)]$ b) $-(R_2/R_1)[(s+1/R_1 C_1)/(s +1/R_2 C_2)]$

   c) $-(C_1/C_2)[(s + 1/R_2 C_2)/(s + 1/R_1 C_1)]$ d) $-(C_1/C_2)[(s + 1/R_1 C_1)/(s + 1/R_2 C_2)]$

2. The transfer function is also known as a:

   a) Lag   b) Lead-Lag   c) Lead   d) Lag-Lead

3. The zero and pole frequencies are most nearly:

   a) 10 and 100   b) 628 and 6,280   c) 15.9 and 159   d) 100 and 10

4. The value of ω at the -3 dB magnitude point is most nearly:

   a) 100   b) 500   c) 5,000   d) 1,000

5. The maximum phase shift occurs when ω is most nearly:

   a) 100   b) 500   c) 1,000   d) 2,000

3.55 Solution:

1. $Z_2(s) = (1 + R_2C_2 s)/sC_2$; $\quad Z_1(s) = (1 + R_1C_1 s)/sC_1$

   $V_{out}/V_{in} = -Z_2(s)/Z_1(s) = -(C_1/C_2)[(1 + R_2C_2 s)/(1 + R_1C_1 s)]$

   $Z_2(s)/Z_1(s) = -(R_2/R_1)[(s + 1/R_2C_2)/(s + 1/R_1C_1)]$

   The correct answer is (a).

2. $V_{out}/V_{in} = -(10)[(s + 1\times 10^2)/(s + 1\times 10^3)]$

   Since $R_2C_2 > R_1C_1$ $\quad f_z < f_p$ thus it is a lead network

   The correct answer is (c).

3. $f_z = 1/(2\pi R_2 C_2) = 15.9$ Hz

   $f_p = 1/(2\pi R_1 C_1) = 159$ Hz

   The correct answer is (c).

4. $-3 = 20\log X$

   $X = 0.707$

   $0.707 \times 10 = 10 \times |(j\omega + 100)/(j\omega + 1,000)|$

   $7.07 = 10[(\omega^2 + 1\times 10^4)^{1/2}/(\omega^2 + 1\times 10^6)^{1/2}]$ at $\omega = 1,000$ rads/sec

   The correct answer is (d).

5. $\theta = \tan^{-1}(\omega/100) - \tan^{-1}(\omega/1,000)$

   at $\omega = 100$ rads/sec $\quad \theta = 45° - 5.7° \approx 39°$

   at $\omega = 500$ rads/sec $\quad \theta = 78.7° - 26.6° \approx 52°$

   at $\omega = 1,000$ rads/sec $\quad \theta = 84.3° - 45° \approx 39°$

   The correct answer is (b).

4.0 COMMUNICATIONS
4.1 Fourier series

Calculate the Fourier series representation of the time domain waveform shown in Figure 4.1a. Also, plot the resulting amplitude line spectrum assuming $T = 1 \times 10^{-3}$ seconds.

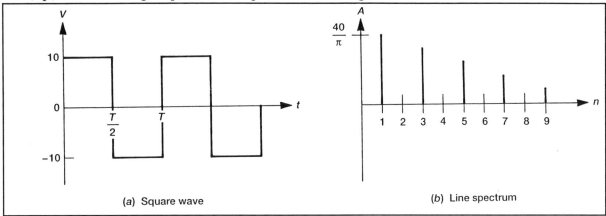

Figure 4.1 Fourier series

Solution:

$$f(t) = \frac{a_0}{2} + \sum_{n=1}^{\infty} [a_n \cos(n\omega_0 t) + b_n \sin(n\omega_0 t)], \qquad \omega_0 = \frac{1}{T} = 1{,}000 \text{ Hz}$$

$a_0 = (1/T)\int_0^T f(t)dt = 1/T[\int_0^{T/2} 10 dt - \int_{T/2}^T 10 dt] = (10/T)[T/2 - 0 - T + T/2] = 0$

$a_n = (1/T)[\int_0^T f(t)\cos(n\omega_0 t)dt]$, let $\omega_0 T = 2\pi$

$a_n = (1/\pi)[\int_0^\pi 10\cos(n\omega_0 t)dt - \int_\pi^{2\pi} 10\cos(n\omega_0 t)dt] = (10/n\pi)([\sin(n\omega_0 t)]_0^\pi - [\sin(n\omega_0 t)]_\pi^{2\pi})$

$a_n = (10/n\pi)[\sin(n\pi) - \sin 0 - \sin(n2\pi) + \sin(n\pi)]$

$a_n = (10/n\pi)[2\sin(n\pi) - \sin(n2\pi)] = 0$, for all n

$b_n = (1/T)[\int_0^T f(t)\sin(n\omega_0 t)dt]$, let $\omega_0 T = 2\pi$

$b_n = (1/\pi)[\int_0^\pi 10\sin(n\omega_0 t)dt - \int_\pi^{2\pi} 10\sin(n\omega_0 t)dt] = (10/n\pi)([-\cos(n\omega_0 t)]_0^\pi + [\cos(n\omega_0 t)]_\pi^{2\pi})$

$b_n = (10/n\pi)[-\cos n\pi + \cos 0 + \cos n2\pi - \cos n\pi] = [10/n\pi][1 - 2\cos n\pi + \cos n2\pi]$

$b_n = (10/n\pi)[1 - 2 + 1] = 0$, for even n

$b_n = (10/n\pi)[1 + 2 + 1] = [40/n\pi]$, for odd n

$f(t) = (40/\pi)[\sin 1{,}000t + (1/3)\sin 3{,}000t + (1/5)\sin 5{,}000t + \cdots + (1/n)\sin(n\omega_0 t)]$, odd n

Due to the (1/n) coefficient, only the first few terms of the equation are significant.

4.2 Fourier integral

Calculate the Fourier integral representation of the time domain pulse shown in Figure 4.2a. Also, plot the resulting amplitude continuous spectrum assuming $\tau = 0.25 \times 10^{-3}$ seconds.

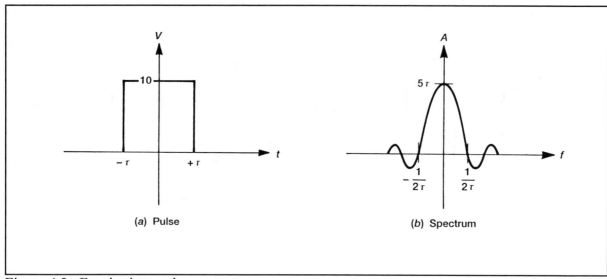

Figure 4.2 Fourier integral

Solution:

$$f(t) = 10 \quad \text{for } -\tau < t < +\tau$$

$$f(t) = 0 \quad \text{elsewhere}$$

$$F(\omega) = \int_{-\infty}^{\infty} f(t) e^{-j\omega t} dt = \int_{-\tau}^{\tau} 10 e^{-j\omega t} dt = -\left(\frac{10}{j\omega}\right) [e^{-j\omega t}]_{-\tau}^{\tau}$$

$$F(\omega) = -(10/j\omega)[e^{-j\omega\tau} - e^{j\omega\tau}]$$

$$F(\omega) = \frac{5}{\omega}\left[\frac{e^{j\omega\tau} - e^{-j\omega\tau}}{2j}\right] = \frac{5}{\omega}(\sin \omega\tau)$$

$$F(\omega) = 5\tau\left[\frac{\sin \omega\tau}{\omega\tau}\right] = 1.25 \times 10^{-3} \times \left(\frac{\sin 0.25 \times 10^{-3}\omega}{0.25 \times 10^{-3}\omega}\right)$$

$$F(f) = 5\tau\left[\frac{\sin(2\pi f\tau)}{(2\pi f\tau)}\right] = 1.25 \times 10^{-3} \times \left(\frac{\sin 0.5 \times 10^{-3}\pi f}{0.5 \times 10^{-3}\pi f}\right)$$

4.3 Convolution

Determine the output response, y(t), of a linear system whose impulse response equals h(t) = u(t) - u(t - 3). The input signal is x(t) = [t - 2(t - 1) + (t - 2)] as shown in Figure 4.3.

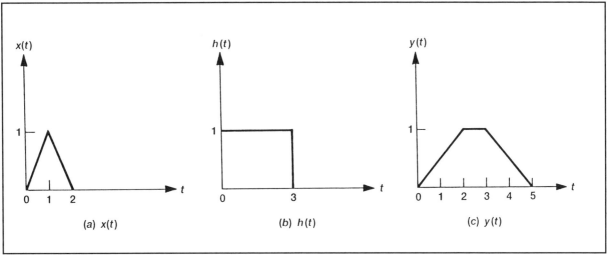

Figure 4.3 Convolution

Solution:

$$y(t) = \int_0^t [h(\tau)x(t - \tau)d\tau]$$

$$y(t) = \int_0^t [\tau - 2(\tau - 1) + (\tau - 2)][u(t - \tau) - u(t - 3 - \tau)]d\tau$$

$$y(t) = \int_0^t [\tau - 2(\tau - 1) + (\tau - 2)]d\tau - \int_0^{t-3} [\tau - 2(\tau - 1) + (\tau - 2)]d\tau$$

$$y(t) = \int_0^t \tau d\tau - 2\int_1^t (\tau - 1)d\tau + \int_2^t (\tau - 2)d\tau - \int_0^{t-3} \tau d\tau + 2\int_1^{t-3} (\tau - 1)d\tau - \int_2^{t-3} (\tau - 2)d\tau$$

$$y(t) = (1/2)t^2 u(t) - (t - 1)^2 u(t - 1) + (1/2)(t - 2)^2 u(t - 2) - (1/2)(t - 3)^2 u(t - 3) +$$

$$(t - 4)^2 u(t - 4) - (1/2)(t - 5)^2 u(t - 5)$$

t	0	1	2	3	4	5
Calculation	0	0.5	2 - 1	4.5 - 4 + 0.5	8 - 9 + 2 - 0.5	12.5 - 16 + 4.5 - 2 + 1
y(t)	0	0.5	1.0	1.0	0.5	0

4.4 Correlation

Determine the autocorrelation function, $R_{xx}(\tau)$, for the time domain signal shown in Figure 4.4a.

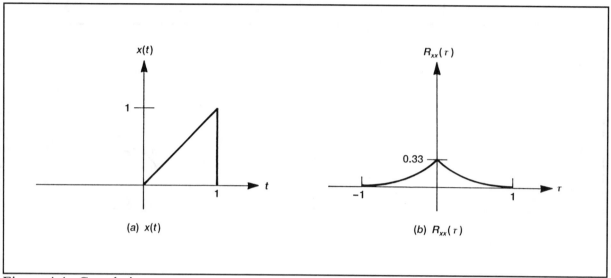

(a) x(t)   (b) $R_{xx}(\tau)$

Figure 4.4   Correlation

Solution:

$$R_{xx}(\tau) = \int_a^b [x(t)x(t+\tau)dt] = \int_0^{1-\tau}[t(t+\tau)dt]$$

$$R_{xx}(\tau) = \int_0^{1-\tau}[t^2 dt] + \int_0^{1-\tau}[\tau t\, dt] = \frac{1}{3}[t^3]\Big|_0^{1-\tau} + \frac{\tau}{2}[t^2]\Big|_0^{1-\tau}$$

$$R_{xx}(\tau) = (1/3)(1-\tau)^3 + (\tau/2)(1-\tau)^2$$

$$R_{xx}(\tau) = (1/3)(1 - 3\tau + 3\tau^2 - \tau^3) + (1/2)(\tau - 2\tau^2 + \tau^3)$$

$$R_{xx}(\tau) = (\tau^3/6 - \tau/2 + 1/3) \qquad 0 \le \tau \le 1$$

And $R_{xx}(\tau) = R_{xx}(-\tau)$

$|\tau|$	0	0.25	0.50	0.75	1.0
$R_{xx}(\tau)$	0.33	0.21	0.10	0.03	0

4.5 Spectral density

Determine the spectral density function, $S_{xx}(f)$, for the autocorrelation function, $R_{xx}(\tau)$, of the time domain signal $x(t) = 1$ as shown in Figure 4.5a.

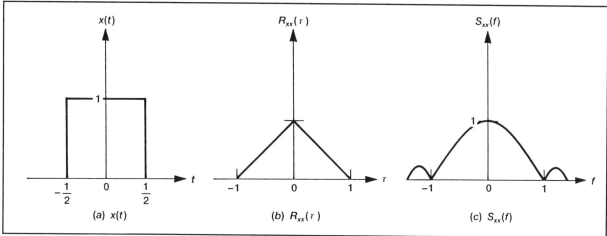

Figure 4.5 Spectral density

Solution:

$x(t) = 1$ for $-1/2 \leq t \leq 1/2$

$$R_{xx}(\tau) = \int_{-1/2}^{1/2 - \tau} [x(t)x(t + \tau)dt] = \int_{-1/2}^{1/2 - \tau} [1 dt]$$

$$R_{xx}(\tau) = [t]_{-1/2}^{1/2 - \tau} = [(1/2 - \tau) - (-1/2)] = (1 - \tau), \quad -1 \leq \tau \leq 1$$

Since $R_{xx}(\tau) = R_{xx}(-\tau)$

$R_{xx}(\tau) = x(\tau) * x(\tau)$

$S_{xx}(f) = X(f)X(f)$

$$X(f) = \int_{-\infty}^{\infty} x(t)e^{-j\omega t} dt = \int_{-1/2}^{1/2} e^{-j2\pi ft} dt$$

$$X(f) = \frac{1}{-j2\pi f}[e^{-j2\pi ft}]_{-1/2}^{1/2} = \frac{e^{-j\pi f} - e^{j\pi f}}{-j2\pi f}$$

$$X(f) = \frac{1}{\pi f}\left(\frac{e^{j\pi f} - e^{-j\pi f}}{2j}\right) = \frac{\sin(\pi f)}{\pi f}$$

$$S_{xx}(f) = X(f)X(f) = X^2(f) = [\sin^2(\pi f)] / (\pi f)^2$$

4.6 Cross correlation

Determine the cross correlation function, $R_{xy}(\tau)$, of the time domain signals, x(t) and y(t), shown in Figure 4.6a.

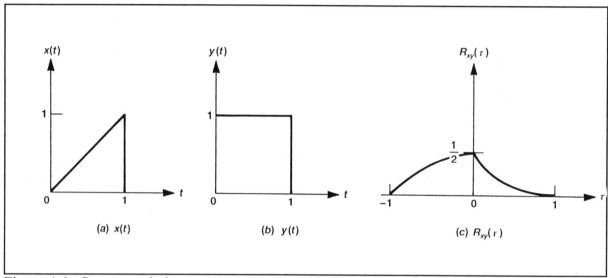

Figure 4.6   Cross correlation

Solution:

$$R_{xy}(\tau) = \int_0^{1-\tau}[x(t)y(t+\tau)dt] = \int_0^{1-\tau}[t\,dt] = \frac{1}{2}[t^2]_0^{1-\tau}$$

$$R_{xy}(\tau) = (1-\tau)^2/2$$

$$R_{xy}(-\tau) = R_{yx}(\tau) = \int_0^{1-\tau}[y(t)x(t+\tau)dt] = \int_0^{1-\tau}(t+\tau)dt = \int_0^{1-\tau}t\,dt = \int_0^{1-\tau}\tau\,dt$$

$$R_{xy}(-\tau) = \frac{1}{2}[t^2]_0^{1-\tau} + \tau[t]_0^{1-\tau} = \frac{(1-\tau)^2}{2} + \tau(1-\tau)$$

$$R_{xy}(-\tau) = (1/2)(1-2\tau+\tau^2) + \tau - \tau^2 = (1/2) - \tau + (1/2)\tau^2 + \tau - \tau^2$$

$$R_{xy}(-\tau) = (1-\tau^2)/2$$

τ	0	0.25	0.50	0.75	1.0
$R_{xy}(\tau)$	0.5	0.28	0.125	0.03	0
$R_{xy}(-\tau)$	0.5	0.46	0.375	0.218	0

4.7 Cross spectral density

Determine the cross spectral density function, $S_{xy}(f)$, for the correlation function shown in Figure 4.7a and defined as $R_{xy}(\tau) = e^{at}/(a+b)$ $t<0$ and $R_{xy}(\tau) = e^{bt}/(a+b)$ $t>0$.

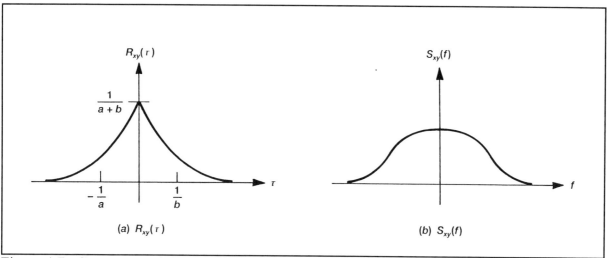

(a) $R_{xy}(\tau)$

(b) $S_{xy}(f)$

Figure 4.7 Cross spectral density

Solution:

$$S_{xy}(f) = \int_{-\infty}^{\infty} R_{xy}(\tau) e^{-j\omega t} dt$$

$$S_{xy}(f) = \frac{1}{a+b}\left[\int_{-\infty}^{0} e^{(a-j\omega)t} dt + \int_{0}^{\infty} e^{-(b+j\omega)t} dt\right]$$

$$S_{xy}(f) = \frac{1}{a+b}\left[\frac{e^{(a-j\omega)t}}{(a-j\omega)}\right]_{-\infty}^{0} + \frac{1}{a+b}\left[\frac{e^{-(b+j\omega)t}}{-(b+j\omega)}\right]_{0}^{\infty}$$

$$S_{xy}(f) = \frac{1}{a+b}\left[\frac{1}{(a-j\omega)} + \frac{1}{(b+j\omega)}\right] = \frac{a-j\omega+b+j\omega}{(a+b)(a-j\omega)(b+j\omega)}$$

$$S_{xy}(f) = \frac{a+b}{(a+b)(a-j\omega)(b+j\omega)} = \frac{1}{(a-j\omega)(b+j\omega)}$$

If $a = b$ then,

$$S_{xy}(f) = 1/(a^2 + \omega^2) = 1/[a^2 + (2\pi f)^2]$$

4.8 Probability

A box contains ten checkers, five red, three black, and two white. One checker is drawn at random.

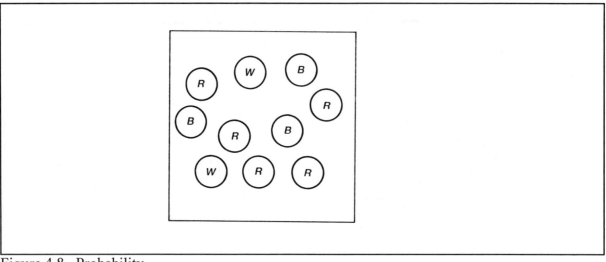

Figure 4.8 Probability

1. The probability of it being red is most nearly:

   a) 3/10   b) 1/5   c) 7/10   d) 1/2

2. The probability of it being black is most nearly:

   a) 1/5   b) 3/10   c) 7/10   d) 1/2

3. The probability of it being white is most nearly:

   a) 1/5   b) 7/10   c) 1/2   d) 3/10

4. The probability of it being not black is most nearly:

   a) 1/2   b) 3/10   c) 7/10   d) 1/5

5. The probability of it being black or white is most nearly:

   a) 1/2   b) 7/10   c) 3/10   d) 1/5

4.8 Solution:

1. P(red) = $5 / (5 + 3 + 2)$ = 5/10 = 1/2

   The correct answer is (d).

2. P(black) = $3 / (5 + 3 + 2)$ = 3/10

   The correct answer is (b).

3. P(white) = $2 / (5 + 3 + 2)$ = 2 / 10 = 1/5

   The correct answer is (a).

4. $P(\overline{black})$ = 1 − P(black) = 1 − 3 / 10 = 7 / 10

   The correct answer is (c).

5. P(black ∪ white) = P(black + white) = $(3 + 2) / (5 + 3 + 2)$ = 5/10 = 1/2

   The correct answer is (a).

4.9 PDF function

A box contains eight bills, two $1, two $2, two $5, and two $10 respectively. Determine the probability density function (PDF) for the dollar sum of two successive random draws, with replacement after each draw, and plot it..

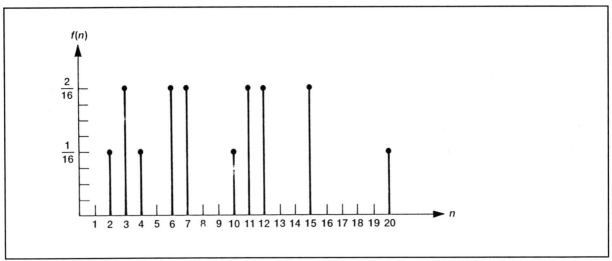

Figure 4.9 PDF function

Solution:

$$P(i \cap j) = P(i)P(j) = (2/8)(2/8) = 1/16$$

VALUE	$1	$2	$5	$10
$1	2	3	6	11
$2	3	4	7	12
$5	6	7	10	15
$10	11	12	15	20

$n = V_i + V_j$,   the sum of two $ values

$f(n) = P(V_i + V_j)$,   the frequency of n

4.10 CDF function

Determine the resulting cumulative distribution function (CDF) for problem 4.9 and plot it. Also, calculate the probability of drawing at least $10 with two random draws.

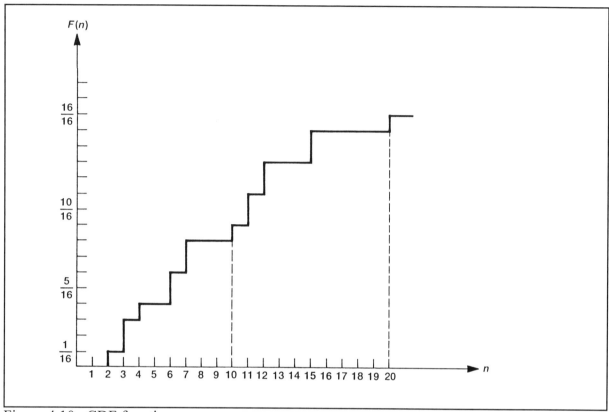

Figure 4.10 CDF function

Solution:

$$F(n) = \sum_{n=1}^{N} f(n)$$

$P(\geq 10) = F(20) - F(10)$

$P(\geq 10) = 16/16 - 9/16 = 7/16$

4.11 Joint probability

Three boxes contain fifteen balls each. Box A contains seven red, five white, and three blue. Box B contains eight red, two white, and five blue. Box C contains five red, six white, and four blue. Calculate (a) the probability that results in three white balls if one ball is drawn from each box, and (b) the probability that at least one of the three balls is white.

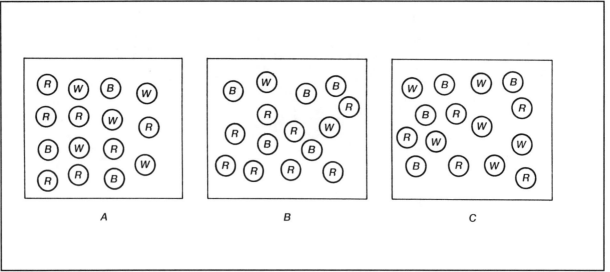

Figure 4.11 Joint probability

Solution:

a) $P(W_A W_B W_C) = P(W_A)P(W_B)P(W_C)$

$P(W_A W_B W_C) = (5/15)(2/15)(6/15) = (4/225)$

b) $P(W) = 1 - P(\overline{W})$

$P(\overline{W}) = \dfrac{10}{15} \times \dfrac{13}{15} \times \dfrac{9}{15} = \dfrac{78}{225}$

$P(W) = 1 - (78/225) = (225/225) - (78/225) = (49/75)$

4.12 Conditional probability

A box contains a total of twenty-four checkers, seven red, nine green, and eight blue. In addition, five of the red, three of the green, and four of the blue are shiny new. A checker is drawn at random from the box. Determine the probability that it (a) will be red, (b) it will be a new red piece, (c) that it will be either new or red.

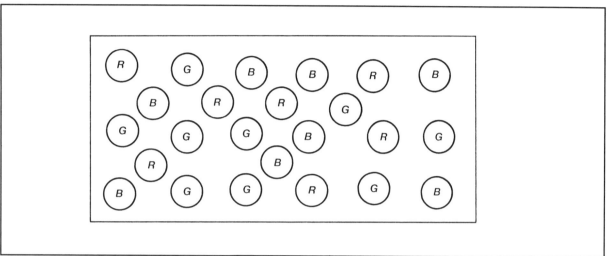

Figure 4.12 Conditional probability

Solution:

COLOR	OLD	NEW	TOTAL
R	2	5	7
G	6	3	9
B	4	4	8
SUM	12	12	24

a) $P(R) = (2+5)/(12+12) = 7/24$

b) $P(N/R) = P(N \cap R)/P(R) = (5/24)/(7/24) = 5/7$

c) $P(N \cup R) = P(N) + P(R) - P(N \cap R) = (12/24) + (7/24) - (5/24) = (7/12)$

4.13 Mean

An exponential PDF is shown in Figure 4.13. Calculate the mean value if $f(x) = 2e^{-2x}$, $x \geq 0$.

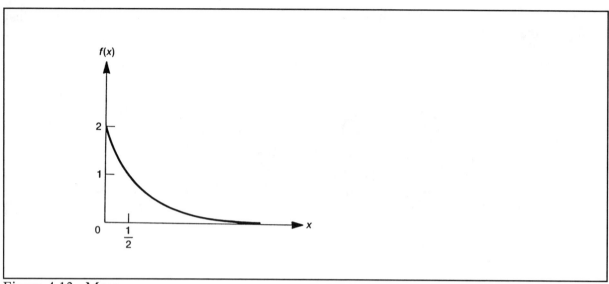

Figure 4.13 Mean

Solution:

$$\text{Mean} = \bar{x} = \int_{-\infty}^{\infty} x f(x) dx$$

$$F(x) = \int_{-\infty}^{\infty} f(x) dx = \int_{-\infty}^{\infty} 2e^{-2x} dx = [e^{-2x}]_0^{\infty} = 0 + 1 = 1$$

$$\bar{x} = \int_0^{\infty} (x) 2e^{-2x} dx = 2\int_0^{\infty} x e^{-2x} dx$$

Using: $\int_0^{\infty} (x^n) e^{-ax} dx = \dfrac{n!}{a^{n+1}}$

$$\bar{x} = 2\left(\dfrac{1}{-2^2}\right) = \dfrac{1}{2}$$

4.14 Mean squared value
A uniform PDF is shown in Figure 4.14. Calculate its mean squared value.

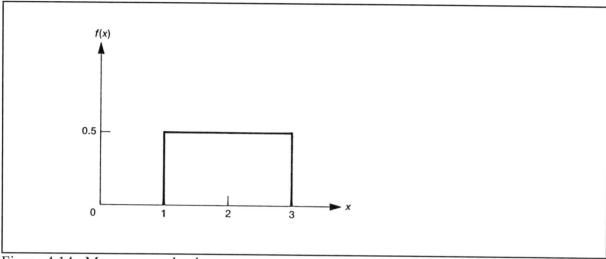

Figure 4.14 Mean squared value

Solution:

Mean squared value $= \overline{x^2} = \int_{-\infty}^{\infty} x^2 f(x)dx$

$F(x) = \int_{-\infty}^{\infty} f(x)dx = \frac{1}{2}\int_{1}^{3} dx = \frac{1}{2}[x]_{1}^{3} = 1$

$\overline{x^2} = \int_{1}^{3} (x)^2 1\, dx = \frac{1}{2}\int_{1}^{3} x^2 dx$

$\overline{x^2} = \frac{1}{6}[x^3]_{1}^{3} = \frac{1}{6}(27 - 1) = \frac{26}{6}$

$\overline{x^2} = \frac{13}{3}$

4.15 Variance

A sawtooth PDF is shown in Figure 4.15. Calculate its variance, σ^2.

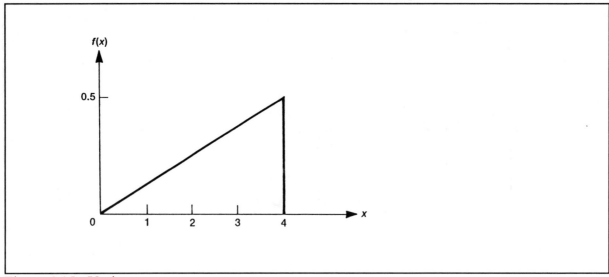

Figure 4.15 Variance

Solution:

$$\text{Variance} = \sigma^2 = \overline{x^2} - (\overline{x})^2$$

$$F(x) = \int_{-\infty}^{\infty} f(x)dx = \frac{1}{8}\int_0^4 x\,dx = \frac{1}{16}[x^2]_0^4 = 1$$

$$\overline{x} = \int_{-\infty}^{\infty} xf(x)dx = \frac{1}{8}\int_0^4 x^2\,dx = \frac{1}{24}[x^3]_0^4 = \frac{64}{24} = \frac{8}{3}$$

$$\overline{x^2} = \int_{-\infty}^{\infty} (x)^2 f(x)dx = \frac{1}{8}\int_0^4 x^3\,dx = \frac{1}{32}[x^4]_0^4 = \frac{256}{32} = 8$$

$$\sigma^2 = \overline{x^2} - (\overline{x})^2 = 8 - \left(\frac{8}{3}\right)^2 = 8 - 7.1 = 0.9$$

4.16 Standard deviation

The PDF corresponding to the possible outcome of a fair die toss is shown in Figure 4.16. Calculate the resulting standard deviation, σ.

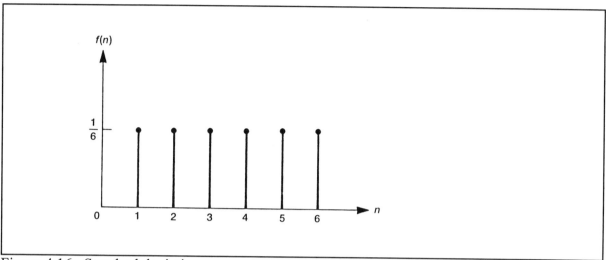

Figure 4.16 Standard deviation

Solution:

Standard deviation $\sigma = \sqrt{\sigma^2}$

$$F(n) = \sum_{i=1}^{N} f(n) = \left[\frac{1}{6} + \frac{1}{6} + \frac{1}{6} + \frac{1}{6} + \frac{1}{6} + \frac{1}{6}\right] = 1$$

$$\bar{n} = \frac{1}{N}\sum_{i=1}^{N} n_i = \frac{1}{6}[1 + 2 + 3 + 4 + 5 + 6] = \frac{21}{6} = \frac{7}{2}$$

$$\overline{n^2} = \frac{1}{N}\sum_{i=1}^{N} (n_i)^2 = \frac{1}{6}[1 + 4 + 9 + 16 + 25 + 36] = \frac{91}{6}$$

$$\sigma^2 = \frac{1}{N}\sum_{i=1}^{N} (n_i - \bar{n})^2 = \overline{n^2} - (\bar{n})^2 = \frac{91}{6} - \frac{49}{4} = 2.916$$

$$\sigma = (\sigma^2)^{1/2} = 1.707$$

4.17 Noise figure

Determine (a) the noise figure, NF, and (b) the output signal-to-noise ratio, $(S/N)_{out}$, for the amplifier circuit model shown in Figure 4.17. The input signal-to-noise ratio, $(S/N)_{in}$, is 10 dB and $K = 1.37 \times 10^{-23}$ J/°K.

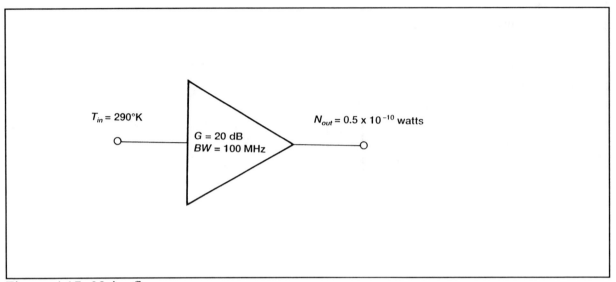

Figure 4.17 Noise figure

Solution:

$$N_{out} = KP(BW)(N_{in} + T_{eff})$$

$$G = 10\log P$$

$$\log P = G/10 = 20/10 = 2$$

$$P = 10^2 = 100$$

$$K = 1.37 \times 10^{-23} \text{ J/°K}$$

$$T_{eff} = \frac{N_{out} - KP(BW)T_{in}}{KP(BW)} = \frac{0.5 \times 10^{-10} - 0.393 \times 10^{-10}}{0.137 \times 10^{-12}} = 75° \text{ K}$$

a) $\quad NF = \left(1 + \dfrac{T_{eff}}{T_{in}}\right) = \left(1 + \dfrac{75}{290}\right) = 1.26$ dB

b) $\quad (S/N)_{out} = (1/NF)(S/N)_{in} = (1/1.26)(10) = 7.93$ dB

4.18 Sensitivity

Determine the sensitivity, S, for the amplifier circuit model shown in Figure 4.18 so that it produces 0.1 watt of output power at room temperature.

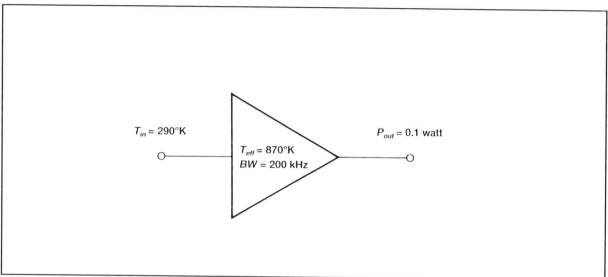

Figure 4.18 Sensitivity

Solution:

$$S = \text{Thermal noise power} + NF + 10\log(BW) + P_{out}$$

$$K = 1.37 \times 10^{-23} \text{ J/}^\circ K, \quad T_{in} = 290 \text{ }^\circ K, \quad BW = 1 \text{ Hz}$$

$$\text{Thermal noise power (dBm)} = 10\log\left(\frac{KT_{in}BW}{1 \times 10^{-3}}\right) = 10\log\left(\frac{3.973 \times 10^{-21}}{1 \times 10^{-3}}\right)$$

$$\text{Thermal noise power (dBm)} = 10 \times (-17.4) = -174 \text{ dBm}$$

$$NF = \left(1 + \frac{T_{eff}}{T_{in}}\right) = \left(1 + \frac{870}{290}\right) = 4 \text{ dB}$$

$$10\log(BW) = 10\log(2 \times 10^5) = 53 \text{ dB}$$

$$\text{Output power (dBm)} = 10\log\left(\frac{P_{out}}{1 \times 10^{-3}}\right) = 10\log\left(\frac{0.1}{1 \times 10^{-3}}\right) = 20 \text{ dBm}$$

$$S = -174 + 4 + 53 + 20 = -174 + 77 = -97 \text{ dBm}$$

4.19 Dynamic range

The amplifier circuit model show in Figure 4.19 exhibits an 1 dB compression of its output power when the input signal equals 100×10^{-3} volts. Find the dynamic range for a noise figure of 4 dB.

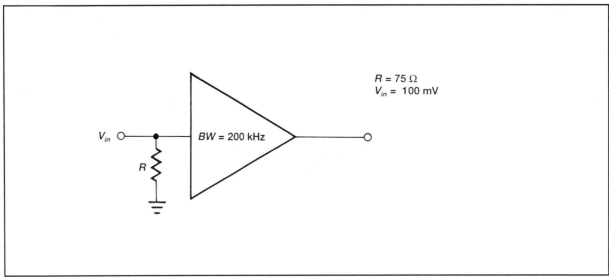

Figure 4.19 Dynamic range

Solution:

Dynamic range = P_{in} (1 dB compression) - MDS

$$P_{in}(dB) = 10\log\left(\frac{V_{in}^2}{R}\right) = 10\log\frac{(100 \times 10^{-3})^2}{75} = 10\log\left(\frac{0.01}{75}\right)$$

P_{in} (dB) = 10(-3.875) = -38.75 dB

P_{in} (dBm) = -38.75 + 30 = -8.75 dBm

MDS = -174 dBm + 4 dB + 10log(BW) + 3 dBm

MDS = -171 dBm + 4 dB + 10log(2×10^5)

MDS = -171 dBm + 4 dB + 53 dB = -114 dBm

Dynamic range = P_{in} (dBm) - MDS

Dynamic range = -8.75 - (-114) = 105.25 dBm

4.20 Intermodulation distortion

An amplifier circuit exhibits a transfer function of $f(x) = a_0 + a_1x + a_2x^2$. Determine the resulting output frequency components for an input signal $x(t) = V\cos\omega_1 t + V\cos\omega_2 t$.
Let $\omega_1 = 691.15 \times 10^6$ and $\omega_2 = 816.81 \times 10^6$.

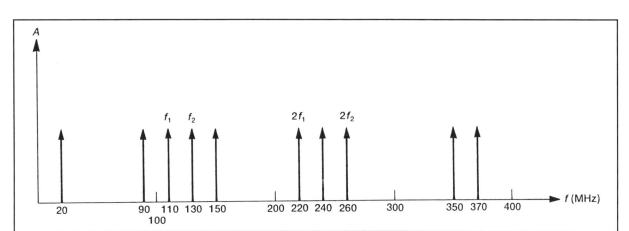

Figure 4.20 Intermodulation distortion

Solution: $y = f(x) = a_0 + a_1x + a_2x^2$

$y(t) = a_0 + a_1(V\cos\omega_1 t + V\cos\omega_2 t) + a_2(V\cos\omega_1 t + V\cos\omega_2 t)^2$

$y(t) = a_0 + a_1V\cos\omega_1 t + a_1V\cos\omega_2 t + a_2(V^2\cos^2\omega_1 t + V^2\cos\omega_1 t \cos\omega_2 t + V^2\cos^2\omega_2 t)$

Using: $\cos^2 A = 1/2 + (1/2)\cos 2A$ and

$\cos A \cos B = (1/2)[\cos(A+B) + \cos(A-B)]$

$y(t) = a_0 + a_2V^2 + a_1V\cos\omega_1 t + a_1V\cos\omega_2 t + (a_2/2)V\cos(\omega_1 + \omega_2)t +$

$\qquad (a_2/2)V\cos(\omega_1 - \omega_2)t + (a_2/2)V^2\cos 2\omega_1 t + (a_2/2)V^2\cos 2\omega_2 t$

$f_1 = \omega_1/(2\pi) = 110$ MHz

$f_2 = \omega_2/(2\pi) = 130$ MHz

The resulting frequency spectrum consists of

$f_1 = 110$ MHz, $\quad f_2 = 130$ MHz,

$2f_1 = 220$ MHz, $\quad 2f_2 = 260$ MHz,

$f_1 + f_2 = 240$ MHz, $\quad 2f_1 + f_2 = 350$ MHz, $\quad 2f_2 + f_1 = 370$ MHz,

$f_2 - f_1 = 20$ MHz, $\quad 2f_1 - f_2 = 90$ MHz, $\quad 2f_2 - f_1 = 150$ MHz

4.21 Intercept point

An amplifier circuit exhibits the input-output power characteristics shown in Figure 4.21. The dotted line depicts the desired transfer function, f(x), of two input signals at f_1 and f_2 respectively. The dashed line shows the undesired $(2f_1 - f_2)$ and $(2f_2 - f_1)$ resulting output, g(x), due to inherent nonlinearities. Determine the third order intercept point.

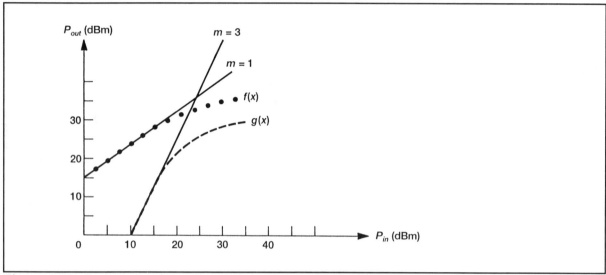

Figure 4.21   Intercept point

Solution:

Projecting the linear portions of f(x) and g(x),

$f(x) = mx + b = P_{in} + 15$

$g(x) = mx + b = 3P_{in} - 30$

At $f(x) = g(x)$

$P_{in} + 15 = 3P_{in} - 30$

$2P_{in} = 45$

$P_{in} = 22.5 \text{ dBm}$

$P_{out} = 22.5 + 15 = 37.5 \text{ dBm}$

4.22 Lowpass filter

Determine (a) the magnitude and (b) phase shift of the lowpass filter circuit shown in Figure 4.22 at its cutoff frequency, f_2.

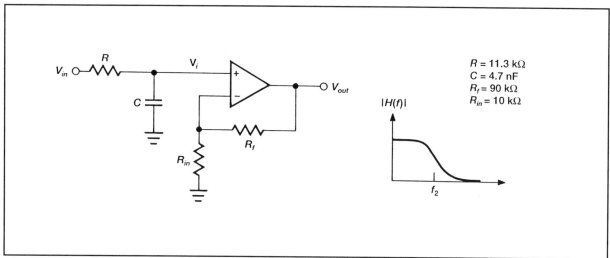

Figure 4.22 Lowpass filter

Solution: $V_{out} = A_v v_i$

$$A_v = \frac{R_f + R_{in}}{R_{in}} = \frac{100 \times 10^3}{10 \times 10^3} = 10$$

$$v_i = V_{in}\left(\frac{Z_2}{Z_1 + Z_2}\right) = V_{in}\left(\frac{1/sC}{R + 1/sC}\right) = V_{in}\left(\frac{1}{1 + sRC}\right)$$

Let $s = j\omega$ $\quad v_i = V_{in}\left(\frac{1}{1 + j\omega RC}\right)$

$$V_{out} = A_v v_i = V_{in}\left(\frac{10}{1 + j\omega RC}\right) = V_{in}\left(\frac{10}{1 + j2\pi f RC}\right)$$

$$H(f) = \frac{V_{out}}{V_{in}} = \frac{10}{1 + j2\pi f RC} \quad \text{and} \quad \theta(f) = \tan^{-1}(-2\pi f RC)$$

$$|H(f)| = \frac{10}{[1 + (2\pi f RC)^2]^{1/2}}$$

a) Let $f_2 = \dfrac{1}{2\pi RC} = \dfrac{1}{2\pi \times 11.3 \times 10^3 \times 4.7 \times 10^{-9}} = 3$ kHz

$$|H(f)| = \frac{10}{\left[1 + \left(\dfrac{f}{f_2}\right)^2\right]^{1/2}} = \frac{10}{[1 + 1]^{1/2}} = \frac{10}{\sqrt{2}} = 7.07$$

b) $\theta(f_2) = \tan^{-1}(-2\pi f RC) = \tan^{-1}-(f/f_2) = \tan^{-1}-(1) = -45°$

4.23 Highpass filter

Determine (a) the magnitude and (b) phase shift of the highpass filter circuit shown in Figure 4.23 at its cutoff frequency, f_1.

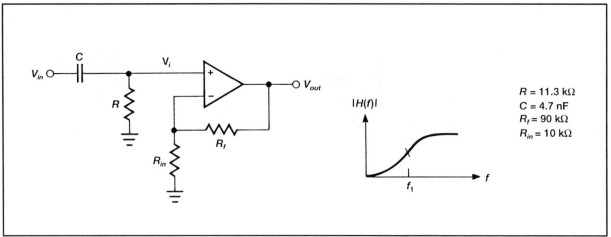

Figure 4.23 Highpass filter

Solution: $V_{out} = A_v v_i$

$$A_v = \frac{R_f + R_{in}}{R_{in}} = \frac{100 \times 10^3}{10 \times 10^3} = 10$$

$$v_i = V_{in}\left(\frac{Z_2}{Z_1 + Z_2}\right) = V_{in}\left(\frac{R}{R + 1/sC}\right) = V_{in}\left(\frac{sRC}{1 + sRC}\right)$$

Let $s = j\omega$
$$v_i = V_{in}\left(\frac{j\omega RC}{1 + j\omega RC}\right) = V_{in}\left(\frac{1}{1 - \frac{j}{\omega RC}}\right)$$

$$V_{out} = A_v v_i = V_{in}\left(\frac{10}{1 - \frac{j}{\omega RC}}\right) = V_{in}\left(\frac{10}{1 - \frac{j}{2\pi fRC}}\right)$$

$$H(f) = \frac{V_{out}}{V_{in}} = \frac{10}{1 - \frac{j}{2\pi fRC}} \quad \text{and} \quad \theta(f) = \tan^{-1}\left(\frac{1}{2\pi fRC}\right)$$

$$|H(f)| = \frac{10}{\left[1 + \left(\frac{1}{2\pi fRC}\right)^2\right]^{1/2}}$$

a) Let $f_1 = \dfrac{1}{2\pi RC} = \dfrac{1}{2\pi \times 11.3 \times 10^3 \times 47 \times 10^{-9}} = 300$ Hz

$$|H(f)| = \frac{10}{\left[1 + \left(\frac{f_1}{f}\right)^2\right]^{1/2}} = \frac{10}{[1 + 1]^{1/2}} = \frac{10}{\sqrt{2}} = 7.07$$

b) $\theta(f_1) = \tan^{-1}[1/(2\pi fRC)] = \tan^{-1}[f_1/f] = \tan^{-1}[1] = 45°$

4.24 Bandpass filter

Determine the magnitude and phase shift of the bandpass filter circuit shown in Figure 4.24 at its cutoff frequencies, f_1 and f_2 respectively.

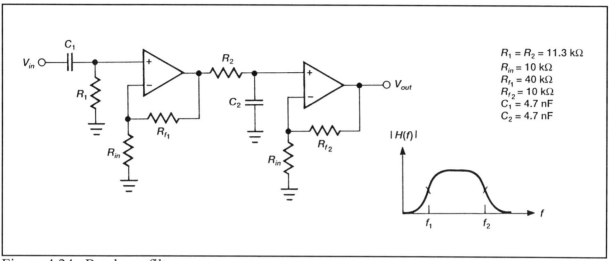

Figure 4.24   Bandpass filter

Solution: $V_{out} / V_{in} = H(f) = H_1(f)H_2(f)$

$$A_{v1} = \frac{40 \times 10^3 + 10 \times 10^3}{10 \times 10^3} = 5 \quad \text{and} \quad A_{v2} = \frac{10 \times 10^3 + 10 \times 10^3}{10 \times 10^3} = 2$$

$$f_1 = \frac{1}{2\pi R_1 C_1} = \frac{1}{2\pi \times 11.3 \times 10^3 \times 47 \times 10^{-9}} = 300 \text{ Hz}$$

$$f_2 = \frac{1}{2\pi R_2 C_2} = \frac{1}{2\pi \times 11.3 \times 10^3 \times 4.7 \times 10^{-9}} = 3 \text{ kHz}$$

$$|H(f)| = |H_1(f)H_2(f)| = \frac{5}{\left[1+\left(\frac{f_1}{f}\right)^2\right]^{1/2}} \times \frac{2}{\left[1+\left(\frac{f}{f_2}\right)^2\right]^{1/2}}$$

$$\theta(f) = \theta_1(f) + \theta_2(f) = \tan^{-1}\left(\frac{f_1}{f}\right) + \tan^{-1}-\left(\frac{f}{f_2}\right)$$

a) $|H(f_1)| = \dfrac{5}{[1+1]^{1/2}} \times \dfrac{2}{[1+(0.1)^2]^{1/2}} = 7.036$

$\theta(f_1) = \tan^{-1}(f_1/f) + \tan^{-1}-(f/f_2) = 45° - 5.7° = 39.3°$

b) $|H(f_2)| = \dfrac{5}{[1+(0.1)^2]^{1/2}} \times \dfrac{2}{[1+1]^{1/2}} = 7.035$

$\theta(f_2) = \tan^{-1}(f_1/f) + \tan^{-1}-(f/f_2) = 5.7° - 45° = -39.3°$

4.25   Bandreject filter

4-26

Determine (a) the magnitude and (b) phase shift of the bandreject filter circuit shown in Figure 4.25 at its cutoff frequencies, f_1 and f_2 respectively.

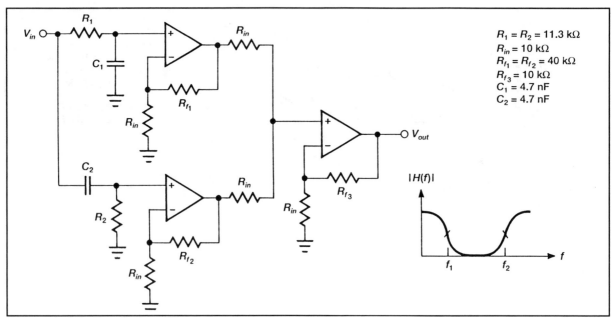

Figure 4.25 Bandreject filter

Solution: $V_{out}/V_{in} = H(f) = H_1(f)A_{v3} + H_2(f)A_{v3}$

$$A_{v1} = A_{v2} = \frac{40 \times 10^3 + 10 \times 10^3}{10 \times 10^3} = 5 \quad \text{and} \quad A_{v3} = \frac{10 \times 10^3 + 10 \times 10^3}{10 \times 10^3} = 2$$

$$f_1 = \frac{1}{2\pi R_1 C_1} = \frac{1}{2\pi \times 11.3 \times 10^3 \times 47 \times 10^{-9}} = 300 \text{ Hz}$$

$$f_2 = \frac{1}{2\pi R_2 C_2} = \frac{1}{2\pi \times 11.3 \times 10^3 \times 4.7 \times 10^{-9}} = 3 \text{ kHz}$$

$$|H(f)| = |H_1(f)|A_{v3} + |H_2(f)|A_{v3} = \frac{10}{\left[1 + \left(\frac{f}{f_1}\right)^2\right]^{1/2}} + \frac{10}{\left[1 + \left(\frac{f_2}{f}\right)^2\right]^{1/2}}$$

a) $|H(f_1)| = \dfrac{10}{[1+1]^{1/2}} + \dfrac{2}{[1+(10)^2]^{1/2}} = 7.07 + 0.995 = 8.08$

$\theta(f_1) \approx \tan^{-1} -(f/f_1) = -45°$

b) $|H(f_2)| = \dfrac{10}{[1+(10)^2]^{1/2}} \times \dfrac{2}{[1+1]^{1/2}} = 0.995 + 7.07 = 8.08$

$\theta(f_2) \approx \tan^{-1}(f_2/f) = 45°$

4.26 Delay line LPF

Determine the frequency response, H(f), of the impulse response function for the delay line lowpass filter (LPF) circuit shown in Figure 4.26. Let $\tau = 0.001$, $f_1 = 0$ Hz, and $f_2 = 1,000$ Hz.

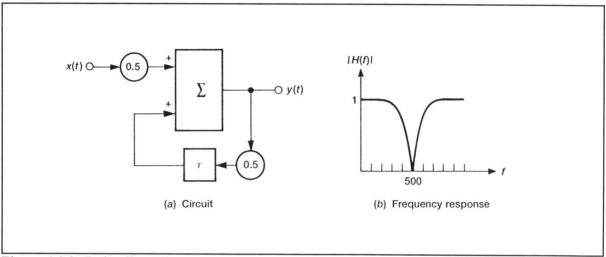

Figure 4.26 Delay line LPF

Solution:

$y(t) = 0.5x(t) + 0.5y(t-\tau)$

$h(t) = 0.5\delta(t) + 0.5\delta(t-\tau)$

$H(\omega) = 0.5(1) + 0.5(e^{-j\omega\tau})$

$|H(\omega)| = [(0.5 + 0.5e^{-j\omega\tau})(0.5 + 0.5e^{j\omega\tau})]^{\frac{1}{2}}$

$|H(\omega)| = [(0.25 + 0.25e^{-j\omega\tau} + 0.25e^{j\omega\tau} + 0.25)]^{\frac{1}{2}}$

$\theta = \omega\tau = 2\pi f\tau$

Using: $\cos\theta = 0.5(e^{j\theta} + e^{-j\theta}) = 0.5(e^{j\omega\tau} + e^{-j\omega\tau}) = 0.5(e^{j2\pi f\tau} + e^{-j2\pi f\tau})$

$|H(\omega)| = [0.5 + 0.5\cos(2\pi f\tau)]^{\frac{1}{2}}$

4.27 Delay line HPF

Determine the frequency response, H(f), of the impulse response function for the delay line highpass filter (HPF) circuit shown in Figure 4.27. Let $\tau = 0.001$, $f_1 = 0$ Hz, and $f_2 = 1,000$ Hz.

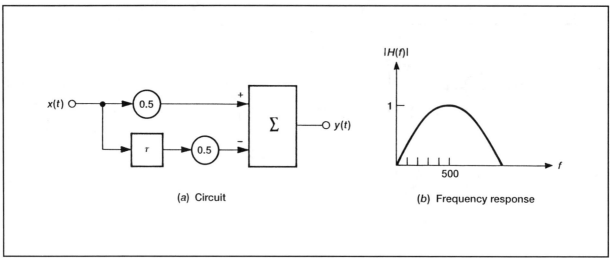

Figure 4.27   Delay line HPF

Solution:

$y(t) = 0.5x(t) - 0.5x(t-\tau)$

$h(t) = 0.5\delta(t) - 0.5\delta(t-\tau)$

$H(\omega) = 0.5(1) - 0.5(e^{-j\omega\tau})$

$|H(\omega)| = [(0.5 - 0.5e^{-j\omega\tau})(0.5 - 0.5e^{j\omega\tau})]^{½}$

$|H(\omega)| = [(0.25 - 0.25e^{-j\omega\tau} - 0.25e^{j\omega\tau} + 0.25)]^{½}$

$\theta = \omega\tau = 2\pi f\tau$

Using: $-\cos\theta = -0.5(e^{j\theta} + e^{-j\theta}) = -0.5(e^{j\omega\tau} + e^{-j\omega\tau}) = -0.5(e^{j2\pi f\tau} + e^{-j2\pi f\tau})$

$|H(\omega)| = [0.5 - 0.5\cos(2\pi f\tau)]^{½}$

4.28 Matched filter

A signal is defined as $x(t) = 4e^{-2t}u(t)$ and is applied to the input of a matched filter, $h(t) = x(t_0 - t)$. Determine the resulting time domain output signal, $y(t)$.

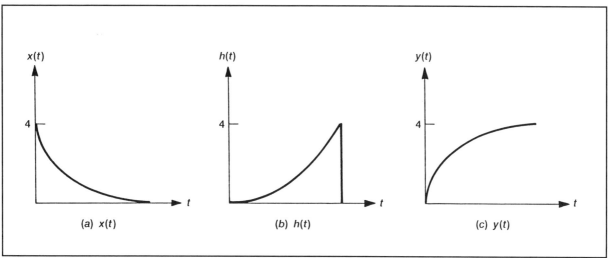

(a) $x(t)$ (b) $h(t)$ (c) $y(t)$

Figure 4.28 Matched filter

Solution:

$$x(t) = 4e^{-2t}u(t)$$

$$X(\omega) = \int_0^\infty [4e^{-2t}u(t)]e^{-j\omega t}dt = 4\int_0^\infty e^{-t(2+j\omega)}dt$$

$$X(\omega) = \frac{4}{-(2+j\omega)}[e^{-t(2+j\omega)}]_0^\infty = \frac{-4}{(2+j\omega)}[0-1] = \frac{4}{(2+j\omega)}$$

Let $H(\omega) = X(-\omega)e^{-j\omega t_0} = \frac{4}{(2-j\omega)}e^{-j\omega t_0}$

$$h(t) = x(t_0 - t) = 4e^{2(t-t_0)} \quad 0 \leq t \leq t_0$$

$$y(t) = \int_{-\infty}^{t_0} x(\tau)h(t_0 - \tau)d\tau = \int_0^{t_0} x(\tau)x(t_0 - \tau)d\tau$$

$$y(t) = \int_0^{t_0} x^2(\tau)d\tau = \int_0^{t_0}(4e^{-2\tau})^2 d\tau$$

$$y(t) = \frac{-16}{4}[e^{-4\tau}]_0^{t_0} = 4(1 - e^{-4t_0})$$

4.29 AM modulation

A 1 MHz sinewave carrier is 50% amplitude modulated using a 1 kHz signal. The resulting modulation envelope is shown in Figure 4.29a. Determine the corresponding frequency spectrum.

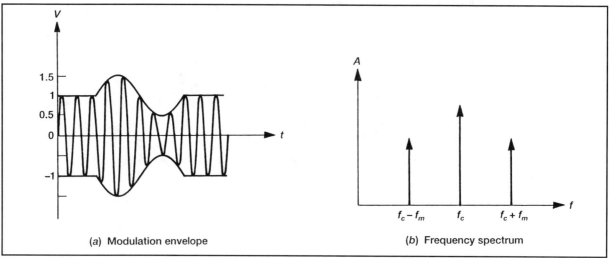

Figure 4.29 AM modulation

Solution:

(Carrier) $V_c(t) = A_c \cos 2\pi f_c t = 1.0 \cos 2\pi (1 \times 10^6)t$

(Signal) $V_m(t) = A_m \cos 2\pi f_m t = 0.5 \cos 2\pi (1 \times 10^3)t$

$V_{AM}(t) = [1 + mV_m(t)]V_c(t) = A_c[1 + mA_m \cos 2\pi f_m t]\cos 2\pi f_c t$

Using $\cos A \cos B = (1/2)[\cos(A+B) + \cos(A-B)]$

$V_{AM}(t) = A_c \cos 2\pi f_c t + [(mA_c A_m)/2]\cos 2\pi (f_c - f_m)t + [(mA_c A_m)/2]\cos 2\pi (f_c + f_m)t$

$$m = \frac{V_{max} - V_{min}}{V_{max} + V_{min}} = \frac{1.5 - 0.5}{1.5 + 0.5} = \frac{1}{2} = 0.5 \quad \text{(modulation index)}$$

$V_{AM}(t) = 1.0 \cos 2\pi (1 \times 10^6)t + 0.125 \cos 2\pi (9.99 \times 10^5)t + 0.125 \cos 2\pi (1.001 \times 10^6)t$

4.30 FM modulation

A 1 MHz sinewave carrier is frequency modulated using a 1 kHz signal. The resulting modulation envelope is shown in Figure 4.30a. Determine the corresponding frequency spectrum for a modulation index, β, of two.

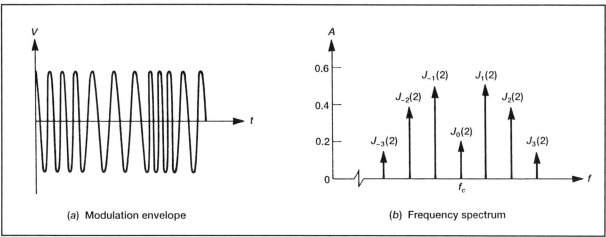

(a) Modulation envelope   (b) Frequency spectrum

Figure 4.30 FM modulation

Solution:

$V_c(t) = A_c \cos 2\pi f_c t = 1.0 \cos 2\pi (1 \times 10^6) t$

$V_m(t) = A_m \cos 2\pi f_m t = 0.5 \cos 2\pi (1 \times 10^3) t$

$V_{FM}(t) = A_c \cos[2\pi f_c t + 2\pi \Delta f \int (A_m \cos 2\pi f_m t \, dt)] = A_c \cos[2\pi f_c t + (A_m \Delta f / f_m) \sin 2\pi f_m t]$

Let $\beta = (A_m \Delta f / f_m) = 2$

$V_{FM}(t) = A_c \cos[2\pi f_c t + \beta \sin 2\pi f_m t]$

Using $\cos(A+B) = (\cos A \cos B - \sin A \sin B)$

$V_{FM}(t) = A_c[\cos 2\pi f_c t \times \cos(\beta \sin 2\pi f_m t) - \sin 2\pi f_c t \times \sin(\beta \sin 2\pi f_m t)]$

$V_{FM}(t) = 1.0[\cos 2\pi (1 \times 10^6) t \times \cos(2 \sin 2\pi (1 \times 10^3) t) - \sin 2\pi (1 \times 10^6) t \times \sin(2 \sin 2\pi (1 \times 10^3) t)]$

Using the Fourier series expansion and simplifying, results in the Bessel functions of the first kind:

$V_{FM}(t) = J_0(\beta) \cos 2\pi f_c t - J_1(\beta)[\cos 2\pi (f_c - f_m) t - \cos 2\pi (f_c + f_m) t] +$
$\quad J_2(\beta)[\cos 2\pi (f_c - 2f_m) t + \cos 2\pi (f_c + 2f_m) t] - J_3(\beta)[\cos 2\pi (f_c - 3f_m) t - \cos 2\pi (f_c + 3f_m) t]$

For $\beta = 2$, the coefficients are:
$J_0(2) = 0.22,$ $J_1(2) = 0.58,$ $J_2(2) = 0.35,$ $J_3(2) = 0.13$

4.31 PM modulation

A 1 MHz sinewave carrier is phase modulated using a 1 kHz signal. The resulting modulation envelope is shown in Figure 4.31a. Determine the corresponding frequency spectrum for a maximum phase deviation, $\Delta\phi \ll \pi$.

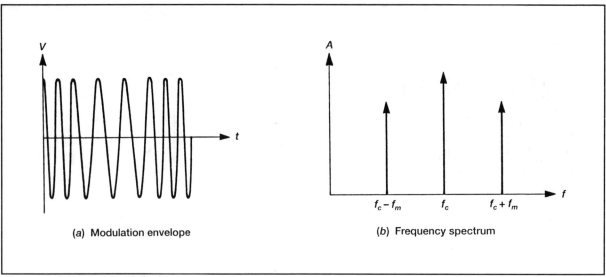

Figure 4.31 PM modulation

Solution:

$$V_c(t) = A_c \cos 2\pi f_c t = 1.0 \cos 2\pi (1 \times 10^6) t$$

$$V_m(t) = A_m \cos 2\pi f_m t = 0.5 \cos 2\pi (1 \times 10^3) t$$

$$V_{FM}(t) = A_c \cos[2\pi f_c t + (\Delta\varphi) A_m \sin 2\pi f_m t] = A_c \cos[2\pi f_c t + \beta \sin 2\pi f_m t]$$

Where $\beta = A_m \Delta\varphi$

Let $x(t) = A_m \sin 2\pi f_m t$, for $\Delta\varphi \ll \pi$,

Using $\cos(A+B) = (\cos A \cos B - \sin A \sin B)$

Let $A = \Delta\varphi x(t)$ and $B = 2\pi f_c t$

$$V_{FM}(t) = A_c [\cos(\Delta\varphi x(t)) \cos 2\pi f_c t - \sin(\Delta\varphi x(t)) \sin 2\pi f_c t]$$

$$V_{FM}(t) \approx A_c [\cos 2\pi f_c t - (\Delta\varphi x(t)) \sin 2\pi f_c t]$$

$$V_{FM}(f) = (A_c/2)[\delta(f - f_c) + \delta(f + f_c)] + j(A_c \Delta\varphi/2)[X(f - f_c) - X(f + f_c)]$$

4.32 nFM modulation

A 1 MHz sinewave carrier is frequency modulated using a 1 kHz signal. Determine the corresponding frequency spectrum for a modulation index, $\beta = 0.01$.

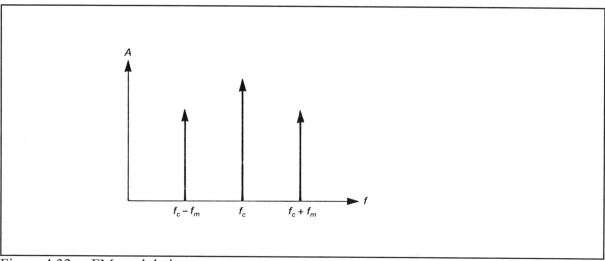

Figure 4.32 nFM modulation

Solution:

$V_{FM}(t) = A_c \cos[2\pi f_c t + \beta \sin 2\pi f_m t\,]$

Using $\cos(A+B) = (\cos A \cos B - \sin A \sin B)$

$V_{FM}(t) = A_c[\cos 2\pi f_c t \times \cos(\beta \sin 2\pi f_m t) - \sin 2\pi f_c t \times \sin(\beta \sin 2\pi f_m t)\,]$

for $\beta \ll 1$

$\cos(\beta \sin 2\pi f_m t) \approx 1$

$\sin(\beta \sin 2\pi f_m t) \approx \beta(\sin 2\pi f_m t)$

$V_{FM}(t) = A_c[\cos 2\pi f_c t - \sin 2\pi f_c t (\beta \sin 2\pi f_m t)\,]$

Using $\sin A \sin B = (1/2)[\cos(A-B) - \cos(A+B)]$

Let $A = 2\pi f_c t$ and $B = 2\pi f_m t$

$V_{FM}(t) = A_c \cos 2\pi f_c t - (\beta A_c/2)[\cos 2\pi(f_c - f_m)t + \cos 2\pi(f_c + f_m)t\,]$

$V_{FM}(t) = 10\cos 2\pi(1\times 10^6)t - (0.05)[\cos 2\pi(9.99\times 10^5)t + \cos 2\pi(1.001\times 10^6)t\,]$

4.33 wFM modulation

A 1 MHz sinewave carrier is frequency modulated using a bandlimited, baseband signal. The required carrier transmission bandwidth, B_T, is 180 kHz. Determine (a) the modulation signal bandwidth, f_m, and (b) peak frequency deviation, Δf, of the carrier for a modulation index, $\beta = 5$.

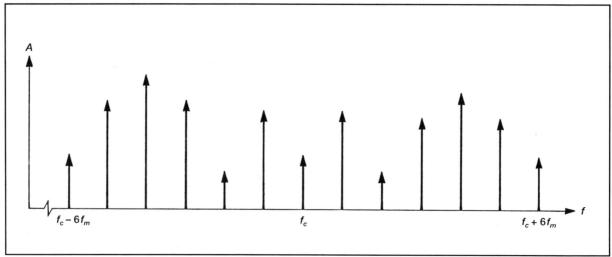

Figure 4.33 wFM modulation

Solution:

$$\beta = \Delta f / f_m$$

$$B_T = 2(\beta + 1)f_m$$

$$\beta = 5 = \Delta f / f_m$$

$$B_T = 180 \times 10^3 = 2(\beta + 1)f_m = 2(5 + 1)f_m$$

a) $f_m = (180 \times 10^3 / 12) = 15 \times 10^3$ Hz

b) $\Delta f = \beta f_m = 5(15 \times 10^3) = 75 \times 10^3$ Hz

$B_T = 2(\beta + 1)f_m = 2(5 + 1)(15 \times 10^3) = 180 \times 10^3$ Hz

4.34 AM/FM modulation

A baseband signal, bandlimited to 10 kHz, is used to AM modulate an 80 kHz sinewave subcarrier. The resulting waveform is used to FM modulate a 10 MHz sinewave carrier. Determine (a) the modulation index, β, and (b) the required carrier transmission bandwidth for a peak carrier deviation, Δf, of 180 kHz.

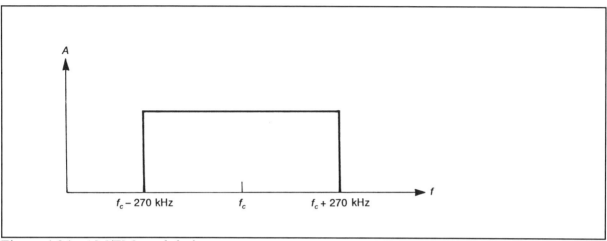

Figure 4.34 AM/FM modulation

Solution:

$$\text{AM Bandwidth} = f_s \pm f_m = 80 \times 10^3 \pm 10 \times 10^3 \text{ Hz}$$

$$\text{(AM)} \ f_{LSB} = f_s - f_m = 80 \times 10^3 - 10 \times 10^3 = 70 \times 10^3 \text{ Hz}$$

$$\text{(AM)} \ f_{USB} = f_s + f_m = 80 \times 10^3 + 10 \times 10^3 = 90 \times 10^3 \text{ Hz}$$

a) $\text{(FM)} \ \beta = \Delta f / f_m' = \Delta f / f_{USB} = (180 \times 10^3) / (90 \times 10^3) = 2$

$\text{(FM)} \ B_T = 2(\beta + 1) f_m' = 2(2+1) f_m' = 2(2+1) f_{USB} = 6 \times 90 \times 10^3 = 540 \times 10^3 \text{ Hz}$

b) $\text{(FM) Transmission bandwidth} = f_c \pm B_T/2 = 10 \times 10^6 \pm 270 \times 10^3 \text{ Hz}$

$\text{(FM)} \ f_{LSB} = f_c - B_T/2 = 10 \times 10^6 - 270 \times 10^3 = 9.730 \times 10^6 \text{ Hz}$

$\text{(FM)} \ f_{USB} = f_c + B_T/2 = 10 \times 10^6 + 270 \times 10^3 = 10.270 \times 10^6 \text{ Hz}$

4.35 FM/AM modulation

A baseband signal, bandlimited to 10 kHz, is used to FM modulate an 80 kHz sinewave subcarrier. The resulting waveform is used to AM modulate a 10 MHz sinewave carrier. Determine the required carrier transmission bandwidth if the peak subcarrier deviation, Δf, is 25 kHz.

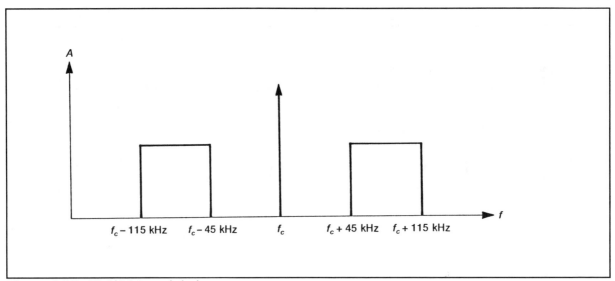

Figure 4.35 FM/AM modulation

Solution:

(FM) $\beta = \Delta f / f_m = (25 \times 10^3) / (10 \times 10^3) = 2.5$

(FM) $B_T = 2(\beta + 1) f_m = 2(2.5 + 1) f_m = 70 \times 10^3$ Hz

(FM) Bandwidth $= f_s \pm B_T/2$ kHz $= 80 \times 10^3 \pm 35 \times 10^3$ Hz

(FM) $f_{LSB} = f_s - B_T/2 = 80 \times 10^3 - 35 \times 10^3 = 45 \times 10^3$ Hz

(FM) $f_{USB} = f_s + B_T/2 = 80 \times 10^3 + 35 \times 10^3 = 115 \times 10^3$ Hz

AM Transmission bandwidth $= f_c \pm f_m' = f_c \pm f_{USB} = 10 \times 10^6 \pm 115 \times 10^3$ Hz

(AM) $f_{LSB} = f_c - f_m = 10 \times 10^6 - 115 \times 10^3 = 9.885 \times 10^6$ Hz

(AM) $f_{USB} = f_c + f_m = 10 \times 10^6 + 115 \times 10^3 = 10.115 \times 10^6$ Hz

4.36 SSB modulation

A 1 MHz sinewave carrier is modulated using a bandlimited 1 kHz signal to generate single-sideband (SSB) modulation. Determine the corresponding frequency spectrum using the sideband filter method.

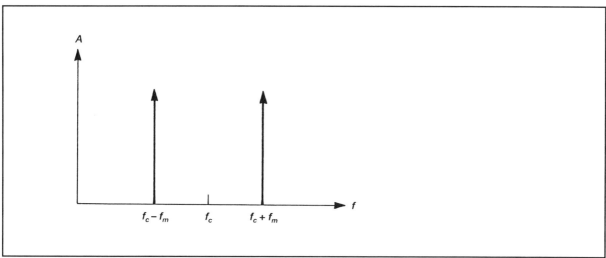

Figure 4.36 SSB modulation

Solution:

(Carrier) $V_c(t) = A_c \cos 2\pi f_c t = 1.0 \cos 2\pi (1 \times 10^6) t$

(Signal) $V_m(t) = A_m \cos 2\pi f_m t = 0.5 \cos 2\pi (1 \times 10^3) t$

$V_{SSB}(t) = A_c [A_m \cos 2\pi f_m t (\cos 2\pi f_c t) \mp A_m \sin 2\pi f_m t (\sin 2\pi f_c t)]$

Using $\cos A \cos B = (1/2)[\cos(A+B) + \cos(A-B)]$

Using $\sin A \sin B = (1/2)[\cos(A-B) - \cos(A+B)]$

$V_{SSB}(t) = (A_m A_c)[\cos 2\pi (f_c \pm f_m) t]$

$V_{SSB}(t) = (0.5)[\cos 2\pi (1 \times 10^6 \pm 1 \times 10^3) t]$

The carrier is eliminated and using either a USB or LSB bandpass filter SSB modulation results.

4.37 QAM modulation

A 1 MHz sinewave carrier is quadrature amplitude modulated (QAM) using a 1 kHz signal in phase and a 2 kHz signal at 90° as shown in Figure 4.37. Determine the resulting frequency spectrum.

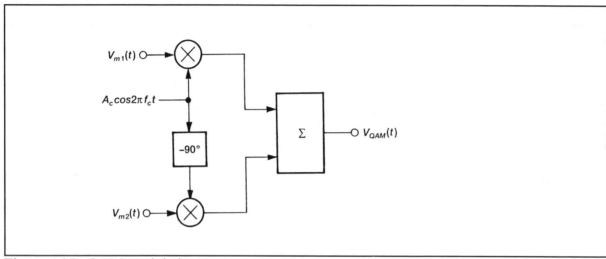

Figure 4.37 QAM modulation

Solution:

(Carrier) $V_c(t) = A_c\cos2\pi f_c t = 1.0\cos2\pi(1\times10^6)t$

(Signal 1) $V_{m1}(t) = A_m\cos2\pi f_{m1}t = 0.5\cos2\pi(1\times10^3)t$

(Signal 2) $V_{m2}(t) = A_m\cos2\pi f_{m2}t = 0.5\cos2\pi(2\times10^3)t$

$V_{QAM}(t) = (A_c\cos2\pi f_c t)V_{m1}(t) + A_c\cos(2\pi f_c t - 90°)V_{m2}(t)$

$V_{QAM}(t) = (A_c\cos2\pi f_c t)V_{m1}(t) + (A_c\sin2\pi f_c t)V_{m2}(t)$

Using $\cos A \cos B = (1/2)[\cos(A+B) + \cos(A-B)]$ where $A = 2\pi f_c t$ and $B = 2\pi f_{m1}t$

Using $\sin A \cos B = (1/2)[\sin(A+B) + \sin(A-B)]$ where $A = 2\pi f_c t$ and $B = 2\pi f_{m2}t$

$V_{QAM}(t) = (A_{m1}A_c/2)[\cos2\pi(f_c \pm f_{m1})t] + (A_{m2}A_c/2)[\sin2\pi(f_c \pm f_{m2})t]$

$V_{QAM}(t) = (0.25)[\cos2\pi(1\times10^6 \pm 1\times10^3)t] + (0.25)[\sin2\pi(1\times10^6 \pm 2\times10^3)t]$

The carrier is eliminated and the USB and LSB of each signal exist in quadrature within the band.

4.38 PAM modulation

A square wave whose period is 1×10^{-6} seconds is used to pulse amplitude modulate (PAM) a 1 kHz signal. Determine the resulting frequency spectrum.

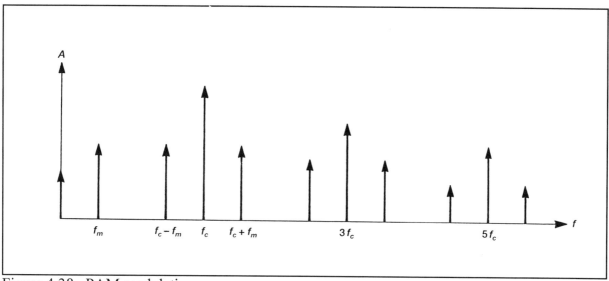

Figure 4.38 PAM modulation

Solution:

$f_c = 1/T = 1 \times 10^6$ Hz

$V_c(t) = A_c[a_0 + a_1\cos 2\pi f_c t + a_3\cos 2\pi(3f_c)t + a_5\cos 2\pi(5f_c)t + ... + a_n\cos 2\pi(nf_c)t]$ (n = odd)

$V_m(t) = A_m\cos 2\pi f_m t = A_m\cos 2\pi(1\times 10^3)t$

$V_{PAM}(t) = [1 + mV_m(t)]V_c(t)$

Using $\cos A \cos B = (1/2)[\cos(A+B) + \cos(A-B)]$

$V_{PAM}(t) = (a_0 A_m A_c) + (ma_0 A_m A_c)[\cos 2\pi f_m t] +$

$\qquad (a_1 A_m A_c)[\cos 2\pi f_c t] + [(ma_1 A_m A_c)/2][(\cos 2\pi(f_c \pm f_m)t] +$

$\qquad (a_3 A_m A_c)[\cos 2\pi(3f_c)t] + [(ma_3 A_m A_c)/2][(\cos 2\pi(3f_c \pm f_m)t] +$

$\qquad (a_5 A_m A_c)[\cos 2\pi(5f_c)t] + [(ma_5 A_m A_c)/2][(\cos 2\pi(5f_c \pm f_m)t] + ... +$

$\qquad (a_n A_m A_c)[\cos 2\pi(nf_c t)t] + [(ma_n A_m A_c)/2][(\cos 2\pi(nf_c \pm f_m)t]$

4.39 PDM modulation

A pulse duration modulation (PDM) signal can be generated using a reference PAM waveform. Draw a block diagram representation, and illustrate the signals at each intermediate point.

Solution:

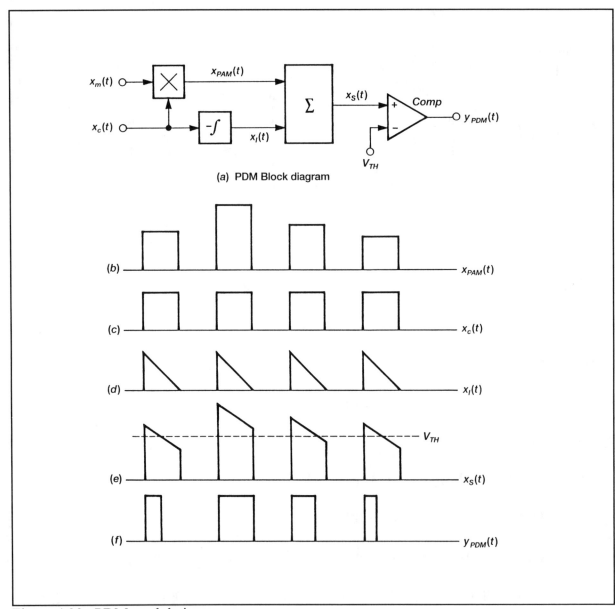

Figure 4.39 PDM modulation

4.40 PPM modulation

A pulse position modulation (PPM) signal can be generated using a reference PDM waveform. Draw a block diagram representation, and illustrate the signals at each intermediate point.

Solution:

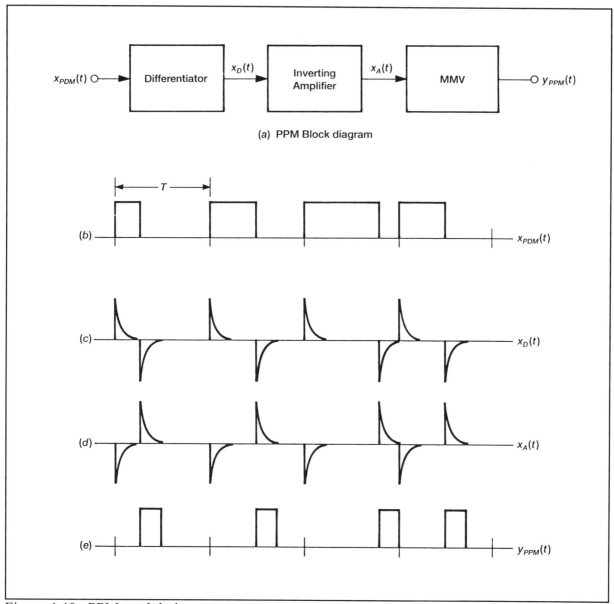

Figure 4.40 PPM modulation

4.41 PCM modulation

A pulse code modulation (PCM) signal can be generated from a PAM waveform by quantizing and encoding it. Draw a block diagram for a four-level PCM, and illustrate the signals at each stage.

Solution:

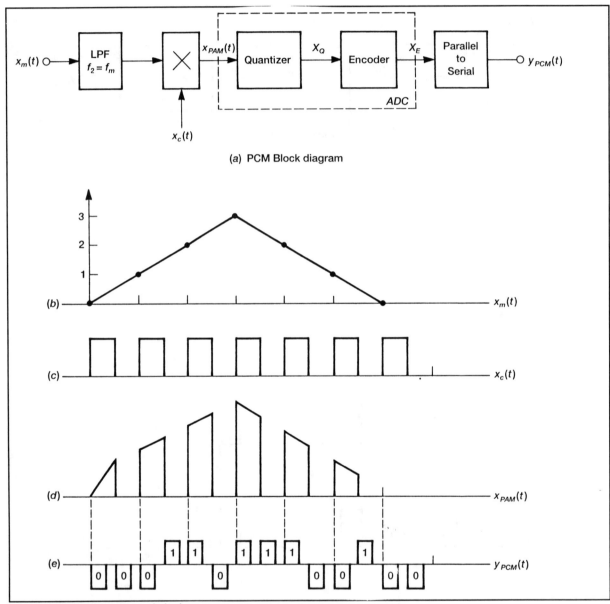

Figure 4.41 PCM modulation

4.42 FSK modulation

Draw a frequency shift keying (FSK) modulation block diagram, using a VCO, and define the output signal. Also, determine (a) the carrier frequency, f_c, and (b) frequency deviation, Δf, for $f(1) = 1{,}700$ Hz and $f(0) = 1{,}300$ Hz. For a modulation voltage, m(t), is ± 5 volts (c) find the VCO gain constant, k_v. Also, draw an FSK demodulator block diagram and (d) define its output signal.

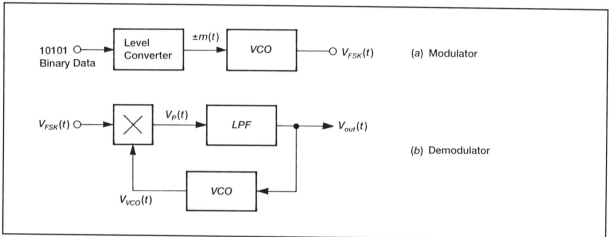

Figure 4.42 FSK modulation

Solution:

$$V_{FSK}(1) = A\sin[2\pi(f_c + f_m)t]$$

$$V_{FSK}(0) = A\sin[2\pi(f_c - f_m)t]$$

a) $f(1) = 1{,}700$ Hz $= (f_c + f_m)$

$f(0) = 1{,}300$ Hz $= (f_c - f_m)$

Solving for $f_c = 1{,}500$ Hz

b) Solving for $f_m = \pm 200$ Hz

c) $f_v(t) = f_c \pm m(t)k_v = 1{,}500 \pm 5 \times k_v$

$k_v = 40$ volts/Hz

d) The demodulator phase detector output is

$$V_p(t) = K_1 \sin[2\pi f_c t \pm \theta_c(t)][\cos(2\pi f_v t + \theta_v(t)]$$

Using $\sin A \cos B = (1/2)[\sin(A+B) + \sin(A-B)]$

The filtered output $V_{out}(t) = K_2 \sin(\pm \theta_c - \theta_v) \approx \pm \theta_c - \theta_v$

4.43 PSK modulation

Draw a phase shift keying (PSK) modulation block diagram and define the output signal. Also, draw a PSK demodulator block diagram and define its output signal.

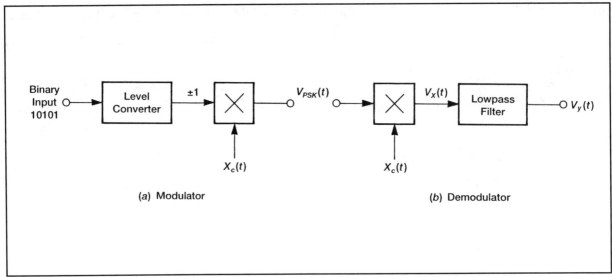

Figure 4.43   PSK modulation

Solution:

$X_c(t) = A_c \cos(2\pi f_c t)$

$V_{PSK}(t) = \pm A_c \cos(2\pi f_c t)$

$V_{PSK}(t) = A_c \cos(2\pi f_c t + \theta)$,   where $\theta = 0°, 180°$

For the demodulator

$V_x(t) = V_{PSK}(t) X_c(t) = [A_c \cos(2\pi f_c t + \theta)][A_c \cos(2\pi f_c t)]$

Using $\cos A \cos B = (1/2)[\cos(A+B) + \cos(A-B)]$

$V_x(t) = (1/2)[\cos(4\pi f_c t + \theta) + \cos(\theta)]$

$V_y(t) = (1/2)\cos(\theta)$

4.44 Balanced modulator

A balanced modulator consists of two AM modulators whose outputs are combined as to eliminate the carrier signal, while retaining the sidebands. Determine the output signal frequency spectrum using a 1 MHz sinewave carrier and a 1 kHz signal.

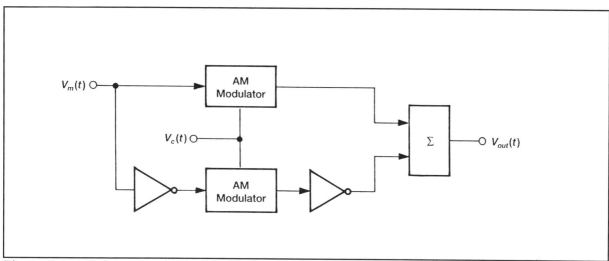

Figure 4.44 Balanced modulator

Solution:

$$V_{out}(t) = V_c(t)[1 + mV_m(t)] - V_c(t)[1 - mV_m(t)]$$

(Carrier) $V_c(t) = A_c \cos 2\pi f_c t = 1.0 \cos 2\pi (1 \times 10^6)t$

(Signal) $V_m(t) = A_m \cos 2\pi f_m t = 0.5 \cos 2\pi (1 \times 10^3)t$

$$V_{out}(t) = [A_c \cos 2\pi f_c t + A_c \cos 2\pi f_c t (mV_m(t)) - A_c \cos 2\pi f_c t + A_c \cos 2\pi f_c t(mV_m(t))]$$

$$V_{out}(t) = [2A_c \cos 2\pi f_c t][mV_m(t)] = [2A_c \cos 2\pi f_c t][mA_m \cos 2\pi f_m t]$$

Using $\cos A \cos B = (1/2)[\cos(A+B) + \cos(A-B)]$

$$V_{out}(t) = A_c m A_m [\cos 2\pi (f_c \pm f_m)t]$$

$$V_{out}(t) = (0.5m)[\cos 2\pi (1 \times 10^6 \pm 1 \times 10^3)t]$$

4.45 Envelope detector

An envelope detector circuit consists of a diode followed by an RC lowpass filter as shown in Figure 4.45. Determine the maximum AM modulation index which will not produce output distortion for a maximum signal bandwidth of 5 kHz.

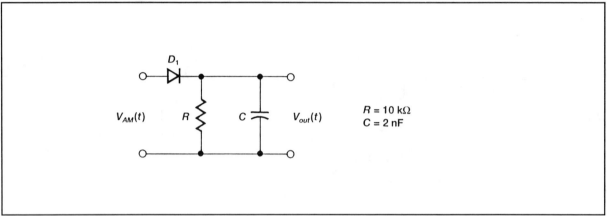

Figure 4.45 Envelope detector

Solution:

$V_{out}(t)$ follows the envelope of $V_{AM}(t)$

$V_{AM}(t) = V_c(t)[1 + mV_m(t)] = V[1 + m\cos 2\pi f_m t]\cos 2\pi f_c t$

$V_{out}(t) = V[1 + m\cos 2\pi f_m t]$

$I_{out}(t) = I_{DC} + I_{AC}[\cos 2\pi f_m t + \theta],$ Where $I_{DC} = V/R$, and $I_{AC} = (mV)/|Z|$

$Z = \dfrac{1}{\dfrac{1}{R} + j\omega C} = \dfrac{R}{1 + j\omega RC}$

$|Z| = \dfrac{R}{[(1 + j\omega RC)(1 - j\omega RC)]^{1/2}} = \dfrac{R}{[1 + (\omega RC)^2]^{1/2}}$

Distortion occurs when $I_{DC} = I_{AC}$, since $I_{DC} = V/R$

$I_{AC} = (mV)/|Z| = (mV/R)[1 + (\omega RC)^2]^{1/2} = mI_{DC}[1 + (\omega RC)^2]^{1/2}$

$I_{AC}/I_{DC} = 1 = m[1 + (\omega RC)^2]^{1/2}$

$\omega = \omega_m = 2\pi f_m = 2\pi(5 \times 10^3) = 31.416 \times 10^3$

$RC = 1 \times 10^4 \times 2 \times 10^{-9} = 2 \times 10^{-5}$

$m \le \dfrac{R}{[1 + (\omega RC)^2]^{1/2}} \le \dfrac{1}{[1 + 0.4]^{1/2}} \le 0.84$

4.46 Square law detector

A square law detector circuit consists of a square law device followed by a lowpass filter as shown in Figure 4.46. Determine the output signal corresponding to an AM modulated 1 MHz carrier with a 1 kHz signal.

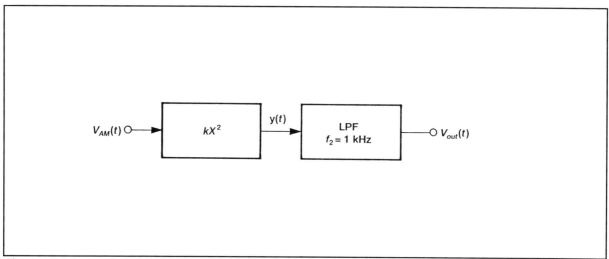

Figure 4.46 Square law detector

Solution:

$$V_{AM}(t) = V_c(t)[1 + mV_m(t)]$$

(Carrier) $V_c(t) = A_c \cos 2\pi f_c t = 1.0 \cos 2\pi (1 \times 10^6) t$

(Signal) $V_m(t) = A_m \cos 2\pi f_m t = 0.5 \cos 2\pi (1 \times 10^3) t$

$$y(t) = k[V_c(t) + mV_c(t)V_m(t)]^2 = k[V_c^2(t) + 2mV_m(t)V_c^2(t) + m^2 V_m^2(t) V_c^2(t)]$$

$$y(t) = kA_c^2 [(\cos 2\pi f_c t)^2 + (2mA_m \cos 2\pi f_m t)(\cos 2\pi f_c t)^2 + m^2 (A_m \cos 2\pi f_m t)^2 (\cos 2\pi f_c)^2]$$

Using $\cos^2 A = [0.5 + (0.5)\cos 2A]$

$$y(t) = kA_c^2 [0.5 + 0.5 \cos 2\pi (2f_c t) + (mA_m \cos 2\pi f_m t)(1 + \cos 2\pi (2f_c t)) +$$

$$m^2 (A_m \cos 2\pi f_m t)^2 (0.5 + 0.5 \cos 2\pi (2f_c t))]$$

Since the LPF cutoff frequency is 1×10^3 Hz, the output becomes

$$V_{out}(t) = kA_c^2 [0.5 + mA_m \cos 2\pi f_m t + (0.5 m^2)(A_m \cos 2\pi f_m t)^2]$$

$$V_{out}(t) = kA_c^2 [0.5 + mV_m(t) + (0.5 m^2) V_m(t)^2]$$

4.47 Synchronous detector

A synchronous detector circuit consists of a multiplier device followed by a lowpass filter as shown in Figure 4.47. Determine the output signal corresponding to an AM modulated 1 MHz carrier with a 1 kHz signal.

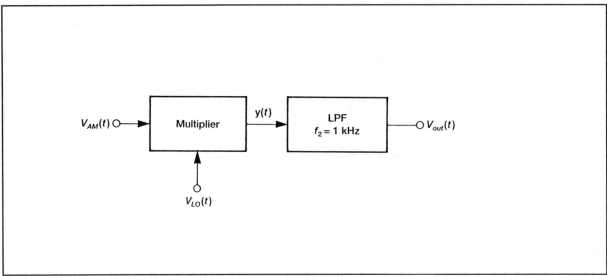

Figure 4.47 Synchronous detector

Solution:

$$V_{AM}(t) = V_c(t)[1 + mV_m(t)]$$

(Carrier) $V_c(t) = A_c \cos 2\pi f_c t = 1.0 \cos 2\pi (1 \times 10^6) t$

(Signal) $V_m(t) = A_m \cos 2\pi f_m t = 0.5 \cos 2\pi (1 \times 10^3) t$

$$V_{LO}(t) = A_{LO}[\cos 2\pi f_c t] = 1.0 \cos 2\pi (1 \times 10^6) t$$

$$y(t) = [V_c(t) + mV_c(t)V_m(t)]V_{LO}(t)$$

$$y(t) = A_c A_{LO}(\cos 2\pi f_c t)^2 + (m A_c A_{LO} A_m)(\cos 2\pi f_c t)^2 (\cos 2\pi f_m t)]$$

Using $\cos^2 A = [0.5 + (0.5)\cos 2A]$

$$y(t) = (0.5 A_c A_{LO})[1 + \cos 2\pi (2f_c t)] + (0.5 m A_m A_c A_{LO})(\cos 2\pi f_m t)[1 + \cos 2\pi (2f_c t)]$$

Since the LPF cutoff frequency is 1×10^3 Hz, the output becomes

$$V_{out}(t) = 0.5 A_c A_{LO}[1 + (mA_m)(\cos 2\pi f_m t)]$$

4.48 Ratio detector

A ratio detector circuit consists of a frequency-to-voltage converter followed by an envelope detector/limiter and a lowpass filter as shown in Figure 4.48. Determine the output signal corresponding to an FM modulated carrier input signal, $X_c(t) = A_c \cos\theta_c(t)$.

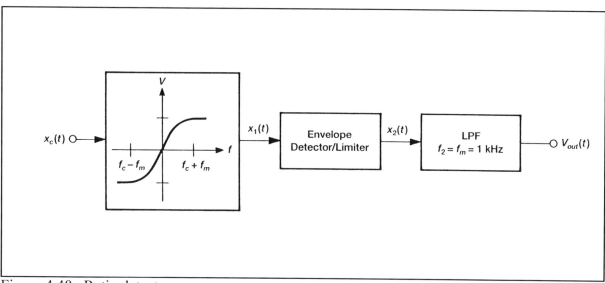

Figure 4.48 Ratio detector

Solution:

Note: $\theta_c(t) = 2\pi f_c t + \phi(t) = [2\pi f_c t + (2\pi\Delta f)\int V_m(t)dt]$

and $\dfrac{d\theta_c(t)}{dt} = 2\pi[f_c + (\Delta f)V_m(t)]$

$X_c(t) = A_c\cos\theta_c(t) = A_c\cos[2\pi f_c t + (2\pi\Delta f)\int V_m(t)dt]$

$x_1(t) = \dfrac{d}{dt}X_c(t) = -A_c \sin\theta_c(t)\left[\dfrac{d\theta_c(t)}{dt}\right]$

$x_1(t) = -A_c\sin\theta_c(t)[2\pi(f_c + (\Delta f)V_m(t)dt)]$

$x_2(t) = 2\pi A_c[f_c + (\Delta f)V_m(t)]$

since $V_m(t) = A_m\cos 2\pi f_m t$

$x_2(t) = 2\pi A_c[f_c + (\Delta f)A_m\cos 2\pi f_m t]$

$V_{out}(t) = (2\pi A_c A_m \Delta f)[\cos 2\pi f_m t]$

4.49 AM receiver

Draw a well labeled AM receiver block diagram, and state its various frequency specifications.

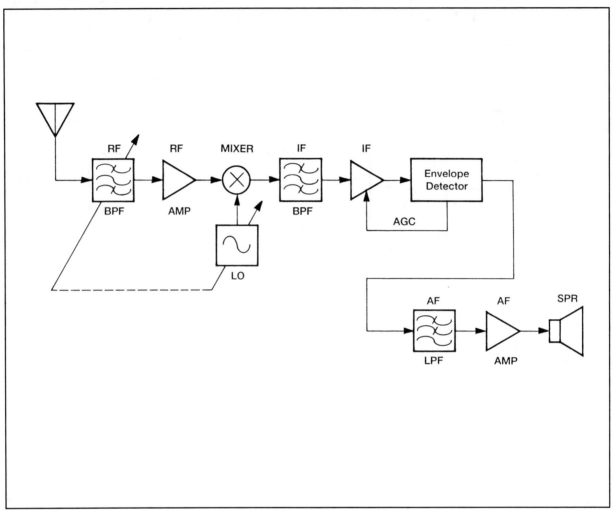

Figure 4.49 AM receiver

Solution:

RF BPF	Tuning range: 535 - 1,605 kHz;	Channel Spacing: 10 kHz
	Bandwidth: 10 kHz < BW_{RF} < 910 kHz	
LO	Tuning range: 990 - 2,060 kHz;	$f_{LO} = f_c + f_{IF}$
IMAGE	Frequency range: 80 - 1,150 kHz;	$f_{IM} = f_c - f_{IF}$
IF BPF	Center frequency: 455 kHz;	$BW_{IF} \leq 10$ kHz
LPF	Cutoff frequency: $f_m = 5$ kHz	

4.50 FM receiver
Draw a well labeled FM receiver block diagram, and state its various frequency specifications.

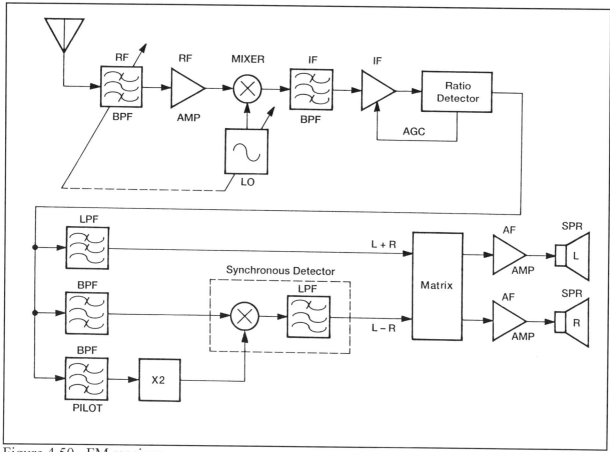

Figure 4.50 FM receiver

Solution:

RF BPF   Tuning range: 88 - 108 MHz;   Channel Spacing: 200 kHz

   Bandwidth: 200 kHz < BW_{RF} < 21.4 MHz

LO   Tuning range: 77.3 - 97.3 MHz; $f_{LO} = f_c - f_{IF}$

IMAGE   Frequency range: 98.7 - 118.7 MHz; $f_{IM} = f_c + f_{IF}$

IF BPF   Center frequency: 10.7 MHz; $BW_{IF} \leq 180$ kHz

LPF   Cutoff frequency: $f_m = 15$ kHz

BPF   Center frequency: 38 kHz;   Bandwidth: 23 - 53 kHz

Pilot BPF   Center frequency: 19 kHz;   Bandwidth: 15 - 23 kHz

4.51 Sample and Hold

The sample and hold circuit shown in Figure 4.51 has a sample pulsewidth, t_s, of 7×10^{-6} seconds, and a hold pulsewidth, t_h, of 50×10^{-6} seconds. The control signal pulse rise and fall times, t_r and t_f, equal 2×10^{-6} seconds. The FET switch R_{on} value equals 10 ohms. The op amp $Z_{in} = 1\times10^6$ ohms, and its slew rate = 5 volts/μsec.

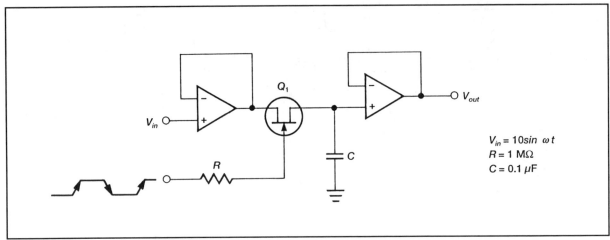

Figure 4.51 Sample and Hold

1. The sample error (%) is most nearly:

   a) 0.3 b) 0.1 c) 1.3 d) 1.7

2. The hold error (%) is most nearly:

   a) 0.5 b) 1.0 c) 0.01 d) 0.05

3. The maximum large input signal frequency is most nearly:

   a) 80×10^3 b) 13×10^3 c) 125×10^3 d) 40×10^3

4. The maximum large input signal sample rate is most nearly:

   a) 40×10^3 b) 250×10^3 c) 80×10^3 d) 160×10^3

5. The maximum small input signal sample rate is most nearly:

   a) 1.43×10^5 b) 5×10^5 c) 2.5×10^5 d) 1.11×10^5

4.51 Solution:

1. $\tau = R_{on}C = 10 \times 0.1 \times 10^{-6} = 1 \times 10^{-6}$ seconds

   $t_s = 7 \times 10^{-6} = 7\tau$

   $V_s = V_{in}(1 - e^{-7}) = 9.99$

   error(%) = $[(10 - 9.99)/10] \times 100 = 0.1\%$

   The correct answer is (b).

2. $I = V_{in} / Z_{in} = 10 / (1 \times 10^6) = 10 \times 10^{-6}$ amps

   $\Delta V = It / C = (10 \times 10^{-6} \times 50 \times 10^{-6}) / (0.1 \times 10^{-6}) = 5 \times 10^{-3}$ volts

   error(%) = $(\Delta V / V) \times 100 = [5 \times 10^{-3} / 10] \times 100 = 0.05\%$

   The correct answer is (d).

3. $f_{max} = SR / 2\pi V_{peak} = (5 \times 10^6) / (2\pi \times 10) \approx 80 \times 10^3$ Hz

   The correct answer is (a).

4. $T = 1/2f_{max} = 1 / (2 \times 80 \times 10^3) = 6.3 \times 10^{-6}$

   $1/T = 160 \times 10^3$ samples per second

   The correct answer is (d).

5. $T = t_{ap} + t_{acq} = t_r + t_f = 4 \times 10^{-6}$ seconds

   $1/T = 2.5 \times 10^5$ samples per second

   The correct answer is (c).

4.52 Noise Power

A simplified front end block diagram of a radio receiver is shown in Figure 4.52. It consists of the antenna, a pre-amplifier, a passive filter, and an RF amplifier. The signal bandwidth is 3×10^5 Hz.

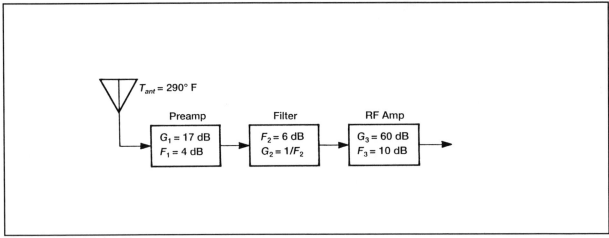

Figure 4.52 Noise Power

1. The available noise power (dBm) at the pre-amplifier input is most nearly:

   a) -174   b) -229   c) -119   d) -148

2. The effective noise temperature of the pre-amplifier is most nearly:

   a) 870   b) 1,160   c) 725   d) 580

3. The pre-amplifier output noise power is most nearly:

   a) 8.16×10^{-14} b) 6.1×10^{-14} c) 1.8×10^{-13} d) 2.4×10^{-13}

4. The overall receiver noise figure (dB) is most nearly:

   a) 3.34   b) 5.24   c) 7.13   d) 3.9

5. The overall receiver noise figure (dB) without the pre-amplifier is most nearly:

   a) 40   b) 16   c) 66   d) 60

4.52 Solution:

1. $N_0 = kTB = 1.38 \times 10^{-23} \times 290 \times 1 = 4 \times 10^{-21}$

   $N_0 \text{ (dBm)} = 10\log(N_0 / 1 \times 10^{-3}) = -174 \text{ dBm}$

   $N_{in} = N_0 + 10\log(B) = -174 + 10\log(3 \times 10^5) = -174 + 55 = -119 \text{ dBm}$

   The correct answer is (c).

2. $NF = (1 + T_e/T_{in})$

   $T_e = T_{in}(NF - 1) = 290°(4 - 1) = 870°K$

   The correct answer is (a).

3. $17 \text{ dB} = 10\log P$

   $P = \log^{-1}(1.7) = 50$

   $N_{out} = kPB(T_{in} + T_e) = 1.38 \times 10^{-23} \times 50 \times 3 \times 10^5 \times 1,160 = 2.4 \times 10^{-13} \text{ watts}$

   The correct answer is (d).

4. $F = F_1 + (F_2 - 1)/G_1 + (F_3 - 1)/G_1G_2$

   $F = 2.5 + (4 - 1)/50 + (10 - 1)/(50 \times 0.25) = 3.34$

   $F \text{ (dB)} = 10\log(3.34) = 5.24 \text{ dB}$

   The correct answer is (b).

5. $F = F_2 + (F_3 - 1)/G_2$

   $F = 4 + (10 - 1)/0.25 = 40$

   $F \text{ (dB)} = 10\log(40) = 16 \text{ dB}$

   The correct answer is (b).

4.53 Channel capacity

The block diagram shown in Figure 4.53 represents a simplified datalink. It consists of a transmitter, a noisy channel, and a receiver.

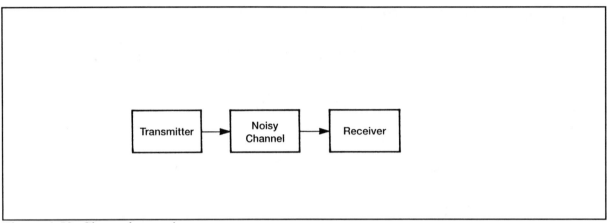

Figure 4.53 Channel capacity

1. The symbol rate, r_s, over the channel having an information rate, R, of 38.4×10^3 bps and an information content, H, of 16 bits per symbol is:

   a) 9,600 b) 153.6×10^3 c) 76.8×10^3 d) 2,400

2. The information rate, R, over a bandlimited 3,000 Hz channel using a 256 level transmission signaling scheme is:

   a) 768×10^3 b) 48×10^3 c) 384×10^3 d) 24×10^3

3. The channel capacity, C, of a 3,000 Hz noisy channel with a signal-to-noise ratio (S/N) of 511 is most nearly:

   a) 27×10^3 b) 8×10^3 c) 19×10^3 d) 13.5×10^3

4. The signal-to-noise ratio (S/N), in dB, of a 2,400 Hz noisy channel with a channel capacity of 27×10^3 bps is most nearly:

   a) 22.5 b) 17 c) 34 d) 11.25

5. The ASCII character rate over a channel, having a capacity of 153×10^3 bps, using one start and two stop bits is most nearly:

   a) 4,000 b) 3,111 c) 3,500 d) 2,800

4.53 Solution:

1. $R = r_s H$

    $r_s = R/H = 38.4 \times 10^3 / 16 = 2,400$ symbols/sec

    The correct answer is (d).

2. $R = \log_2(M) \times 2B = \log_2(256) \times 2 \times 3,000 = 48 \times 10^3$ bps

    The correct answer is (b).

3. $C = B\log_2(1 + S/N) = 3,000\log_2(1 + 511) = 3,000 \times 9 = 27,000$ bps

    The correct answer is (a).

4. $C/B = 27,000 / 2,400 = 11.25$

    $S/N = 2^{C/B} - 1 = 2^{11.25} - 1 \approx 2,434$

    S/N (dB) $= 10\log(2,434) \approx 34$ dB

    The correct answer is (c).

5. $t_{bit} = 1/R = 1/28,000 = 35.7 \times 10^{-6}$ seconds

    $H = \log_2(128) = 7$ bits/character

    $T_{char} = 1 + 7 + 2 = 10\ t_{bit}$

    $r_s = 1/T_{char} = 1/(357 \times 10^{-6}) = 2,800$ characters/sec

    The correct answer is (d).

4.54 DAC

The simplified schematic diagram of an eight-bit Digital-to-Analog (DAC) converter is shown in Figure 4.54. It consists of a weighted resistor network, and an op amp.

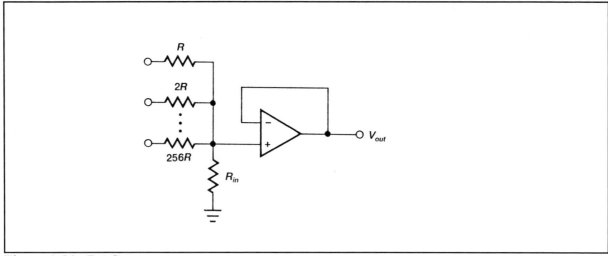

Figure 4.54 DAC

1. The LSB and FS output values, using a 10 volt reference voltage, are most nearly:

   a) 0.08 and 9.2   b) 0.02 and 9.8

   c) 1.25 and 8.75   d) 0.04 and 9.96

2. The LSB and ± FS output values, using a ± 5 volt reference voltage, are most nearly:

   a) 0.04 and 4.96, -5   b) 0.625 and 4.375, -5

   c) 0.08 and 4.92, -5   d) 0.02 and 4.98, -5

3. The DAC resolution (%) is most nearly:

   a) 0.2   b) 0.6   c) 0.4   d) 0.8

4. The DAC dynamic range (dB) is most nearly:

   a) 48   b) 18   c) 9   d) 24

5. The DAC output value, using a 10 volt reference, for an input of 10101101_2 is most nearly:

   a) 0.363   b) 6.75   c) 0.725   d) 3.375

4.54 Solution:

1. LSB = $V_{ref}/2^n$ = 10/256 = 0.04 volts

   FS = $V_{ref}(1 - LSB)$ = 10 - 0.04 = 9.96 volts

   The correct answer is (d).

2. LSB = $|V_{ref}|/2^{n-1}$ = 5/128 = 0.04 volts

   FS = $V_{ref}(1 - LSB)$ = 5 - 0.04 = 4.96 volts

   -FS = $-V_{ref}$ = -5 volts

   The correct answer is (a).

3. Resolution ($\%V_{ref}$) = $(1/2^n) \times 100$ = $(1/256) \times 100$ = 0.4 %

   The correct answer is (c).

4. Dynamic range (dB) = $20\log(2^n)$ ≈ 48 dB

   The correct answer is (a).

5. $V_{out} = V_{ref}[\sum a_i b_i]$

   V_{out} = 10[1/2 + 0 + 1/8 + 0 + 1/32 + 1/64 + 0 + 1/256] = 10[0.675] = 6.75 volts

   The correct answer is (b).

4.55 ADC

The simplified block diagram shown in Figure 4.55 represents an eight-bit successive approximation Analog-to-Digital (ADC) converter. It consists of a comparator, a successive approximation register (SAR), and a DAC. The module clock input frequency is 1×10^6 Hz, and $v_{in} = 10\sin\omega t$.

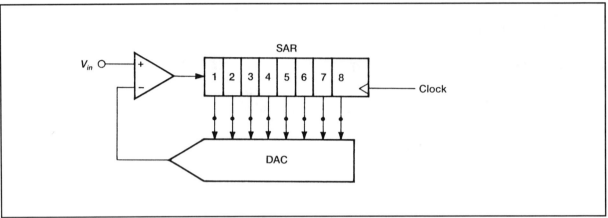

Figure 4.55 ADC

1. The conversion rate per second is most nearly:

   a) 8×10^6 b) 1×10^6 c) 2×10^6 d) 1.25×10^5

2. The maximum input signal frequency, for ± 1 LSB accuracy, using a sample and hold circuit is most nearly:

   a) 1.25×10^5 b) 6.25×10^4 c) 2.5×10^5 d) 5×10^5

3. The maximum input signal frequency, for ± 1 LSB accuracy, without a sample and hold circuit is most nearly:

   a) 15.62×10^3 b) 77.5   c) 155   d) 488

4. The RMS quantization noise of the converter is most nearly:

   a) 22×10^{-3} b) 40×10^{-3} c) 80×10^{-3} d) 11×10^{-3}

5. Converter dynamic range, in dB, is most nearly:

   a) 48   b) 42   c) 39   d) 45

4.55 Solution:

1. $T = 1/f = 1 \times 10^{-6}$ seconds

   $T_c = nT = 8T = 8 \times 10^{-6}$ seconds

   $R = 1/T_c = 1.25 \times 10^5$ conversions per second

   The correct answer is (d).

2. $f_{max} = 1/(2T_c) = 1/(16 \times 10^{-6}) = 6.25 \times 10^4$ Hz

   The correct answer is (b).

3. $f_{max} = 1/(2^n \pi T_c) = 1/(256 \times \pi \times 8 \times 10^{-6}) = 155$ Hz

   The correct answer is (c).

4. $q = (V_{max} - V_{min})/2^n = (2V_{peak})/2^n = 20/256 = 0.078$ volts

   RMS noise $= q/\sqrt{12} = 22 \times 10^{-3}$ volts

   The correct answer is (a).

5. Dynamic Range (dB) $= 6.02n - 3.01 = (6.02 \times 8) - 3.01 \approx 45$ dB

   The correct answer is (d).

NOTES:

5.0 ELECTROMAGNETICS

5.1 Resistance

Determine the resistance of a solid copper conductor at 100 MHz as shown in Figure 5.1. The conductivity of copper, σ, equals 5.8×10^7 siemens/meter.

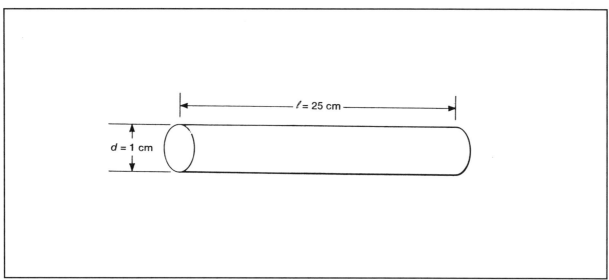

Figure 5.1 Resistance

Solution:

$$R = \left(\frac{R_s}{2\pi r}\right) l$$

$$R_s = [(\pi f \mu) / \sigma]^{1/2}$$

$\mu = 4\pi \times 10^{-7}$ henries/meter

$\sigma = 5.8 \times 10^7$ siemens/meter

$$R_s = \sqrt{\frac{\pi \times 1 \times 10^8 \times 4\pi \times 10^{-7}}{5.8 \times 10^7}} = 2.61 \times 10^{-3} \text{ ohms / meter}$$

$r = d / 2 = 0.5$ cm

$l = 25$ cm

$$R = \frac{2.61 \times 10^{-3}}{2\pi \times 0.005} \times (0.25) = 2.078 \times 10^{-2} \text{ ohms}$$

5.2 Inductance

Determine the inductance of a copper coaxial cable as shown in Figure 5.2. The permeability of copper, μ, equals $4\pi \times 10^{-7}$ henries/meter.

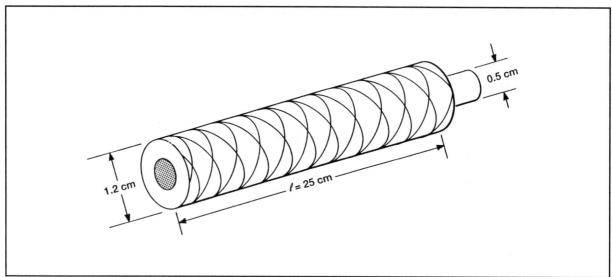

Figure 5.2 Inductance

Solution:

$$L = \frac{\mu}{2\pi} \ln\left(\frac{b}{a}\right) l$$

b = outer radius = 1.2/2 = 0.6 cm

a = inner radius = 0.5/2 = 0.25 cm

l = 25 cm

μ = $4\pi \times 10^{-7}$ henries/meter

$$L = \frac{4\pi \times 10^{-7}}{2\pi} \times \ln\left(\frac{0.6 \times 10^{-2}}{0.25 \times 10^{-2}}\right) \times (0.25)$$

$L = (2 \times 10^{-7})(0.875) \times (0.25)$

$L = 0.4375 \times 10^{-7}$ henries

5.3 Capacitance

Two pennies are held 0.1 cm apart as shown in Figure 5.3. Determine the resulting capacitance. The permittivity of free space, ϵ_0, equals 8.854×10^{-12} farads/meter.

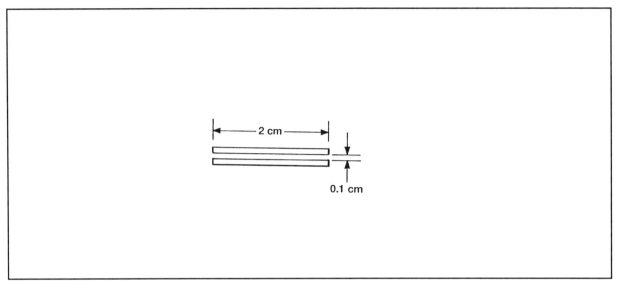

Figure 5.3 Capacitance

Solution:

$$C = \epsilon_0 \left(\frac{A}{d} \right)$$

$\epsilon_0 = 8.854 \times 10^{-12}$ farads/meter

$A = \pi r^2 = \pi (1 \times 10^{-2})^2$

$A = \pi \times 10^{-4}$

$C = \epsilon_0 (\pi \times 10^{-4}) / (1 \times 10^{-3}) = \epsilon_0 (\pi \times 10^{-1})$

$C = 8.854 \times 10^{-12} \times (\pi \times 10^{-1}) = 2.78 \times 10^{-12}$ farads

5.4 Parallel plate capacitor

Two capacitors are formed using a pair of pennies separated by two spacers with different dielectric constant. Figure 5.4a shows a capacitor formed with the dielectric spacers in parallel. Figure 5.4b shows a capacitor formed with the dielectric spacers in series. Determine the resulting capacitances using dielectric constant values, k_1, k_2, of 2.1 and 3.4 respectively.

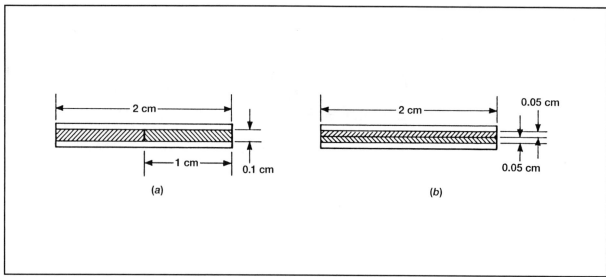

Figure 5.4 Parallel plate capacitor

Solution:

a) $C = \dfrac{\epsilon_1 A_1 + \epsilon_2 A_2}{d} = \dfrac{\epsilon_1 A + \epsilon_2 A}{2d}$

$A = \pi r^2 = \pi(1\times 10^{-2})^2 = \pi(1\times 10^{-4})$

$\epsilon_0 = 8.854\times 10^{-12}$ farads/meter

$\epsilon_1 = k_1 \epsilon_0 = 2.1\epsilon_0$

$\epsilon_2 = k_2 \epsilon_0 = 3.4\epsilon_0$

$C = 7.65\times 10^{-11}$ farads

b) $C = \dfrac{A}{\dfrac{d_1}{\epsilon_1} + \dfrac{d_2}{\epsilon_2}} = \dfrac{\pi \times 1 \times 10^{-4}}{\dfrac{0.05 \times 10^{-12}}{18.6 \times 10^{-12}} + \dfrac{0.05 \times 10^{-12}}{30.1 \times 10^{-12}}}$

$C = 7.22\times 10^{-12}$ farads

5.5 Spherical capacitor

A capacitor is formed using two concentric copper spheres separated by a material with a dielectric constant of two. Determine the resulting capacitance using the dimensions shown in Figure 5.5.

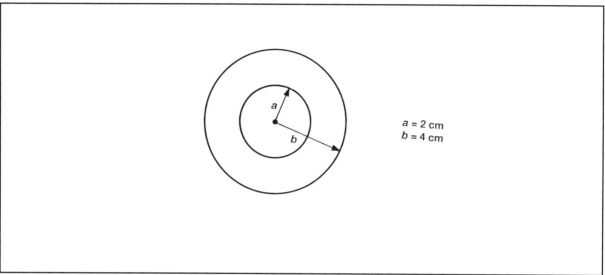

Figure 5.5 Spherical capacitor

Solution:

$$C = 4\pi\epsilon_1 \left(\frac{ab}{b - a} \right)$$

$\epsilon_0 = 8.854 \times 10^{-12}$ farads/meter

$\epsilon_1 = k_1 \epsilon_0 = 2\epsilon_0$

$$C = 2.225 \times 10^{-10} \left(\frac{2 \times 10^{-2} \times 4 \times 10^{-2}}{4 \times 10^{-2} - 2 \times 10^{-2}} \right) = 2.225 \times 10^{-10} \left(\frac{8 \times 10^{-4}}{2 \times 10^{-2}} \right)$$

$C = 2.225 \times 10^{-10} (4 \times 10^{-2})$

$C = 8.9 \times 10^{-12}$ farads

5.6 Cylindrical capacitor

A capacitor is formed using two concentric copper cylinders separated by a material with a dielectric constant, k_1, of two. Determine the resulting capacitance using the dimensions shown in Figure 5.6.

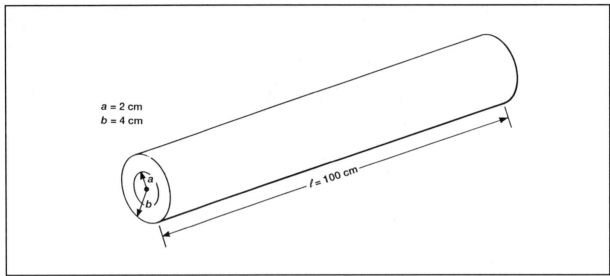

Figure 5.6 Cylindrical capacitor

Solution:

$$C = \frac{2\pi\epsilon_1 l}{\ln\left(\dfrac{b}{a}\right)}$$

$\epsilon_0 = 8.854 \times 10^{-12}$ farads/meter

$\epsilon_1 = k_1 \epsilon_0 = 2\epsilon_0$

$C = [1.1126 \times 10^{-10}] / \ln[(4 \times 10^{-2}) / (2 \times 10^{-2})]$

$C = [1.1126 \times 10^{-10}] / [0.693]$

$C = 160.55 \times 10^{-12}$ farads

5.7 Electric force

Two electric point charges are placed 25 cm apart as shown in Figure 5.7. The value of q_1 equals 50×10^{-6} coulombs, and q_2 equals 25×10^{-6} coulombs. Determine the electric force exerted by q_1 on q_2.

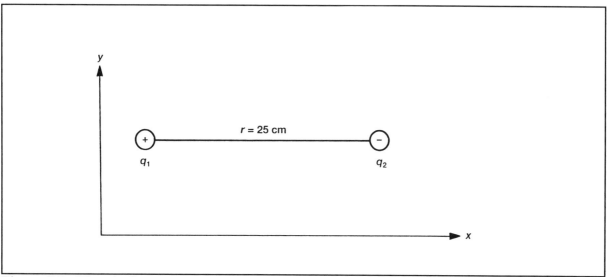

Figure 5.7 Electric force

Solution:

$$F_{12} = \kappa \left(\frac{q_1 q_2}{r^2} \right)$$

$$\kappa = 1/(4\pi\epsilon_0)$$

$\epsilon_0 = 8.85 \times 10^{-12}$ farads/meter (permittivity of free space)

$$F_{12} = 9 \times 10^9 \left(\frac{50 \times 10^{-6} \times 25 \times 10^{-6}}{0.25 \times 0.25} \right)$$

$$F_{12} = 9 \times 10^9 \left(\frac{1.25 \times 10^{-9}}{0.0625} \right) = 180 \text{ newtons}$$

5.8 Electric field

A stationery electric point charge has a value of $q_1 = 50 \times 10^{-6}$ coulombs as shown in Figure 5.8. Determine (a) the magnitude and (b) direction of the electric field, **E**, at a point p(x,y,z) where x = 30 cm, y = 40 cm, and z = 50 cm.

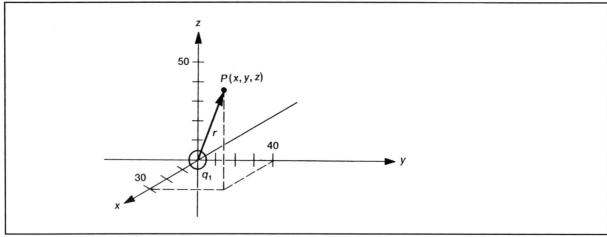

Figure 5.8 Electric field

Solution:

a) $\mathbf{E} = \kappa [q_1 / r^2]$

$\kappa = 1/(4\pi\epsilon_0) = 9 \times 10^9$

$r^2 = x^2 + y^2 + z^2 = (0.3)^2 + (0.4)^2 + (0.5)^2$

$r^2 = (0.09) + (0.16) + (0.25) = 0.5$ meter^2

$E = 9 \times 10^9 \left(\dfrac{50 \times 10^{-6}}{0.5} \right) = 9 \times 10^5$ volts / meter

$r = (0.5)^{\frac{1}{2}} = 0.707$ meter

b) $\cos\alpha = x/r = 0.3/0.707 = 0.42$

$\cos\beta = y/r = 0.4/0.707 = 0.56$

$\cos\gamma = z/r = 0.5/0.707 = 0.707$

$\alpha = \cos^{-1}(0.42) = 65°$
$\beta = \cos^{-1}(0.56) = 56°$
$\gamma = \cos^{-1}(0.707) = 45°$

5.9 Electric field strength

A long wire has a uniform charge distribution, λ, measured in coulombs/meter. Determine the electric field strength, **E**, at a distance of b meters at point p(x,y,z) as shown in Figure 5.9.

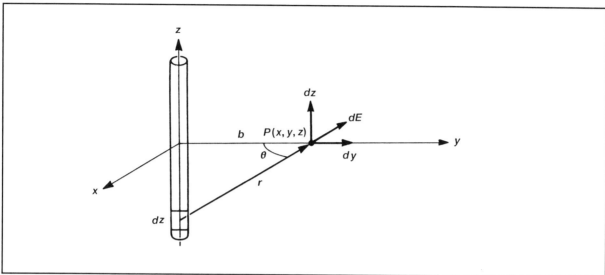

Figure 5.9 Electric field strength

Solution:

$$dE = \kappa \left[\frac{dq}{r^2} \right]$$

$z/b = \tan\theta,$ $z = b\tan\theta,$ $dz = b\sec^2\theta\, d\theta$
$b/r = \cos\theta,$ $r = b/\cos\theta = b\sec\theta,$ $r^2 = (b\sec\theta)^2$
$z/r = \sin\theta,$ $z = r\sin\theta$

$$dE = \kappa \left[\frac{\lambda dz}{r^2} \right] = \frac{\kappa\lambda(b\sec^2\theta\, d\theta)}{(b\sec\theta)^2} = \frac{\kappa\lambda}{b}(d\theta)$$

$dE_y = dE\cos\theta\, d\theta$

$$E_y = \frac{\kappa\lambda}{b}\int_{-\pi/2}^{\pi/2}\cos\theta\, d\theta = \frac{\kappa\lambda}{b}[\sin\theta]_{-\pi/2}^{\pi/2} = \frac{2\kappa\lambda}{b}$$

$\kappa = 1/(4\pi\epsilon_0) = 9\times 10^9$

$\mathbf{E}_y = \lambda/[2\pi\epsilon_0 b]$ volts/meter

5.10 Electric potential

A long wire has a uniform charge distribution, λ, measured in coulombs/meter. Determine the electric potential between two points, a and b, as shown in Figure 5.10.

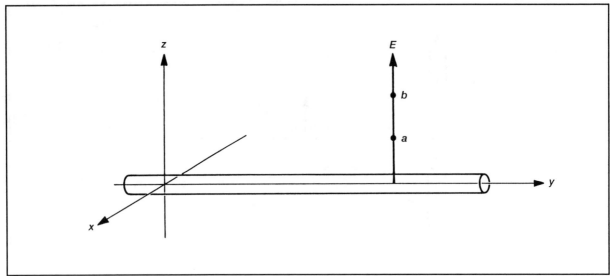

Figure 5.10   Electric potential

Solution:

$$V_{ab} = V_b - V_a = -\int_a^b E \cdot dl$$

$$E = 2\kappa(\lambda / z)$$

$$\kappa = 1/(4\pi\epsilon_0) = 9 \times 10^9$$

$$E = \lambda / (2\pi\epsilon_0 z) \quad \text{volts/meter}$$

$$V_{ab} = -\int_a^b \left[\frac{\lambda}{2\pi\epsilon_0 z}\right] dz = -\frac{\lambda}{2\pi\epsilon_0} \int_a^b \frac{dz}{z}$$

$$V_{ab} = -\frac{\lambda}{2\pi\epsilon_0} [\ln z]_a^b$$

$$V_{ab} = -\left[\frac{\lambda}{2\pi\epsilon_0}\right][\ln b - \ln a] = \left[\frac{\lambda}{2\pi\epsilon_0}\right]\ln\left[\frac{a}{b}\right]$$

5.11   Electric flux density

A parallel plate capacitor has a potential of 100 volts across its plates as shown in Figure 5.11. Each plate has a surface area of 160 cm² and separated by a distance of 0.1 cm with a spacer. Determine the electric flux density, **D**, if the spacer has a dielectric constant, k, of 2.5.

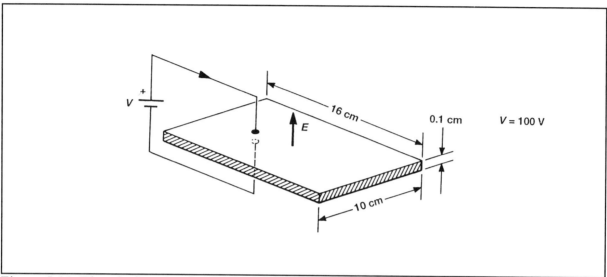

Figure 5.11 Electric flux density

Solution:

$$V = -\int_0^l E \cdot dl = -El$$

$\mathbf{E} \;=\; V/l \;=\; 100/(1\times10^{-3}) \;=\; (1\times10^2)/(1\times10^{-3}) \;=\; 1\times10^5 \quad \text{volts/meter}$

$Q \;=\; \epsilon EA \;=\; (k\epsilon_0)EA \;=\; DA$

$\epsilon_0 \;=\; 8.854\times10^{-12} \quad \text{farads/meter}$

$Q \;=\; k\epsilon_0 EA \;=\; (2.5)(8.854\times10^{-12})(1\times10^5)(0.016) \;=\; 35.41\times10^{-9} \quad \text{coulombs}$

$\mathbf{D} \;=\; \epsilon E \;=\; Q/A \;=\; (35.41\times10^{-9})/(0.016) \;=\; 2.2\times10^{-6} \quad \text{coulombs/meter}^2$

5.12 Electric energy

Determine the electric energy stored in the parallel plate capacitor of problem 5.11. Figure 5.12 illustrates the equivalent circuit.

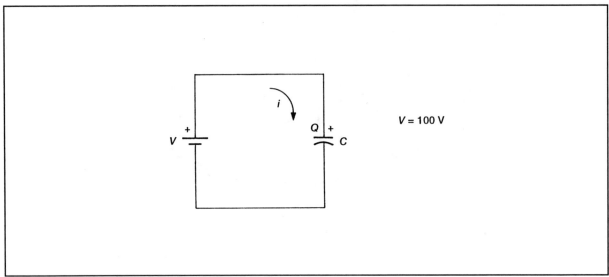

Figure 5.12 Electric energy

Solution:

$$C = \epsilon(A/d)$$

$$\epsilon = k\epsilon_0 = (2.5)(8.854 \times 10^{-12}) = 22.13 \times 10^{-12} \quad \text{farads/meter}$$

$$C = (22.13 \times 10^{-12})(0.016/0.001) = 3.541 \times 10^{-10} \quad \text{farads}$$

$$Q = CV$$

$$dW = VdQ$$

$$dQ = CdV$$

$$dW = CVdV$$

$$W = C\int_0^V VdV = \frac{1}{2}CV^2$$

$$W = (1/2)CV^2 = (1/2)(3.541 \times 10^{-10})(100)^2 = 1.77 \times 10^{-6} \quad \text{joules}$$

5.13 Electric deflection

An electron beam enters at the midpoint of two parallel metal plates with a horizontal velocity, v_x, of 2.5×10^7 meters/second as shown in figure 5.13. Determine (a) the vertical velocity, v_y, (b) the deflection angle, θ, and (c) the vertical distance traveled as it exits the plates.

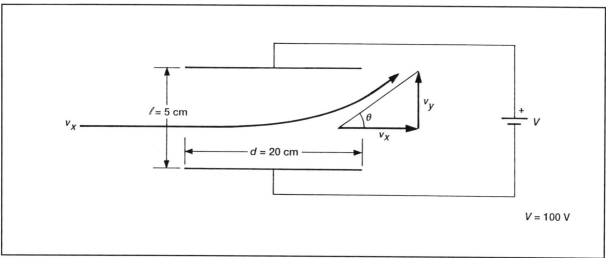

Figure 5.13 Electric deflection

Solution:

$$V = -\int_0^l E_y dy = -El$$

$E = V/l = 100 / 5 \times 10^{-2} = 2 \times 10^3$ volts/meter

$F = ma = qE$

$m = 9.1 \times 10^{-31}$ kg

$q = 1.6 \times 10^{-19}$ coulomb

$a = (qE)/m = (1.6 \times 10^{-19} \times 2 \times 10^3) / (9.1 \times 10^{-31}) = 3.52 \times 10^{14}$ meters/sec²

$t = d/v_x = (20 \times 10^{-2}) / (2.5 \times 10^7) = 8 \times 10^{-9}$ sec

a) $v_y = at = (3.52 \times 10^{14})(8 \times 10^{-9}) = 2.8 \times 10^6$ meters/sec

b) $\theta = \tan^{-1}(v_y/v_x) = \tan^{-1}[(2.8 \times 10^6) / (2.5 \times 10^7)] = 6.4°$

c) $dy/dt = at$

$y = a\int t\,dt = (1/2)at^2 = 1.126 \times 10^{-2}$ meter

5.14 Magnetic force

The opposite poles of two bar magnets are placed 25 cm apart as shown in Figure 5.14. The magnetic pole strengths are $m_1 = 50 \times 10^{-6}$ amperes-meters² and $m_2 = 25 \times 10^{-6}$ amperes-meters². Determine the magnetic force exerted by m_1 on m_2.

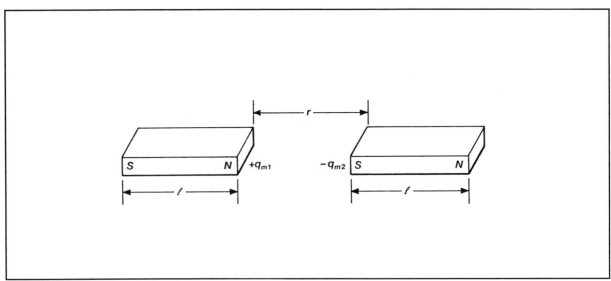

Figure 5.14 Magnetic force

Solution:

$$F_{12} = \kappa \left(\frac{m_1 m_2}{r^2} \right)$$

$$\kappa = 1/(4\pi\mu_0)$$

$\mu_0 = 1.257 \times 10^{-6}$ henries/meter (permeability of free space)

$m_1 = l q_{m1} = 50 \times 10^{-6}$ amperes-meters²

$m_2 = l q_{m2} = 25 \times 10^{-6}$ amperes-meters²

$$F_{12} = (6.33 \times 10^4) \frac{50 \times 10^{-6} \times 25 \times 10^{-6}}{(0.25)^2}$$

$$F_{12} = (6.33 \times 10^4) \frac{1.25 \times 10^{-9}}{6.25 \times 10^{-2}} = 1.26 \times 10^{-3} \text{ newtons}$$

5.15 Magnetic field

The magnetic pole strength of a bar magnet has a value of $m_1 = 50 \times 10^{-6}$ amperes-meters² as shown in Figure 5.15. Determine (a) the magnitude and (b) direction of the magnetic field, **H**, at a point p(x,y,z) where x = 30 cm, y = 40 cm, and z = 50 cm.

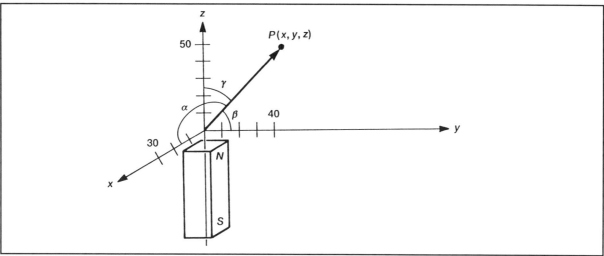

Figure 5.15 Magnetic field

Solution:

a) $\mathbf{H} = \kappa [m_1 / r^2]$

$\kappa = 1/(4\pi\mu_0) = 6.33 \times 10^4$

$r^2 = x^2 + y^2 + z^2 = (0.3)^2 + (0.4)^2 + (0.5)^2$

$r^2 = (0.09) + (0.16) + (0.25) = 0.5$ meter²

$H = (6.33 \times 10^4) \dfrac{50 \times 10^{-6}}{0.5} = 6.33$ amperes/meter

$r = (0.5)^{1/2} = 0.707$ meter

b) $\alpha = \cos^{-1}(x/r) = \cos^{-1}(0.42) = 65°$

$\beta = \cos^{-1}(y/r) = \cos^{-1}(0.56) = 56°$

$\gamma = \cos^{-1}(z/r) = \cos^{-1}(0.707) = 45°$

5.16 Magnetic field strength

A long wire carries a constant current, i, measured in amperes as shown in Figure 5.16. Determine the magnetic field strength, **H**, at a distance of b meters at point p(x,y,z).

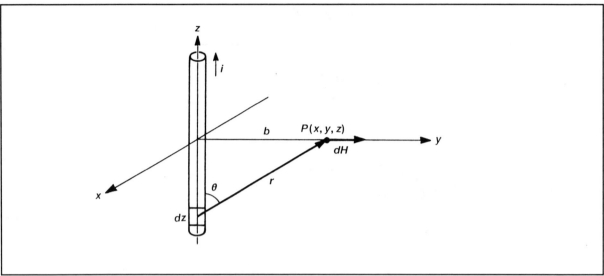

Figure 5.16 Magnetic field strength

Solution:

$$H = \frac{1}{4\pi}\int\left(\frac{i\sin\theta}{r^2}\right)dz \quad \text{amperes/meter}$$

$r = b\csc\theta$

$z = -b\cot\theta$

$dz = b\csc^2\theta\, d\theta$

$$H = \frac{i}{4\pi}\int_0^\pi \frac{\sin\theta}{(b\csc\theta)^2}(b\csc^2\theta\, d\theta)$$

$$H = \frac{i}{4\pi b}\int_0^\pi \sin\theta\, d\theta$$

$$H = \frac{i}{4\pi b}[-\cos\theta]_0^\pi = \frac{i}{2\pi b}$$

5.17 Magnetic potential
A one meter long wire carries a constant current of two amperes as shown in Figure 5.17. Determine (a) the scalar and (b) vector magnetic potentials at a point p(x,y,z) where x = 30 cm, y = 40 cm, and z = 50 cm.

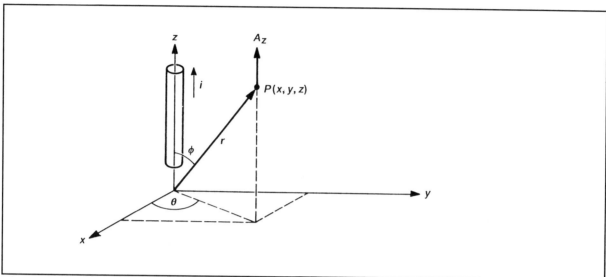

Figure 5.17 Magnetic potential

Solution:

a) $U(\text{scalar}) = -(i/2\pi)\phi$

$\phi = \tan^{-1}(x^2 + y^2)^{1/2}/z = \tan^{-1}(0.25)^{1/2}/0.5 = \tan^{-1}(1) = 45°$

$U = -[45°/360°] \times (2) = -0.25$ amperes

b) $A_z = (\mu_0/4\pi)[(il)/r]$

$\mu_0 = 1.257 \times 10^{-6}$ henries/meter

$r = (x^2 + y^2 + z^2)^{1/2} = [(0.3)^2 + (0.4)^2 + (0.5)^2]^{1/2} = 0.707$ meter

$$A_z = \frac{1 \times 10^{-7} \times 2 \times 1}{0.707} = 2.8 \times 10^{-7}$$ webers / meter

$\theta = \tan^{-1}(y/x) = \tan^{-1}(0.4/0.3) = 53°$

$A_r = A_z \cos\theta = (0.6)A_z$

$A_\theta = -A_z \sin\theta = -(0.8)A_z$

5.18 Magnetic flux density

A ten turn coil has a current flow of 1.25 amperes. It is wound on a rectangular iron bar as shown in Figure 5.18. Determine the magnetic flux density if the relative permeability of the bar is 110.

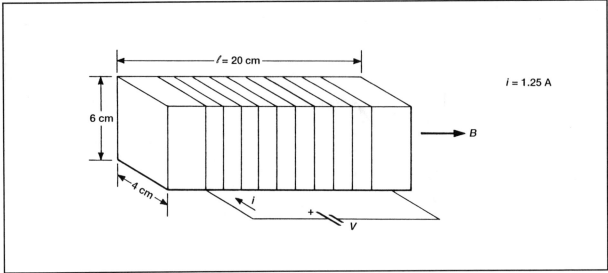

Figure 5.18   Magnetic flux density

Solution:

$$H = \frac{Ni}{l} = \frac{10 \times 1.25}{0.2} = 62.5 \text{ amperes/meter}$$

$$\Phi = \mu H A = (\mu_r \mu_0) H A = BA$$

$$\mu_0 = 1.257 \times 10^{-6} \text{ henries/meter}$$

$$\Phi = (\mu_r \mu_0) H A = (110 \times 1.257 \times 10^{-6})(62.5)(0.06 \times 0.04) = 2.07 \times 10^{-5} \text{ webers}$$

$$B = \mu H = \Phi/A = (2.07 \times 10^{-5}) / (0.0024) = 8.635 \times 10^{-3} \text{ webers/meter}^2$$

5.19 Magnetic energy

Determine the magnetic energy stored in the iron coil of problem 5.18. Figure 5.19 illustrates the equivalent circuit.

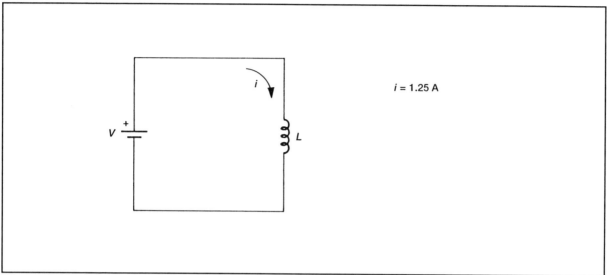

Figure 5.19 Magnetic energy

Solution:

$$L = \frac{N\Phi}{i} = \frac{N(BA)}{i} = \frac{N(\mu HA)}{i}$$

$$\mu = \mu_r\mu_0 = 110 \times 1.257 \times 10^{-6} \quad \text{henries/meter}$$

$$L = \frac{N}{i}\left(\frac{\mu NiA}{l}\right) = \left[\frac{\mu N^2 A}{l}\right] = 1.6 \times 10^{-4} \quad \text{henries}$$

$$dW = L(idi)$$

$$W = L\int_0^i idi = \frac{1}{2}Li^2$$

$$W = (1/2)Li^2 = (1/2)(1.6 \times 10^{-4})(1.25)^2 = 1.3 \times 10^{-4} \quad \text{joules}$$

5.20 Magnetic deflection

An electron beam enters a uniform magnetic field at a right angle as shown in Figure 5.20. The beam velocity, v_x, equals 2.5×10^7 meters/second. Determine (a) the radius of curvature, r, and (b) the deflection angle, θ, of the beam as it exits the field.

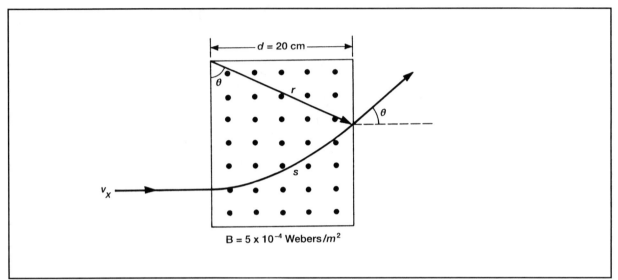

Figure 5.20   Magnetic deflection

Solution:

a) $F = Bqv\sin\phi$

$\phi = 90°$

$F = Bqv = ma = (mv^2)/r$

$r = (m/q)(v/B)$

$m = 9.1\times10^{-31}$ kg

$q = 1.6\times10^{-19}$ coulomb

$r = (m/q)(v_x/B) = (5.69\times10^{-12})[(2.5\times10^7)/(5\times10^{-4})] = 28.45\times10^{-2}$ meters

$r = 28.45$ cm

b) $\theta = \sin^{-1}(d/r) = \sin^{-1}[(0.2)/(0.2845)] = 44.66°$

And $s = (\theta/360°)(2\pi r) = (44.66°/360°)(2\pi\times28.45) = 22.16$ cm

5.21 Laplace's equation

A pair of parallel conductors is 50 cm apart as shown in Figure 5.21. The lower conductor has a potential of 100 volts, the upper conductor has a potential of 150 volts. Determine the potential between the conductors assuming the charge density, ρ, equals zero.

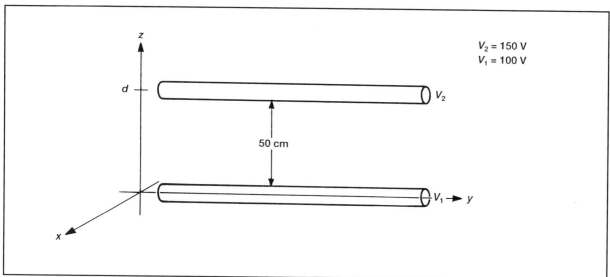

Figure 5.21 Laplace's equation

Solution:

Using the symmetry about the x,y axes

$$\nabla^2 V = \frac{d^2 V(z)}{dz^2} = -\frac{\rho}{\epsilon_0} = 0$$

$$\frac{dV(z)}{dz} = k_1$$

$$V(z) = k_1 z + k_2$$

at z(0) $V = V_1$
at z(d) $V = V_2$

$V_1 = k_1 \times 0 + k_2$
$V_2 = k_1 \times d + k_2$

$k_1 = (V_2 - V_1) / d, \qquad k_2 = V_1$

$V(z) = k_1 z + k_2 = [(V_2 - V_1)/d]z + V_1 = [(150-100)/0.5]z + 100$

$V(z) = 100z + 100 \qquad$ For $0 < z < 0.5$ meter

5.22 Ampere's law

A one meter long conductor carries a constant current of two amperes as shown in Figure 5.22. Determine the expression for the magnetic field around it using Ampere's law.

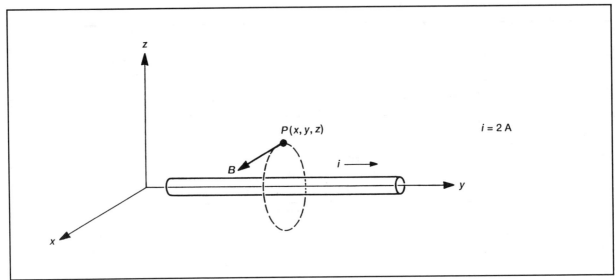

Figure 5.22 Ampere's law

Solution:

$$\oint \mathbf{B} \cdot d\mathbf{l} = \mu_0 I$$

The magnetic field, **B**, is concentric around the y-axis conductor.
The line integral around this path equals the circumference of a circle with radius,

$$r = (x^2 + z^2)^{1/2}.$$

$$\oint \mathbf{B} \cdot d\mathbf{l} = \mu_0 I = \mathbf{B}(2\pi r)$$

$$\mu_0 = 4\pi \times 10^{-7} \text{ henries/meter}$$

$$B = \frac{\mu_0 I}{2\pi r} = \left(\frac{4\pi \times 10^{-7}}{2\pi}\right)\left(\frac{I}{r}\right)$$

$$\mathbf{B} = (2 \times 10^{-7} \times 2) / r$$

$$\mathbf{B} = (4 \times 10^{-7}) / (x^2 + z^2)^{1/2} \quad \text{webers/meter}^2$$

5.23 Faraday's law

A one meter long helical conductor carries a sinusoidal current as shown in Figure 5.23. The resulting magnetic flux, Φ, equals $(16 \times 10^{-6})\sin 377t$ webers. Determine the expression for the resulting electric field around it using Faraday's law.

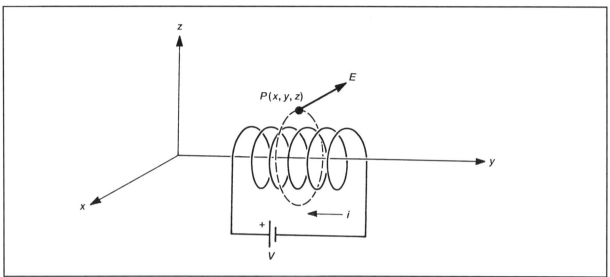

Figure 5.23 Faraday's law

Solution:

$$\oint E \cdot dl = -\frac{d\Phi}{dt}$$

The electric field, **E**, is concentric around the y-axis conductor.
The line integral around this path equals the circumference of a circle with radius,

$$r = (x^2 + z^2)^{1/2}.$$

$$\Phi = (16 \times 10^{-6})\sin 377t$$

$$\oint E \cdot dl = E(2\pi r) = -\frac{d\Phi}{dt} = (16 \times 10^{-6})\cos 377t$$

$$E = -\left(\frac{1}{2\pi r}\right)\frac{d\Phi}{dt}$$

$$E = \frac{1}{2\pi(x^2 + z^2)^{1/2}}(16 \times 10^{-6})\cos 377t \quad \text{volts / meter}$$

5.24 Gauss's law

A long wire has a uniform charge distribution, λ, measured in coulombs/meter as shown in Figure 5.24. Find the electric field strength at a distance of b meters at point p(x,y,z) using Gauss's law.

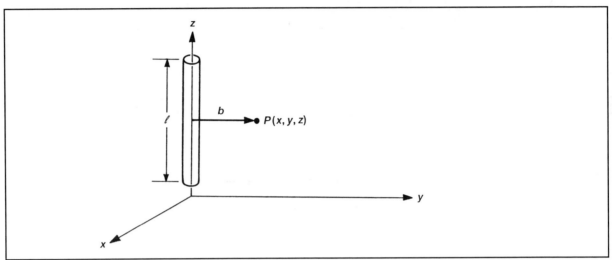

Figure 5.24 Gauss's law

Solution:

$$\oint \mathbf{E} \cdot \mathbf{dS} = q/\epsilon_0$$

The surface area around the wire is defined by a cylinder of radius b.

$$\oint \mathbf{E} \cdot \mathbf{dS} = q/\epsilon_0 = (\lambda l)/\epsilon_0$$

$$\mathbf{E}(2\pi b l) = (\lambda l)/\epsilon_0$$

$$\mathbf{E} = \lambda/(2\pi\epsilon_0 b)$$

$$\kappa = 1/(4\pi\epsilon_0)$$

$$\mathbf{E} = (2\kappa\lambda)/b \quad \text{volts/meter}$$

5.25 Maxwell's law

A circular wire loop carries a current of two amperes as shown in Figure 5.25. Determine the corresponding magnetic flux, Φ, inside the loop for a radius of 50 cm.

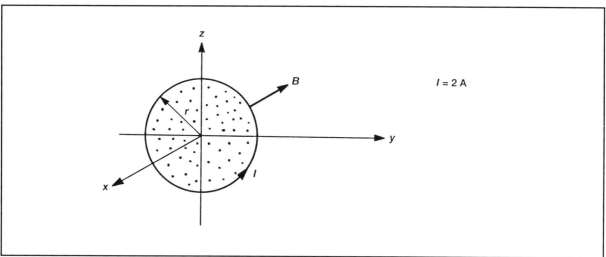

Figure 5.25 Maxwell's law

Solution:

$$\oint \mathbf{B \cdot dS} = 0$$

The radial magnetic flux, Φ, of the entire loop surface equals zero.
Using the Biot-Savart law, the axial magnetic flux density component is

$$d\mathbf{B} = [(\mu_0 \, Idl) / (4\pi r^2)]\sin\theta$$

$$\theta = 90°$$

$$B = \int dB = \frac{\mu_0 I}{4\pi r^2}\int dl = \frac{\mu_0 I}{4\pi r^2}[2\pi r] = \frac{\mu_0 I}{2r}$$

The axial magnetic flux, Φ, of the loop cross section equals

$$\Phi = \oint \mathbf{B \cdot dS} = BA = B(\pi r^2)$$

$$\Phi = [(\mu_0 I) / (2r)] \times (\pi r^2) = 0.5\pi\mu_0 \, Ir$$

$$\Phi = 0.5\pi\mu_0 (2 \times 0.5) = 0.5\pi\mu_0$$

$$\mu_0 = 4\pi \times 10^{-7} \text{ henries/meter}$$

$$\Phi = 19.7 \times 10^{-7} \text{ webers}$$

5.26 Toroid

A toroidal coil consists of twenty turns and has a rectangular cross section as shown in Figure 5.26. Calculate its self inductance assuming it is wound on a nonmagnetic material.

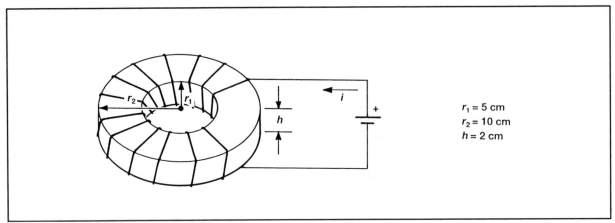

Figure 5.26 Toroid

Solution:

$h = 2$ cm, $r_1 = 5$ cm, $r_2 = 10$ cm, $N = 20$ turns, $I = Ni$ net current

$$\oint \mathbf{B} \cdot dl = \mu_0 I$$

$$\mathbf{B}(2\pi r) = \mu_0 (Ni)$$

$$\mathbf{B} = (\mu_0 / 2\pi)[(Ni)/r] = (\mu_0 Ni)/(2\pi r)$$

$$L = (N\Phi)/i$$

The magnetic flux, Φ, of the toroid cross section equals

$$\Phi = \int B \cdot dS = \frac{\mu_0 Ni}{2\pi} \int_{r_1}^{r_2} h \frac{dr}{r} = \frac{\mu_0 Nih}{2\pi} [\ln r]_{r_1}^{r_2}$$

$$\Phi = [(\mu_0 Nih)/(2\pi)] \ln(r_2/r_1)$$

$$L = (N\Phi)/i = [(\mu_0 N^2 h)/(2\pi)] \ln(r_2/r_1)$$

$$L = \frac{N\Phi}{i} = \left(\frac{4\pi \times 10^{-7} \times 4 \times 10^2 \times 0.02}{2\pi}\right) \ln\left(\frac{10 \times 10^2}{5 \times 10^2}\right) = (2 \times 10^{-7} \times 8) \times \ln(2)$$

$$L = (1.6 \times 10^{-6}) \times \ln(2) = 1.11 \times 10^{-6} \quad \text{henries}$$

5.27 Solenoid

A solenoid coil is 10 cm long and is wound around a paper tube with a 2 cm diameter as shown in Figure 5.27. Determine the resulting self inductance of the coil for a turns ratio of 200 turns/meter.

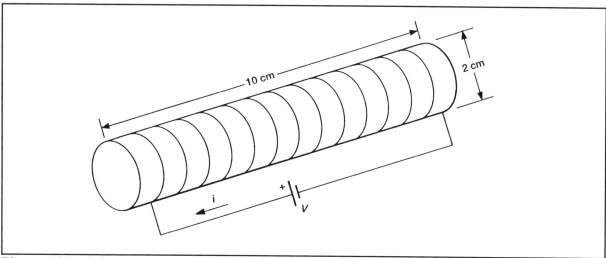

Figure 5.27 Solenoid

Solution:

$N = nl = 200$ turns/meter $\times$ 0.1 meter = 20 turns

$I = iN = I(nl)$ the net current

$$\oint \mathbf{B} \cdot dl = \mu_0 I$$

$\mathbf{B}l = \mu_0 iN = \mu_0 i(nl)$

$\mathbf{B} = \mu_0 ni$

The magnetic flux, Φ, of the solenoid cross section equals $\Phi = \mathbf{B}A$

$$L = \frac{N\Phi}{i} = \left(\frac{nl}{i}\right)(BA) = \left(\frac{nl}{i}\right)(\mu_0 niA)$$

$L = \mu_0 n^2 l A = \mu_0 n^2 l (\pi r^2)$

$L = (4\pi \times 10^{-7}) \times (200)^2 \times (0.1) \times (\pi) \times (0.01)^2 = 1.58 \times 10^{-6}$ henries

5.28 Electric dipole

Two equal stationery charges, q = 50×10⁻⁶ coulombs, of opposite polarity are placed apart at a distance, a, of 10 cm as shown in Figure 5.28. Determine the electric field strength at two points p_1 (0, 0, 100) and p_2 (0, 100, 0) cm away respectively.

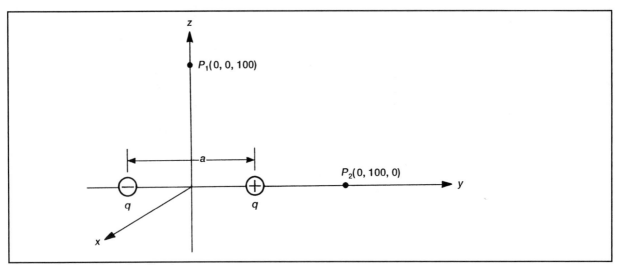

Figure 5.28 Electric dipole

Solution:

Electric dipole moment = qa coulombs-meters

ϵ_0 = 8.85×10⁻¹² farads/meter

$$E_z = \frac{qa}{4\pi\epsilon_0 z^3} = \frac{50 \times 10^{-6} \times 0.1}{1.11 \times 10^{-10} \times (1)^3}$$

$$E_z = \frac{5 \times 10^{-6}}{1.11 \times 10^{-10}} = 4.5 \times 10^4 \text{ volts / meter}$$

$$E_y = \frac{2qa}{4\pi\epsilon_0 y^3} = \frac{100 \times 10^{-6} \times 0.1}{1.11 \times 10^{-10} \times (1)^3}$$

$$E_y = \frac{10 \times 10^{-6}}{1.11 \times 10^{-10}} = 9.0 \times 10^4 \text{ volts / meter}$$

5.29 Electric quadrupole

Four equal stationery charges, q = 50×10⁻⁶ coulombs, are used to form an electric quadrupole as shown in Figure 5.29. Determine the resulting electric field strength at a point p(0, 100, 0) cm away.

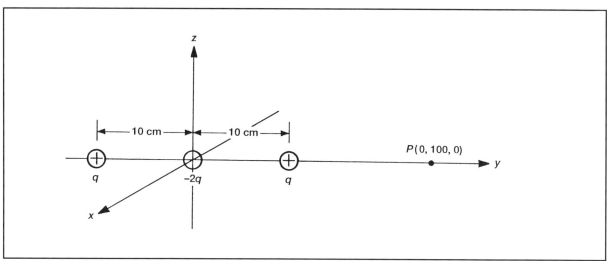

Figure 5.29 Electric quadrupole

Solution:

Electric quadrupole moment $Q = 2qa^2$ coulombs-meters²

$$Q = 2qa^2 = 2\times(50\times10^{-6})\times(0.1)^2 = 1\times10^{-6}$$

$$\epsilon_0 = 8.85\times10^{-12} \quad \text{farads/meter}$$

$$E_y = \frac{3Q}{4\pi\epsilon_0 y^4} = \frac{3\times10^{-6}}{1.11\times10^{-10}\times(1)^4}$$

$$E_y = \frac{3\times10^{-6}}{1.11\times10^{-10}} = 2.7\times10^4 \quad \text{volts/meter}$$

5.30 Magnetic dipole

A magnetic dipole is formed using a 10 cm long permanent magnet, with a magnetic charge $q_m = 5\times10^{-4}$ ampere-meters, as shown in Figure 5.30. Determine the magnetic field strength at two points $p_1(0, 0, 100)$ and $p_2(0, 100, 0)$ cm away respectively.

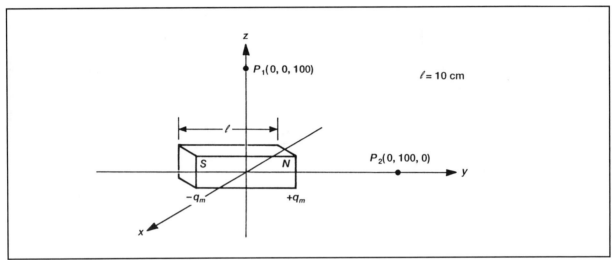

Figure 5.30 Magnetic dipole

Solution:

Magnetic dipole moment = $q_m l$ amperes-meters^2

$\mu_0 = 4\pi\times10^{-7}$ henries/meter

$$B_z = \frac{\mu_0}{4\pi}\left[\frac{q_m l}{z^3}\right] = \frac{4\pi\times10^{-7}}{4\pi}\left[\frac{5\times10^{-4}\times0.1}{(1)^3}\right]$$

$\mathbf{B}_z = (1\times10^{-7})\times(5\times10^{-5}) = 5\times10^{-12}$ webers/meters^2

$$B_y = \frac{\mu_0}{4\pi}\left[\frac{q_m l}{y^3}\right] = \frac{4\pi\times10^{-7}}{4\pi}\left[\frac{10\times10^{-4}\times0.1}{(1)^3}\right]$$

$\mathbf{B}_y = (1\times10^{-7})\times(1\times10^{-4}) = 10\times10^{-12}$ webers/meters^2

5.31 Magnetic quadrupole

A magnetic quadrupole is formed using two 10 cm long permanent magnets, with a magnetic charge $q_m = 5 \times 10^{-4}$ ampere-meters, as shown in Figure 5.31. Determine the magnetic field strength at a point $p(0, 100, 0)$ cm away.

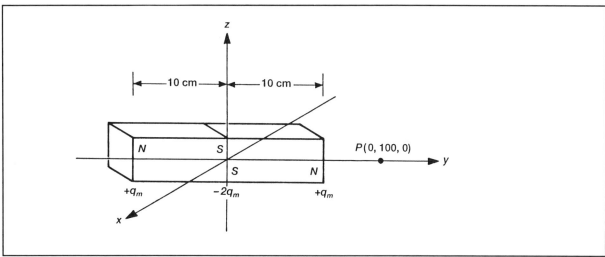

Figure 5.31 Magnetic quadrupole

Solution:

Magnetic quadrupole moment $M = 2q_m l^2$ amperes-meters^3

$$M = 2q_m l^2 = 2 \times (5 \times 10^{-4}) \times (0.1)^2 = 1 \times 10^{-5}$$

$$\mu_0 = 4\pi \times 10^{-7} \text{ henries/meter}$$

$$B_y = \frac{\mu_0}{4\pi}\left[\frac{3M}{y^4}\right] = \frac{4\pi \times 10^{-7}}{4\pi}\left[\frac{3 \times 1 \times 10^{-5}}{(1)^4}\right]$$

$$\mathbf{B}_y = (1 \times 10^{-7}) \times (3 \times 10^{-5}) = 3 \times 10^{-12} \text{ webers/meters}^2$$

5.32 Antenna radiation

A half-wave dipole antenna is driven with a 100 MHz sinewave signal as shown in Figure 5.32. Determine (a) the antenna length in meters. Also calculate (b) the velocity of the radiating electromagnetic wave, and (c) the impedance of free space.

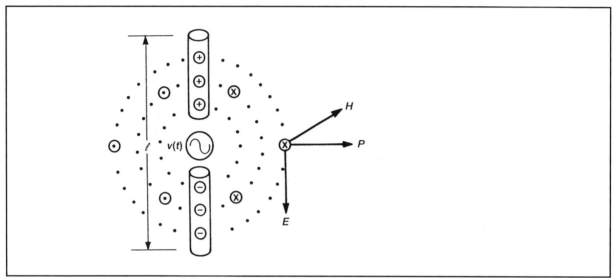

Figure 5.32   Antenna radiation

Solution:

$$c = f\lambda = 3 \times 10^8 \text{ meters/second}$$

$$\lambda = c/f = (3 \times 10^8)/(100 \times 10^6) = 3 \text{ meters}$$

a) $l = \lambda/2 = 3/2 = 1.5$ meters

$$\mu_0 = 4\pi \times 10^{-7} \text{ henries/meter}$$

$$\epsilon_0 = 8.854 \times 10^{-12} \text{ farads/meter}$$

b) $v_0 = \dfrac{1}{\sqrt{\mu_0 \epsilon_0}} = \dfrac{1}{\sqrt{1.257 \times 10^{-6} \times 8.854 \times 10^{-12}}} \approx 3 \times 10^8$ meters/second

c) $Z_0 = \sqrt{\dfrac{\mu_0}{\epsilon_0}} = \sqrt{\dfrac{1.257 \times 10^{-6}}{8.854 \times 10^{-12}}} = 377$ ohms

5.33 Near field

A vertical half-wave dipole antenna is driven with a 100 MHz sinewave signal. A small square wire loop is located 10 meters away as shown in figure 5.33. Determine the induced voltage in the loop assuming a maximum antenna current of 10 amperes.

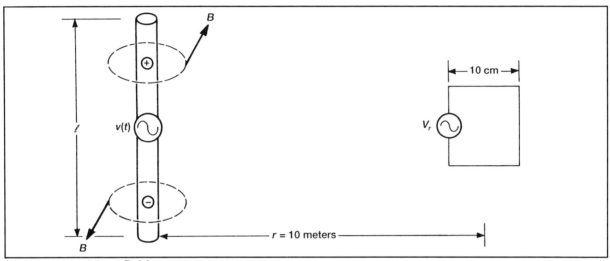

Figure 5.33   Near field

Solution:

$$\lambda = c/f = (3\times 10^8)/(100\times 10^6) = 3 \text{ meters}$$

$$l = \lambda/2 = 3/2 = 1.5 \text{ meters}$$

$$\Phi = BA$$

$$\mu_0 = 4\pi \times 10^{-7} \text{ henries/meter}$$

$$B = \frac{\mu_0 I}{2\pi r}[\sin 2\pi ft] = \frac{40\pi \times 10^{-7}}{2\pi \times 10}[\sin 2\pi ft] = 2\times 10^{-7}[\sin 2\pi ft] \text{ webers/meter}^2$$

$$\Phi = (2\times 10^{-7})(0.1)^2 = (2\times 10^{-9})\sin 2\pi ft$$

$$v_r = \frac{d\Phi}{dt} = 4\pi \times 10^{-9} \times f[\cos 2\pi ft]$$

$$v_r = (4\pi \times 10^{-9} \times 100 \times 10^6)\cos 2\pi ft$$

$$V_r(\max) = (4\pi \times 10^{-9} \times 1 \times 10^8) = 1.256 \text{ volts}$$

5.34   Far field

A vertical half-wave dipole antenna is driven with a 100 MHz sinewave signal. A small wire antenna is located 1,000 meters away as shown in figure 5.34. Determine the induced voltage in the receiving antenna if the transmitting antenna has a maximum current of 10 amperes.

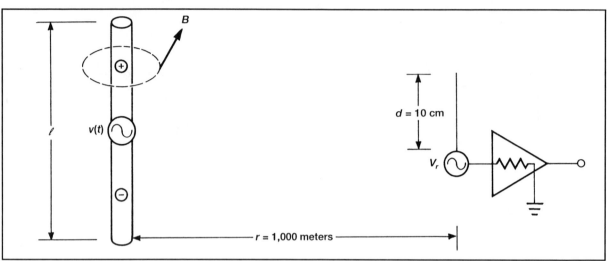

Figure 5.34   Far field

Solution:

$\lambda = c / f = (3 \times 10^8) / (100 \times 10^6) = 3$ meters

$l = \lambda / 2 = 3/2 = 1.5$ meters

$R_s = 73$ ohms

$P_t = I^2 R_s = (10)^2 (73) = 7{,}300$ watts

$E = (49.2 P_t)^{1/2} / r$ volts/meter

$r = 1{,}000$ meters

$E = 0.6$ volts/meter

$V_r = E \cdot d = 0.6 \times 0.1 = 0.06$ volts

$v_r = V_r \cos 2\pi f t = 0.06 \times \cos 2\pi f t$ volts

5.35 Vertical antenna

A vertical quarter-wave Marconi antenna is driven with a 365 volt, 100 MHz sinewave signal as shown in Figure 5.35. Calculate the electric field strength ten wavelengths away. Assume that the effective antenna height equals one-half the actual height.

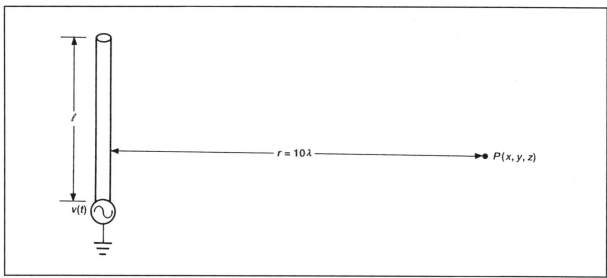

Figure 5.35 Vertical antenna

Solution:

$\lambda = c/f = (3 \times 10^8)/(100 \times 10^6) = 3$ meters

$l = \lambda/4 = 3/4 = 0.75$ meter

$h_e = (1/2)l = 0.375$ meter

$R_s = 36.5$ ohms

$I = V/R_s = 365/36.5 = 10$ amperes

$r = 10\lambda = 30$ meters

$\mathbf{E} = (120\pi I h_e)/(\lambda r)$ volts/meter

$\mathbf{E} = (120 \times \pi \times 10 \times 0.375)/(3 \times 30) = 15.7$ volts/meter

5.36 Dipole antenna

A folded half-wave dipole antenna is driven with a 365 volt, 100 MHz sinewave signal as shown in Figure 5.36. Calculate the transmitted power, P_t.

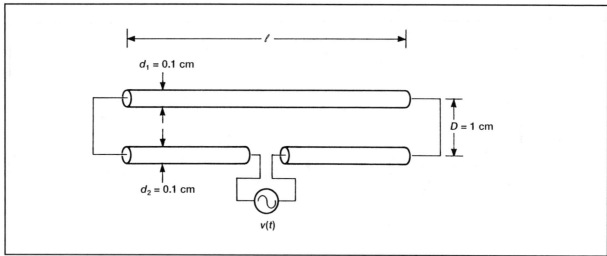

Figure 5.36   Dipole antenna

Solution:

$$\lambda = c/f = (3 \times 10^8)/(100 \times 10^6) = 3 \text{ meters}$$

$$l = \lambda/2 = 3/2 = 1.5 \text{ meters}$$

$$R_s = \left(1 + \frac{\log \frac{2D}{d_2}}{\log \frac{2D}{d_1}}\right)^2 \times 73$$

$$R_s = (1+1)^2 \times (73) = 292 \text{ ohms}$$

$$P_t = V^2/R_s = (365)^2/292 = 465.25 \text{ watts}$$

5.37 Loop antenna

A circular loop antenna has a radius of 10 cm as shown in Figure 5.37. Determine the corresponding radiation resistance at 100 MHz. Also, calculate the induced voltage resulting from an incident electric field of 10 volts/meter placed in the loop plane and oscillating at 100 MHz.

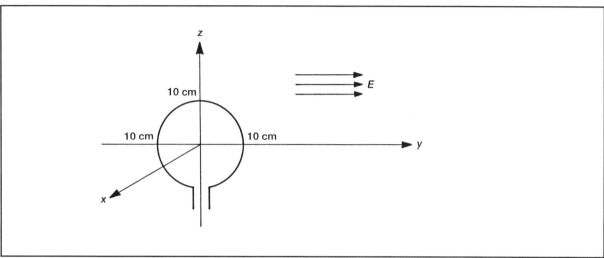

Figure 5.37 Loop antenna

Solution:

$$\lambda = c/f = (3 \times 10^8)/(100 \times 10^6) = 3 \text{ meters}$$

$$R_s = 197 \times \left(\frac{2\pi r}{\lambda}\right)^4 = 197 \times \left(\frac{0.628}{3}\right)^4$$

$$R_s = 0.378 \text{ ohms}$$

$$V_r = (2\pi E A)/\lambda$$

$$A = \pi r^2 = 0.0314 \text{ meter}$$

$$E = 10 \text{ volts/meter}$$

$$V_r = \frac{2\pi \times 10 \times 0.0314}{3} = 0.657 \text{ volts}$$

5.38 Transmission line

A 50 meter long transmission line has a series inductance of 5×10^{-6} henries/meter and a shunt capacitance of 50×10^{-12} farads/meter as shown in Figure 5.38. Determine (a) its characteristic impedance, Z_0, and (b) the propagation delay, τ_p, assuming it is properly terminated.

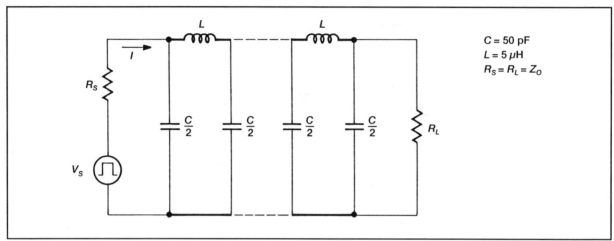

Figure 5.38 Transmission line

Solution:

$I\tau = CV$

$\tau = C(V/I)$

$V\tau = LI$

$\tau = L(I/V)$

$\tau^2 = L(I/V) \times C(V/I) = LC$

$\tau = (LC)^{\frac{1}{2}} = [(5 \times 10^{-6})(50 \times 10^{-12})]^{\frac{1}{2}} = 1.58 \times 10^{-8}$ seconds/meter

a) $Z_0 = V/I = (L/C)^{\frac{1}{2}} = [(5 \times 10^{-6})/(50 \times 10^{-12})]^{\frac{1}{2}} = 316.23$ ohms/meter

$R_s = R_L = Z_0$

b) $\tau_p = 50 \times \tau = 7.9 \times 10^{-7}$ seconds

5.39 Coaxial cable

A coaxial cable has a characteristic impedance, Z_0, of 75 ohms as shown in Figure 5.39. Determine the value of the dielectric constant for the spacer material.

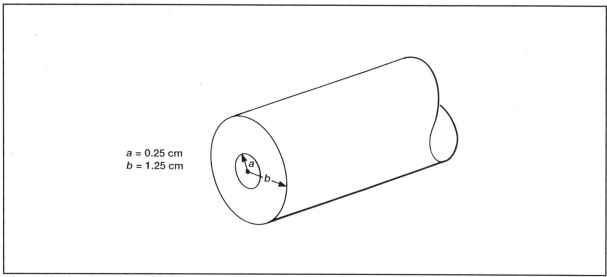

Figure 5.39 Coaxial cable

Solution:

$$Z_0 = \frac{60}{\sqrt{k}} \ln\left(\frac{b}{a}\right)$$

$Z_0 = 75 = [60/\sqrt{k}]\ln(1.25 / 0.25) = [60/\sqrt{k}]\ln(5)$

$\sqrt{k} = [60/75]\ln(5) = [60/75] \times 1.61$

$k = (1.2875)^2 = 1.65$

5.40 Parallel conductors

A two-wire transmission line has a characteristic impedance, Z_0, of 300 ohms as shown in Figure 5.40. Determine the conductor diameter for a spacer material dielectric constant of 2.5.

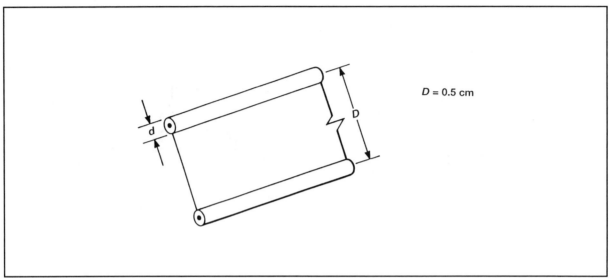

Figure 5.40 Parallel conductors

Solution:

$$Z_0 = \frac{120}{\sqrt{k}} \ln\left(\frac{2D}{d}\right)$$

$2D = 2 \times 0.5 = 1 \text{ cm}$

$Z_0 = 300 = [120/\sqrt{k}]\ln(1/d)$

$k = 2.5$

$(2.5)^{1/2} \times [300/120] = \ln(1/d)$

$3.95 = \ln(1/d)$

$e^{3.95} = (1/d)$

$d = (1/e^{3.95}) = 0.02 \text{ cm}$

5.41 Quarter wave stub

A 100 MHz signal is transmitted over a transmission line which has a characteristic impedance, Z_1, of 300 ohms. A quarter wave stub is used to connect the 300 ohm line to a second transmission line which has a characteristic impedance, Z_3, of 75 ohms as shown in Figure 5.41. Determine (a) the length, l, and (b) characteristic impedance, Z_2, of the impedance matching stub.

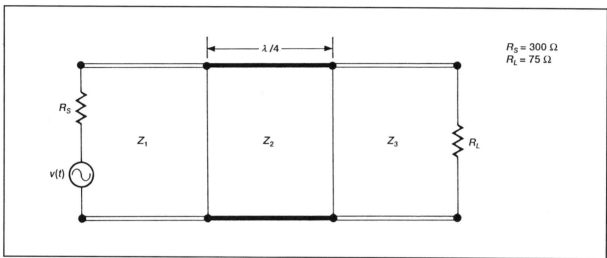

Figure 5.41 Quarter wave stub

Solution:

a) R_s = 300 ohms

   R_L = 75 ohms

   $Z_2 / Z_1 = Z_3 / Z_2$

   $(Z_2)^2 = Z_1 Z_3$

   $Z_2 = (Z_1 Z_3)^{1/2} = (300 \times 75)^{1/2}$

   Z_2 = 150 ohms

b) $\lambda = c / f = (3 \times 10^8) / (100 \times 10^6) = 3$ meters

   $l = \lambda / 4 = 3/4 = 0.75$ meter

5.42 Reflections (R<Z)

A transmission line has a characteristic impedance, Z_0, of 300 ohms and is terminated with a resistive load of 75 ohms as shown in Figure 5.42. Calculate (a) the current reflection ratio, ρ_i, (b) the transmission ratio, τ_i, and (c) the voltage standing wave ratio (VSWR).

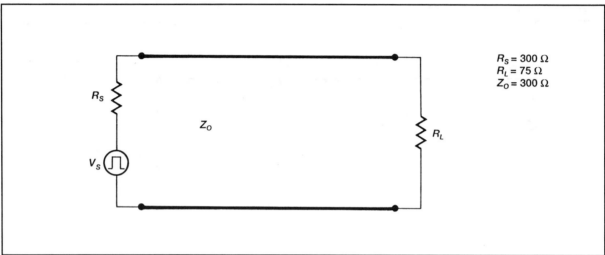

Figure 5.42   Reflections (R<Z)

Solution:

V_{min} at R_L

$R_s = Z_0 = 300$ ohms

$R_L = 75$ ohms

a) $\rho_i = I_r / I_f$

$(I_f + I_r)R_L = (I_f - I_r)Z_0$

$$\rho_i = \frac{Z_0 - R_L}{Z_0 + R_L} = \frac{300 - 75}{300 + 75} = \frac{225}{375} = \frac{3}{5}$$

b) $\tau_i = \dfrac{2Z_0}{Z_0 + R_L} = \dfrac{600}{375} = \rho_i + 1 = \dfrac{8}{5}$

c) $\text{VSWR} = \dfrac{I_{max}}{I_{min}} = \dfrac{1 + \rho_i}{1 - \rho_i} = \dfrac{1 + 0.6}{1 - 0.6} = 4$

$\rho_i = \dfrac{VSWR - 1}{VSWR + 1} = \dfrac{4 - 1}{4 + 1} = \dfrac{3}{5}$

5.43 Reflections (R>Z)

A transmission line has a characteristic impedance, Z_0, of 75 ohms and is terminated with a resistive load of 300 ohms as shown in Figure 5.43. Calculate (a) the voltage reflection ratio, ρ_v, (b) the transmission ratio, τ_v, and (c) the voltage standing wave ratio (VSWR).

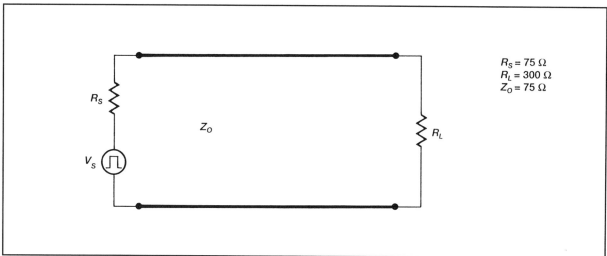

Figure 5.43   Reflections (R>Z)

Solution:

V_{min} at $\lambda/4$ from R_L

$R_s = Z_0 = 75$ ohms

$R_L = 300$ ohms

a) $\rho_v = V_r / V_f$

$(V_f - V_r) / Z_0 = (V_f + V_r) / R_L$

$$\rho_v = \frac{R_L - Z_0}{R_L + Z_0} = \frac{300 - 75}{300 + 75} = \frac{225}{375} = \frac{3}{5}$$

b) $\tau_v = \dfrac{2R_L}{R_L + Z_0} = \dfrac{600}{375} = \rho_v + 1 = \dfrac{8}{5}$

c) $\text{VSWR} = \dfrac{V_{max}}{V_{min}} = \dfrac{1 + \rho_v}{1 - \rho_v} = \dfrac{1 + 0.6}{1 - 0.6} = 4$

$$\rho_v = \frac{VSWR - 1}{VSWR + 1} = \frac{4 - 1}{4 + 1} = \frac{3}{5}$$

5.44 Rectangular wave guide

A rectangular waveguide is 4 cm wide and 2 cm high as shown Figure 5.44. Determine (a) its cutoff wavelength and (b) frequency for the TE_{10} and TE_{11} modes respectively.

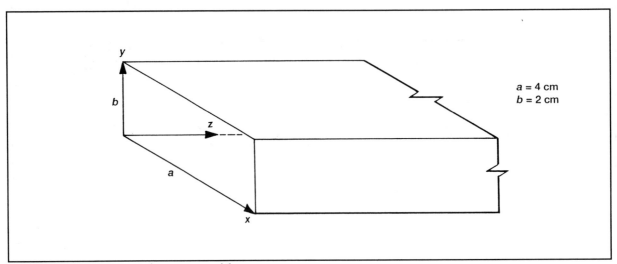

Figure 5.44 Rectangular wave guide

Solution:

$$k_c = [(m\pi/a)^2 + (n\pi/b)^2]^{1/2}$$

$$\lambda_c = 2\pi/k_c = 2/[(m/a)^2 + (n/b)^2]^{1/2}$$

$$f_c = \frac{c}{\lambda_c} = \frac{1}{\sqrt{\mu_0 \epsilon_0}} \times \frac{\left[\left(\frac{m}{a}\right)^2 + \left(\frac{n}{b}\right)^2\right]^{1/2}}{2}$$

a) For TE_{10},

$$\lambda_c = 2a = 2 \times (0.04) = 0.08 \text{ meters}$$

$$f_c = c/\lambda_c = (3 \times 10^8)/(0.08) = 3.750 \times 10^9 \text{ Hz}$$

b) For TE_{11},

$$\lambda_c = 2/[(1/0.04)^2 + (1/0.02)^2]^{1/2} = 0.0358 \text{ meters}$$

$$f_c = c/\lambda_c = (3 \times 10^8)/(0.0358) = 8.385 \times 10^9 \text{ Hz}$$

5.45 Circular wave guide

A circular waveguide has a radius of 4 cm as shown Figure 5.45. Determine (a) its cutoff wavelength and (b) frequency for the TM_{01} and TM_{11} modes respectively.

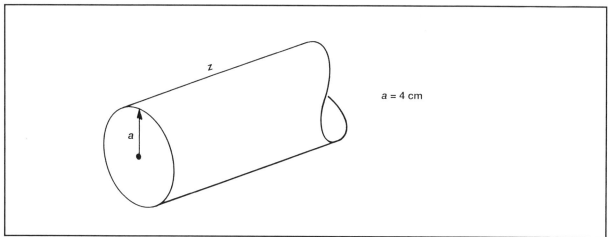

Figure 5.45 Circular wave guide

Solution:

$k_c = P_{mn} / a$
$P_{01} = 2.405$
$P_{11} = 3.83$

$\lambda_c = 2\pi / k_c = (2\pi a) / P_{mn}$

$f_c = \dfrac{c}{\lambda_c} = \dfrac{1}{\sqrt{\mu_0 \epsilon_0}} \times \dfrac{P_{mn}}{2\pi a}$

a) For TM_{01},

$\lambda_c = (2\pi a) / P_{01} = (2\pi a) / 2.405 = 2.61a = 2.61 \times 0.04 = 0.1$ meter

$f_c = c / \lambda_c = 3 \times 10^8 [2.405 / (2\pi a)]$

$f_c = 3 \times 10^8 \times [0.383 / a] = 3 \times 10^8 \times [0.383 / 0.04] = 2.872 \times 10^9$ Hz

b) For TM_{11},

$\lambda_c = (2\pi a) / P_{11} = (2\pi a) / 3.83 = 1.64a = 1.64 \times 0.04 = 0.065$ meter

$f_c = c / \lambda_c = 3 \times 10^8 \times [3.83 / (2\pi a)]$

$f_c = 3\times10^8\times[0.609/a] = 3\times10^8\times[0.609/0.04] = 4.567\times10^9$ Hz

5.46 TM wave

A circular waveguide with a radius of 4 cm is used to propagate an electromagnetic wave in the TM_{01} mode as shown Figure 5.46. Determine (a) the wave impedance, Z_{TM}, (b) phase velocity, v_p, and (c) group velocity, v_g, of the waveguide for a wavelength of 8 cm.

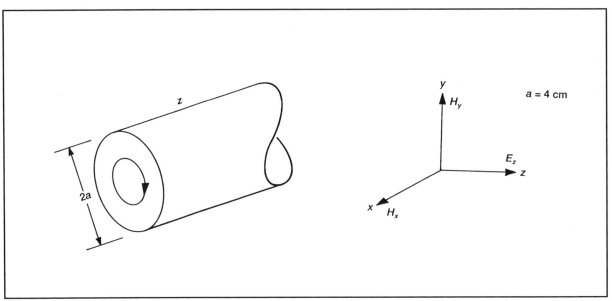

Figure 5.46 TM wave

Solution:

a) $Z_{TM} = (\mu_0/\epsilon_0)^{1/2}[1 - (\lambda/\lambda_c)^2]^{1/2}$

$(\mu_0/\epsilon_0)^{1/2} = 377$ ohms

$1/(\mu_0\epsilon_0)^{1/2} = 3\times10^8$ meters/second

$\lambda_c = 2.61a = 2.61\times0.04 = 0.1$ meter

$Z_{TM} = 377\left[1 - \left(\frac{0.08}{0.1}\right)^2\right]^{1/2} = 377 \times (0.36)^{1/2} = 377 \times 0.6 = 226.2$ ohms

b) $v_p = 1/[(\mu_0\epsilon_0)^{1/2} \times (1 - (\lambda/\lambda_c)^2)^{1/2}]$

$v_p = [3\times10^8/0.6] = 5\times10^8$ meters/second

c) $v_g = [1/(\mu_0\epsilon_0)^{1/2}]\times[1 - (\lambda/\lambda_c)^2]^{1/2}$

$v_g = [3\times10^8]\times[0.6] = 1.8\times10^8$ meters/second

$$c = (v_p v_g)^{1/2} = (5\times10^8 \times 1.8\times10^8)^{1/2} = 3\times10^8 \text{ meters/second}$$

5.47 TE wave

A rectangular waveguide with a width of 4 cm and height of 2 cm is used to propagate an electromagmetic wave in the TE_{10} mode as shown Figure 5.47. Determine (a) the wave impedance, Z_{TE}, (b) phase velocity, v_p, and (c) group velocity, v_g, of the waveguide for a wavelength of 6 cm.

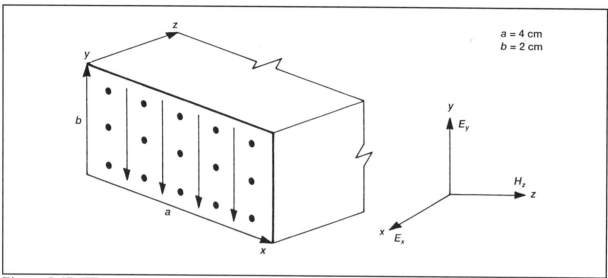

Figure 5.47 TE wave

Solution:

a) $Z_{TE} = [\mu_0/\epsilon_0]^{1/2} / [1 - (\lambda/\lambda_c)^2]^{1/2}$

$(\mu_0/\epsilon_0)^{1/2} = 377$ ohms

$1/(\mu_0\epsilon_0)^{1/2} = 3\times10^8$ meters/second

$\lambda_c = 2a = 2\times0.04 = 0.08$ meter

$$Z_{TE} = \frac{377}{\left[1 - \left(\frac{0.06}{0.08}\right)^2\right]^{1/2}} = \frac{377}{0.66} = 570 \text{ ohms}$$

b) $v_p = 1/[(\mu_0\epsilon_0)^{1/2} \times (1 - (\lambda/\lambda_c)^2)^{1/2}]$

$v_p = [3\times10^8 / 0.66] = 4.54\times10^8$ meters/second

c) $v_g = [1/(\mu_0\epsilon_0)^{1/2}] \times [1 - (\lambda/\lambda_c)^2]^{1/2}$

$v_g = [3\times10^8] \times [0.66] = 1.98\times10^8$ meters/second

$$c = (v_p v_g)^{\frac{1}{2}} = (4.54 \times 10^8 \times 1.98 \times 10^8)^{\frac{1}{2}} = 3 \times 10^8 \text{ meters/second}$$

5.48 TEM wave

A coaxial transmission line is used to propagate a TEM electromagnetic wave as shown in Figure 5.48. Determine the expressions for (a) the electric field intensity, E_x and (b) magnetic field intensity, H_y. And calculate (c) the line impedance, Z_0, (d) the wave impedance Z_{TEM}, (e) the phase velocity, v_p, and (f) the group velocity, v_g, for a spacer dielectric constant of four.

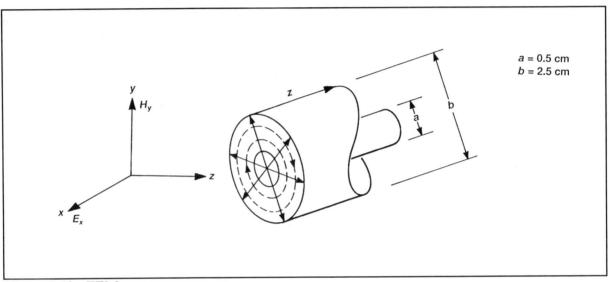

Figure 5.48 TEM wave

Solution:

a) $E_x(z) = E_0 e^{\pm j(\omega t - z/\lambda)}$ volts/meter

b) $\lambda = v/f = (2\pi)/[\omega(\mu_0 \epsilon_0)^{1/2}]$

$H_y(z) = [E_0 / (\mu_0 / k\epsilon_0)^{1/2}] e^{\pm j(\omega t - z/\lambda)}$ amperes/meter

c) $Z_0 = \dfrac{60}{\sqrt{k}} \ln\left(\dfrac{b}{a}\right) = \dfrac{60}{\sqrt{4}} \ln\left(\dfrac{2.5 \times 10^{-2}}{0.5 \times 10^{-2}}\right) = 48$ ohms

d) $Z_{TEM} = (\mu_0 / k\epsilon_0)^{1/2} = 377/\sqrt{k} = 377/\sqrt{4} = 188.5$ ohms

e) $v_p = \sqrt{k} / (\mu_0 \epsilon_0)^{1/2} = \sqrt{4} \times (3 \times 10^8) = 6 \times 10^8$ meters/second

f) $v_g = 1 / (k\mu_0 \epsilon_0)^{1/2} = (3 \times 10^8)/\sqrt{4} = 1.5 \times 10^8$ meters/second

$c = (v_p v_g)^{1/2} = (6 \times 10^8 \times 1.5 \times 10^8)^{1/2} = 3 \times 10^8$ meters/second

5.49 Rectangular cavity

A rectangular resonant cavity is 4 cm wide, 6 cm long, and 2 cm high as shown in Figure 5.49. Determine (a) the resonant wavelength, λ, and (b) resonant frequency, f, for the TE_{10} mode of oscillation.

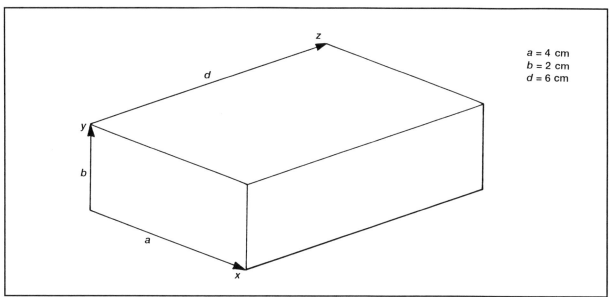

Figure 5.49 Rectangular cavity

Solution:

The condition for oscillation occurs when $d = \lambda_g/2$

$\lambda_g = \lambda / [1 - (\lambda/\lambda_c)^2]^{1/2}$

$\lambda_c = 2a$

$d = [\lambda/2][1 - (\lambda/2a)^2]^{1/2}$

a) $\lambda = [2ad] / [(a)^2 + (d)^2]^{1/2}$

$$\lambda = \frac{[2 \times 0.04 \times 0.06]}{\left[(0.04)^2 + (0.06)^2\right]^{1/2}}$$

$\lambda = [0.0048] / [0.0721] = 0.0665$ meter

b) $f = c/\lambda = (3 \times 10^8) / (0.0665) = 4.51 \times 10^9$ Hz

5.50 Cylindrical cavity

A cylindrical resonant cavity has a radius of 4 cm and is 6 cm long as shown in Figure 5.50. Determine (a) the resonant wavelength, λ, and (b) resonant frequency, f, for the TM_{01} mode of oscillation.

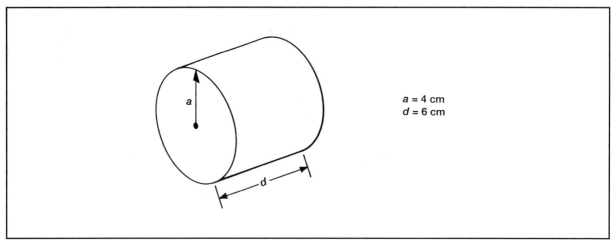

Figure 5.50 Cylindrical cavity

Solution:

The condition for oscillation occurs when $d = \lambda_g/2$

$\lambda_g = \lambda / [1 - (\lambda/\lambda_c)^2]^{1/2}$

$\lambda_c = 2.61a$

$d = [\lambda/2][1 - (\lambda/2.61a)^2]^{1/2}$

a) $\lambda = [3.23ad] / [2.61(a)^2 + 1.53(d)^2]^{1/2}$

$$\lambda = \frac{[3.23 \times 0.04 \times 0.06]}{[2.61 \times (0.04)^2 + 1.53 \times (0.06)^2]^{1/2}}$$

$\lambda = [0.00752] / [0.0984] = 0.0787$ meter

b) $f = c/\lambda = (3 \times 10^8)/(0.0787) = 3.81 \times 10^9$ Hz

5.51 Line parameters

A transmission line is connected between a voltage source, v(t) = Vsinωt, and a load as shown in Figure 5.51. The per unit length electrical parameters are R = 3×10⁻³ ohms/m, L = 5×10⁻⁶ henries/m, G = 1×10⁻⁹ siemens/m, and C = 50×10⁻¹² farads/m.

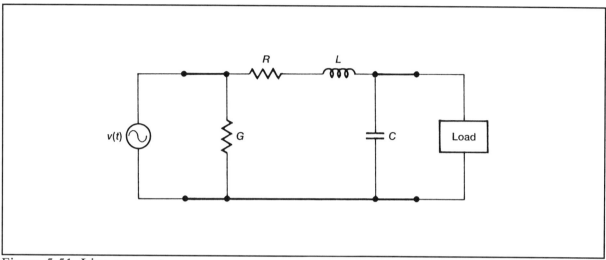

Figure 5.51 Line parameters

1. The line characteristic impedance at f = 400 Hz is most nearly:

   a) 3×10^{-3} b) 2.5×10^{3} c) 3.2×10^{3} d) 1.25×10^{-2}

2. The line phase constant at f = 400 Hz is most nearly:

   a) 4.9×10^{-5} b) 3×10^{-3} c) 7×10^{-3} d) 4×10^{-4}

3. The line phase velocity at f = 400 Hz is most nearly:

   a) 6.28×10^{6} b) 1.57×10^{3} c) 3.25×10^{3} d) 1×10^{6}

4. The line characteristic impedance at f = 400×10⁶ Hz is most nearly:

   a) 1.7×10^{3} b) 270   c) 316   d) 2×10^{3}

5. The line phase velocity at at f = 400×10⁶ Hz is most nearly:

   a) 1×10^{7} b) 4.8×10^{7} c) 3.6×10^{7} d) 6.32×10^{7}

5.51 Solution:

1. $z = R + j\omega L = 3\times 10^{-3} + j1.25\times 10^{-2} = 1.28\angle 76.8°$

   $y = G + j\omega C = 1\times 10^{-9} + j1.25\times 10^{-7} = 1.25\times 10^{-7}\angle 89.5°$

   $Z_0 = (z/y)^{\frac{1}{2}} = 3.2\times 10^3 \angle -6.5°$

   The correct answer is (c).

2. $\gamma = (zy)^{\frac{1}{2}} = 4\times 10^{-4}\angle 83°$

   $\gamma = \alpha + j\beta = 4.9\times 10^{-5} + j4\times 10^{-4}$

   $\alpha = 4.9\times 10^{-5}$ nepers/m

   $\beta = 4.\times 10^{-4}$ radians/m

   The correct answer is (d).

3. $v = \omega / \beta = (2\pi \times 400)/(4\times 10^{-4}) = 6.28\times 10^6$ meters/second

   The correct answer is (a).

4. At $f = 400\times 10^6$, $\omega L \gg R$ and $\omega C \gg G$

   $Z_0 = (L/C)^{\frac{1}{2}} = 316$ ohms

   The correct answer is (c).

5. $\gamma = \alpha + j\beta \approx j\beta$

   $\beta = \omega (LC)^{\frac{1}{2}}$

   $v = \omega / \beta = 1/(LC)^{\frac{1}{2}} = 6.32\times 10^7$ meters/second

   The correct answer is (d).

5.52 Impedance matching

Two transmission line impedance matching transformers are shown in Figure 5.52. The source signal frequency is 500×10^6 Hz.

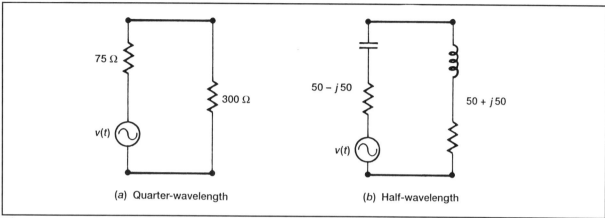

Figure 5.52 Impedance matching

1. The characteristic impedance of the quarter-wavelength matching stub is most nearly:

   a) 187 b) 600 c) 150 c) 225

2. The minimum length of the half-wavelength matching stub at which the reactances cancel out is most nearly:

   a) 0.3 b) 0.15 c) 0.6 d) 0.9

3. The optimum characteristic impedance of the half-wavelength matching stub is most nearly:

   a) 50 b) 100 c) 140 d) 70

4. The voltage standing wave ratio (VSWR) of the quarter-wavelength matching stub is most nearly:

   a) 0.5 b) 2 c) 0.33 d) 3

5. The L and C values of the input and output impedances of the quarter-wavelegth matching stub are most nearly:

   a) 1.6×10^{-8} and 6.4×10^{-12} b) 10×10^{-8} and 1×10^{-12}

   c) 1×10^{-12} and 4×10^{-11} d) 1×10^{-12} and 10×10^{-8}

5.52 Solution:

1. $Z_0 = (Z_{in} Z_{out})^{1/2} = (75 \times 300)^{1/2} = 150$ ohms

   The correct answer is (c).

2. At $l = \lambda/2$, $Z_{in} = Z_{out}$

   $Z = Z_{in} + Z_{out} = (50 - j50) + (50 + j50) = 100$ ohms

   $\lambda = c/f = 3\times 10^8 / 5\times 10^8 = 0.6$ meters

   $\lambda/2 = 0.3$ meters

   The correct answer is (a).

3. For minimum VSWR $Z_0 = |Z_{out}| = (R^2 + X_L^2)^{1/2} = 70$ ohms

   The correct answer is (d).

4. $\rho_v = (Z_{out} - Z_{in})/(Z_{out} + Z_{in}) = 150/450 = 1/3$

   VSWR $= (1 + \rho_v)/(1 - \rho_v) = 2$

   The correct answer is (b).

5. $L = X_L/(2\pi f) = 1.6 \times 10^{-8}$ henries

   $C = 1/(2\pi f X_C) = 6.4 \times 10^{-12}$ farads

   The correct answer is (a).

5.53 Electric interference

A half-wave dipole antenna is driven with a 1×10^3 volt square wave signal as shown in Figure 5.53. An amplifier circuit is located 30 meters away.

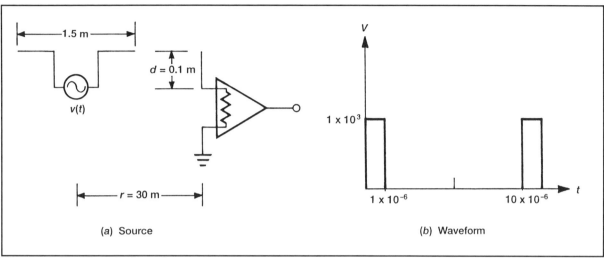

Figure 5.53 Electric interference

1. The dipole antenna resonance frequency is most nearly:

   a) 1×10^8 b) 5×10^7 c) 4×10^8 d) 2×10^8

2. The peak amplitude of the fifth harmonic is most nearly:

   a) 63.7   b) 200   c) 127.4   d) 100

3. The peak radiated power of the fifth harmonic is most nearly:

   a) 550   b) 137   c) 106   d) 222

4. The peak electric field strength, of the fifth harmonic, at r = 30 meters is most nearly:

   a) 29   b) 5.23   c) 10.5   d) 15

5. The peak induced voltage, of the fifth harmonic, in the amplifier circuit is most nearly:

   a) 1.06   b) 0.26   c) 0.6   d) 0.53

5.53 Solution:

1. $l = \lambda/2 = 1.5$ meters

   $\lambda = 2 \times l = 3$ meters

   $f_r = c/\lambda = (3 \times 10^8)/3 = 1 \times 10^8$ Hz

   The correct answer is (a).

2. $V_5 = 2A\tau[(\sin 5\pi\tau)/(5\pi\tau)]$

   $\tau = (1 \times 10^{-6})/(10 \times 10^{-6}) = 0.1$

   $V_5 = [2 \times 1 \times 10^3 \times 0.1] \times [1/1.57] = 127.4$ volts

   The correct answer is (c).

3. $P_t = (V_5)^2/R_s = (16.2 \times 10^3)/73 = 222.1$ watts

   The correct answer is (d).

4. $E = (49.2 P_t)^{1/2}/r$ volts/meter

   $r = 30$ m

   $E = 5.23$ volts/meter

   The correct answer is (b).

5. $V_r = E \cdot d = 5.23 \times 0.1 = 0.53$ volts

   The correct answer is (d).

5.54 Magnetic interference
A conductor is carrying a 20 ampere square wave signal and is located two meters away from a rectangular wire loop as shown in Figure 5.54.

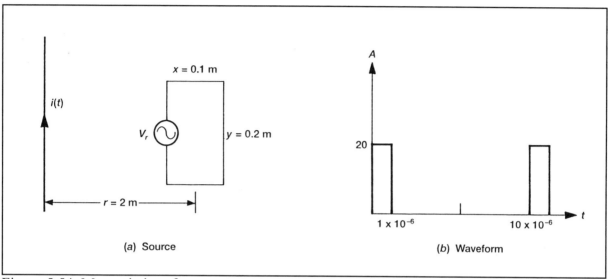

Figure 5.54 Magnetic interference

1. The peak amplitude of the fifth harmonic is most nearly:

   a) 5.1 b) 4 c) 2 d) 2.55

2. The peak magnetic field strength, of the fifth harmonic, at r = 2 meters is most nearly:

   a) 4×10^{-7} b) 2.55×10^{-7} c) 8×10^{-7} d) 11.1×10^{-7}

3. The peak magnetic flux, of the fifth harmonic, at r = 2 meters is most nearly:

   a) 2.22×10^{-8} b) 8×10^{-9} c) 5.1×10^{-9} d) 1.6×10^{-8}

4. The rectangular loop resonant frequency is most nearly:

   a) 5×10^{8} b) 2.5×10^{8} c) 1.5×10^{10} d) 10×10^{8}

5. The peak induced voltage, of the fifth harmonic, in the rectangular loop most nearly:

   a) 0.8 b) 3.2 c) 0.64 d) 0.16

5.54 Solution:

1. $I_5 = 2A\tau[(\sin 5\pi\tau)/(5\pi\tau)]$

   $\tau = 1\times 10^{-6} / 10\times 10^{-6} = 0.1$

   $I_5 = [2\times 20\times 0.1]\times[1/1.57] = 2.55$ amps

   The correct answer is (d).

2. $\mu_0 = 4\pi\times 10^{-7}$ henries/meter

   r = 2 meters

   $\mathbf{B} = (\mu_0 I_5)/(2\pi r) = (4\pi\times 10^{-7}\times 2.55)/(4\pi) = 2.55\times 10^{-7}$ webers/meter²

   The correct answer is (b).

3. $\Phi = \mathbf{B}A = (2.55\times 10^{-7})\times(0.1\times 0.2) = 5.1\times 10^{-9}$ webers

   The correct answer is (c).

4. $\lambda = 2(x+y) = 2(0.1 + 0.2) = 0.6$ meters

   $f = c/\lambda = 3\times 10^8 / 0.6 = 5\times 10^8$ Hz

   The correct answer is (a).

5. $v_r = -d\Phi/dt = -d/dt[(5.1\times 10^{-9})\sin 5\omega t] = -d/dt[(5.1\times 10^{-9})\sin 10\pi ft]$

   $v_r = -(5.1\times 10^{-9}\times 10\pi\times f)\cos 10\pi ft$

   $f_r = 1/\tau = 1/1\times 10^{-6} = 1\times 10^6$ Hz

   $5f_r = 5\times 10^6$ Hz

   $V_r = -(5.1\times 10^{-9}\times 10\pi\times 1\times 10^6) = -0.16$ volts

   The correct answer is (d).

5.55 Radar
A simplified block diagram of a radar system is shown in Figure 5.55.

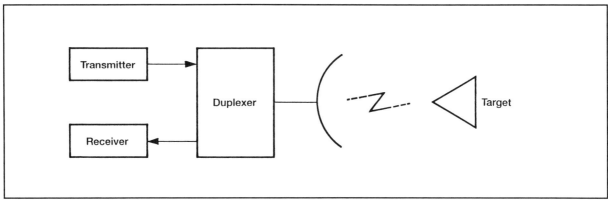

Figure 5.55 Radar

1. The target distance for a 10×10^{-6} seconds time difference between the T_x and R_x pulses is most nearly:

   a) 3×10^3 b) 6×10^3 c) 1×10^3 d) 1.5×10^3

2. The maximum pulse repetition frequency to detect a target at an unambiguous distance of 10×10^3 meters is most nearly:

   a) 60×10^3 b) 15×10^3 c) 7.5×10^3 d) 30×10^3

3. The doppler frequency for a CW radar operating at 10×10^9 Hz for a target velocity of 360 km/hr is most nearly:

   a) 6.6×10^3 b) 1.65×10^3 c) 3.3×10^3 d) 3.6×10^3

4. A triangular pulse FM-CW radar, modulated at 300 Hz rate over a range of 100×10^6 Hz, with a 400 kHz beat frequency has a target range of most nearly:

   a) 2×10^3 b) 5×10^2 c) 1×10^3 d) 4×10^3

5. A pulsed MTI radar operating at 5×10^9 Hz with a pulse repetition frequency of 300×10^3 Hz, and a single delay line canceler, has a minimum blind speed of most nearly:

   a) 1.8×10^4 b) 4.5×10^3 c) 3.6×10^4 d) 9×10^3

5.55 Solution:

1. $d = (c\Delta t)/2 = (3\times10^8 \times 10\times10^{-6})/2 = 1.5\times10^3$ meters

   The correct answer is (d).

2. $f_r = c/(2d) = (3\times10^8)/(2\times10\times10^3) = 15\times10^3$ Hz

   The correct answer is (a).

3. $\lambda = c/f_0 = (3\times10^8)/(10\times10^9) = 0.03$ meters

   $v = (360\times1{,}000)/(60\times60) = 100$ m/sec

   $f_d = (2v)/\lambda = (2\times100)/0.03 = 6.6\times10^3$ Hz

   The correct answer is (a).

4. $f_r = 400\times10^3$ Hz

   $f_m = 300$ Hz

   $\Delta f = 100\times10^6$ Hz,

   $d = (cf_r)/(4f_m\Delta f) = 1\times10^3$ meters

   The correct answer is (c).

5. $\lambda = c/f_0 = (3\times10^8)/(5\times10^9) = 0.06$ m

   $v_1 = (\lambda f_r)/2 = (0.06\times300\times10^3)/2 = 9\times10^3$ m/sec

   The correct answer is (d).

6.0 POWER

6.1 Line resistance

A solid round wire conductor, made of annealed copper, has a length of five kilometers as shown in Figure 6.1. The wire diameter is 0.25 cm and its resistivity constant, ρ, equals 1.7241×10^{-8}, meter-ohms at 20°C. Determine the conductor dc resistance at 20°C and 40°C using a zero resistance temperature of -234.5°C.

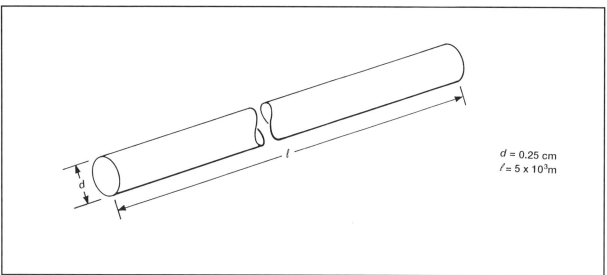

Figure 6.1 Line resistance

Solution:

$$\rho = 1.7241 \times 10^{-8} \text{ meter-ohms at } 20°C$$

$$R = \rho\left(\frac{l}{A}\right) = \rho\left(\frac{l}{\pi r^2}\right) = \frac{1.724 \times 10^{-8} \times 5 \times 10^3}{\pi \times (0.125 \times 10^{-2})^2}$$

$$R = 17.56 \text{ ohms at } 20°C$$

$$R_1 = R\left(\frac{234.5 + T_1}{234.5 + T}\right)$$

$$R_1 = R\left(\frac{234.5 + 40}{234.5 + 20}\right)$$

$$R_1 = 18.94 \text{ ohms at } 40°C$$

6.2 Line inductance

A two conductor solid copper wire transmission line extends for five kilometers. Each wire diameter is 0.25 cm and are spaced at a uniform distance of 30 cm apart as shown in Figure 6.2. Determine the line inductance, L_T.

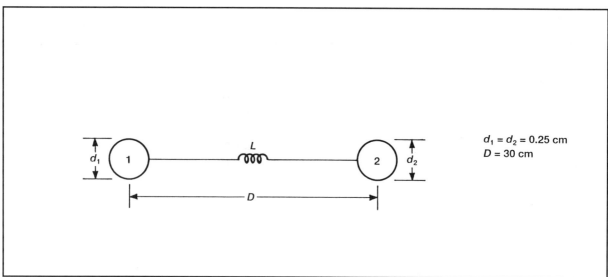

Figure 6.2 Line inductance

Solution:

$$L_1 = L_{int} + L_{ext} = \frac{\mu}{8\pi} + \frac{\mu}{2\pi}\ln\left(\frac{D}{r_1}\right)$$

$$\mu = \mu_r \mu_0$$

$$\mu_r = 1, \quad \mu_0 = 4\pi \times 10^{-7}$$

$$L_1 = L_{int} + L_{ext} = [(4\pi \times 10^{-7})/(8\pi)] + [(4\pi \times 10^{-7})/(2\pi)]\ln(D/r_1)$$

$$L_1 = 2 \times 10^{-7}[(1/4) + \ln(D/r_1)] \quad \text{henries/meter}$$

$$L = L_1 + L_2$$

Due to symmetry $L_1 = L_2$

$$L = 2L_1 = 4 \times 10^{-7} \ln[(1.284D)/r_1]$$

$$L = 4 \times 10^{-7} \ln\left[\frac{1.284 \times 30}{0.125}\right] = 4 \times 10^{-7} \times 5.73 = 2.3 \times 10^{-6} \quad \text{henries}$$

$$L_T = L \times l = 2.3 \times 10^{-6} \times 5 \times 10^{3} = 11.5 \times 10^{-3} \quad \text{henries}$$

6.3 Line capacitance

A two conductor solid copper wire transmission line extends for five kilometers. Each wire diameter is 0.25 cm and are spaced at a uniform distance of 30 cm apart as shown in Figure 6.3. Determine the line capacitance, C_T.

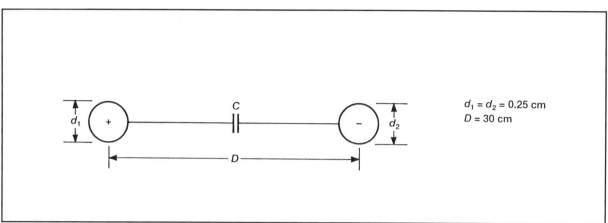

Figure 6.3 Line capacitance

Solution:

$$C = 2\pi\epsilon \left[\frac{1}{\ln \frac{D}{r_1}} - \frac{1}{\ln \frac{r_2}{D}} \right]$$

$$C = (2\pi\epsilon)[1/\ln(D/r_1) + 1/\ln(D/r_2)]$$

$$C = (2\pi\epsilon)[1/\ln(D^2/(r_1 r_2))]$$

$$\epsilon = \epsilon_r \epsilon_0$$

$$\epsilon_r = 1, \quad \epsilon_0 = 8.854 \times 10^{-12}$$

For $r_1 = r_2 = r$

$$C = 2\pi\epsilon \left[\frac{1}{2\ln \frac{D}{r}} \right] = \pi\epsilon \left[\frac{1}{\ln \frac{D}{r}} \right]$$

$$C = [\pi \times 8.854 \times 10^{-1}] \times [1/\ln(30/0.125)]$$

$$C = 27.8 \times 10^{-12} \times [1/5.48] = 5.07 \times 10^{-12} \quad \text{farads/meter}$$

$$C_T = C \times l = 5.07 \times 10^{-12} \times 5 \times 10^3 = 25.35 \times 10^{-9} \quad \text{farads}$$

6.4 Short line

A five kilometer long transmission line is used to supply a resistive load with 500 kwatts at 12.5 kvolts, 60 Hz as shown in Figure 6.4. The line parameters are r = 3.5 ohms/km and z = 2.3×10⁻³ henries/km. Calculate the required source voltage, V_s.

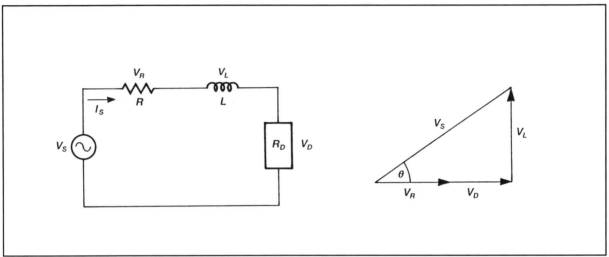

Figure 6.4   Short line

Solution: $I = I_s = I_D$

$V_s = V_D + IZ = V_D + I(R^2 + X_L^2)^{½}$

$R = 2lr = 2 \times 5 \times 3.5 = 35$ ohms

$L = lz = 5 \times 2.3 \times 10^{-3} = 11.5 \times 10^{-3}$ henries

$X_L = 2\pi f L = 377L = 377 \times 11.5 \times 10^{-3} = 4.335$ ohms

$V_s = V_D + IZ = I(R_D + Z) + I[R_D + (R + jX_L)]$

$I = P_D / V_D = (500 \times 10^3) / (12.5 \times 10^3) = 40\angle 0°$ amperes

$R_D = P_D / I^2 = (500 \times 10^3) / (1.6 \times 10^3) = 312.5$ ohms

$V_s = I[R_D + (R + jX_L)] = 40 \times [312.5 + (35 + j4.335)] = [12.5 \times 10^3 + (1.4 \times 10^3 + j173.42)]$

$V_s = [13.9 \times 10^3 + j173.42] = 13.9 \times 10^3 \angle \theta$

$\theta = \tan^{-1}[V_L / (V_D + V_R)] = \tan^{-1}[173.42 / 13.9 \times 10^3] = \tan^{-1}[1.246 \times 10^{-2}] = 0.71°$

$V_s = 13.9 \times 10^3 \angle 0.71°$

6.5 Medium line

A 100 kilometer long transmission line is used to supply a resistive load with 500 kwatts at 12.5 kvolts, 60 Hz as shown in Figure 6.5. The total line impedance $Z = 50 + j50$ ohms with a total line capacitance of $C = 1.5 \times 10^{-6}$ farads. Calculate the required source voltage, V_s.

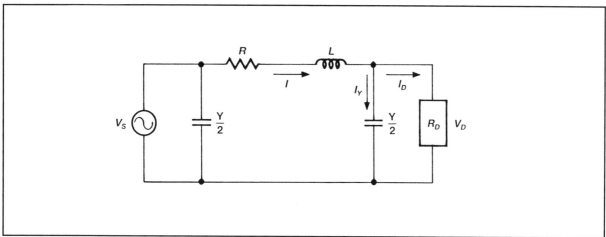

Figure 6.5 Medium line

Solution:

$Y = 1/X_C = 2\pi fC = 377 \times 1.5 \times 10^{-6} = 565.5 \times 10^{-6} \angle 90°$ siemens

$V_s = IZ + V_D$

$I = I_D + I_Y = I_D + V_D(Y/2)$

$V_s = [I_D + V_D(Y/2)]Z + V_D = I_D Z + V_D(Y/2)Z + V_D = I_D Z + V_D[(YZ/2) + 1]$

$I_D = P_D / V_D = (500 \times 10^3) / (12.5 \times 10^3) = 40 \angle 0°$ amperes

$Z = (R^2 + X_L^2)^{\frac{1}{2}} = [(50)^2 + (50)^2]^{\frac{1}{2}} = 70.7 \angle 45°$ ohms

$V_s = [(40 \angle 0°) \times (70.7 \angle 45°)] + [12.5 \times 10^3 \angle 0°][(YZ/2) + 1]$

$YZ/2 = [(565 \times 10^{-6} \angle 90°) \times (70.7 \angle 45°)] / 2 = 0.02 \angle 135°$

$V_s = [2,828 \angle 45° + 250 \angle 135° + 12.5 \times 10^3 \angle 0°]$

$V_s = [1,999.4 + j1,999.4 - 176.8 + j176.8 + 12.5 \times 10^3]$

$V_s = 14,332.6 + j2,176 = 14,486.95 \angle \theta$

$\theta = \tan^{-1}[2,176/14,322.6] = \tan^{-1}[0.151] = 8.64°$

$V_s = 14,486.95 \angle 8.64°$

6.6 Long line

A 200 kilometer long transmission line is used to supply a resistive load with 500 kwatts at 12.5 kvolts, 60 Hz as shown in Figure 6.6. The line parameters are R = 0.125 ohms/km, L = 2×10⁻³ henries/km, and C = 0.015×10⁻⁶ farads/km. Calculate the required source voltage, V_s.

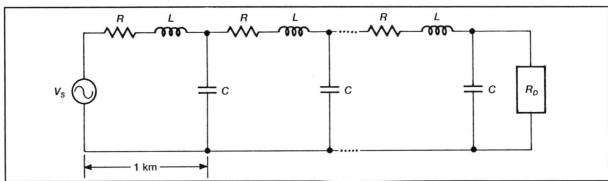

Figure 6.6 Long line

Solution:

$z = R + j2\pi fL = 0.125 + j377 \times 2 \times 10^{-3} = 0.125 + j0.754 = 0.764 \angle 80.6°$ ohms/km

$y = 1/R + jB = G + j2\pi fC = 0 + 377 \times 0.015 \times 10^{-6} = 5.65 \times 10^{-6} \angle 90°$ siemens/km

$Z = (z/y)^{1/2} = [0.764 / (5.65 \times 10^{-6})]^{1/2} \angle [(80.6° - 90°)/2] = 367.7 \angle -4.7°$ ohms

$\gamma = (zy)^{1/2} = [0.764 \times 5.65 \times 10^{-6}]^{1/2} \angle [(80.6° + 90°)/2] = 2.077 \times 10^{-3} \angle 85.3°$

$\gamma = \alpha + j\beta = 0.17 \times 10^{-3} + j2.068 \times 10^{-3}$

$\gamma l = \gamma \times 200 = \alpha l + j\beta l = 0.034 + j0.4136$

$I_D = P_D / V_D = (500 \times 10^3) / (12.5 \times 10^3) = 40 \angle 0°$

$I_D Z = (40 \angle 0°) \times (367.7 \angle -4.7°) = 14,710 \angle -4.7°$

$V_s = [V_D/2 + (I_D Z/2)]e^{\alpha l} e^{j\beta l} + [V_D/2 - (I_D Z/2)]e^{-\alpha l} e^{-j\beta l} = V_f e^{\alpha l} e^{j\beta l} + V_r e^{-\alpha l} e^{-j\beta l}$

$V_f = [6,250 + (7,342 - j602)] = 13,592 - j602 = 13,605 \angle -2.54°$

$V_r = [6,250 - (7,342 + j602)] = -1,092 + j602 = 1,247 \angle 151.14°$

$e^{\pm j\theta} = (\cos\theta \pm j\sin\theta)$ and $\theta° = 57.3° \times \theta = 57.3° \times 0.4136 = 23.7°$

$e^{\alpha l} e^{j\beta l} = e^{0.034} e^{j0.4136} = e^{0.034}(\cos 23.7° + j\sin 23.7°) = 1.035(1) \angle 23.7°$

$e^{-\alpha l} e^{-j\beta l} = e^{-0.034} e^{-j0.4136} = e^{-0.034}(\cos 23.7° - j\sin 23.7°) = 0.966(1) \angle -23.7°$

$V_s = V_f \times 1.035 \angle 23.7° + V_r \times 0.966 \angle -23.7° = 14,081 \angle 21.16° + 1,204 \angle 127.4° = 13,784 \angle 25.9°$

6.7 Power factor

Two reactive loads are connected in parallel across 120 volts ac as shown in Figure 6.7. Determine (a) the resulting real power, P_T, (b) the reactive power, Q_T, and (c) the power factor, PF_T.

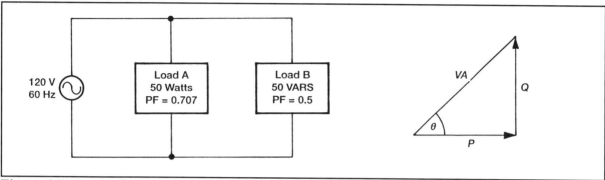

Figure 6.7 Power factor

Solution:

$P_A = 50$ watts

$PF = \cos\theta = 0.707$

$\theta = \cos^{-1}(0.707) = 45°$

$VA_A = P_A / \cos\theta = 70.72$

$Q_A = VA_A \sin\theta = 70.72 \times \sin 45° = 50$ VARs

$Q_B = 50$ VARs

$PF = \cos\theta = 0.5$

$\theta = \cos^{-1}(0.5) = 60°$

$VA_B = Q_B / \sin\theta = 57.7$

$P_B = VA\cos\theta = 57.7 \times \cos 60° = 28.86$ watts

$VA_T = [(P_T)^2 + (Q_T)^2]^{½} = [(50 + 28.6)^2 + (50 + 50)^2]^{½} = 127.7$

a) $P_T = 50 + 28.6 = 78.6$ watts

b) $Q_T = 50 + 50 = 100$ VARs

c) $PF_T = \cos\theta = P_T / VA_T = (78.6/127.7) = 0.617$

$\theta = \cos^{-1}(0.617) = 51.86°$

6.8 Maximum power transfer

Figure 6.8 illustrates two series reactive circuits coupled via an ideal transformer.

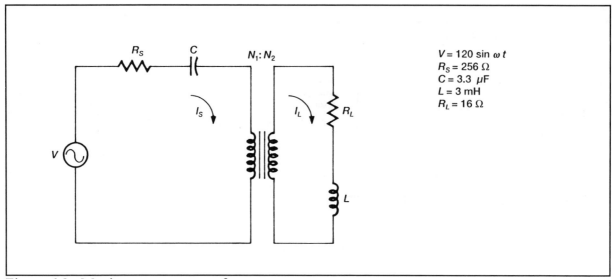

Figure 6.8 Maximum power transfer

1. The transformer turns ratio, $\alpha = N_1/N_2$, is most nearly:

    a) 16   b) 0.25   c) 0.5   d) 4

2. The maximum power transfer frequency, f_0, is most nearly:

    a) 200   b) 400   c) 600   d) 100

3. The load current at f_0, is most nearly:

    a) 3.76   b) 0.47   c) 0.94   d) 1.88

4. The load power dissipated at f_0 is most nearly:

    a) 14   b) 28   c) 30   d) 15

6.8 Solution:

1. $R_s = \alpha^2 R_L$

   $\alpha = (R_s / R_L)^{\frac{1}{2}} = (256/16)^{\frac{1}{2}} = 4$

   The correct answer is (d).

2. When $1/(2\pi fC) = \alpha^2(2\pi fL)$

   $f = \dfrac{1}{2\pi a\sqrt{LC}} = \dfrac{1}{8\pi\sqrt{9.949 \times 10^{-5}}} = 399.89$ Hz

   The correct answer is (b).

3. $I_L = aI_S = a\left(\dfrac{V}{R_s + a^2 R_L}\right) = a\left(\dfrac{V}{2R_s}\right) = 4 \times \left(\dfrac{120}{512}\right) = 0.94$ ampere

   The correct answer is (c).

4. $P = I_L^2 R_L$

   $P = \left(\dfrac{aV}{2R_s}\right)^2 R_L = (0.88)R_L \doteq 14$ watts

   The correct answer is (a).

6.9 Self inductance

A coil consists of twenty turns and it is wound around a rectangular iron bar with a relative permeability, μ, of 120. Determine the coil inductance using the dimensions shown in Figure 6.9.

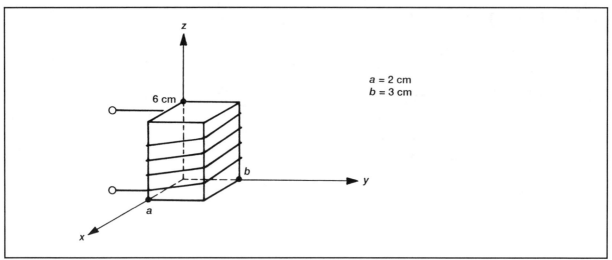

Figure 6.9 Self inductance

Solution:

$$\mu = \mu_r \mu_0$$

$$\mu_r = 120$$

$$\mu_0 = 4\pi \times 10^{-7} \quad \text{henries/meter}$$

$$\mu = 120 \times 4\pi \times 10^{-7}$$

$$B = \mu H = \mu[(NI)/z]$$

$$L\,[(N\Phi)/I] = N[(BA)/I] = \mu N[(NIA)/(Iz)]$$

$$L = \mu\left(\frac{N^2 A}{z}\right) = 120 \times 4\pi \times 10^{-7}\left(\frac{400 \times 0.02 \times 0.03}{0.06}\right)$$

$$L = 480\pi \times 10^{-7} \times [400 \times 0.01] = 6.032 \times 10^{-4} \quad \text{henries}$$

6.10 Mutual inductance

Four coils are connected in a series-parallel arrangement as shown in Figure 6.10. The coefficient of coupling between L_1 and L_2 is 25 %. The coefficient of coupling between L_3 and L_4 is -40 %. Determine the circuit inductance.

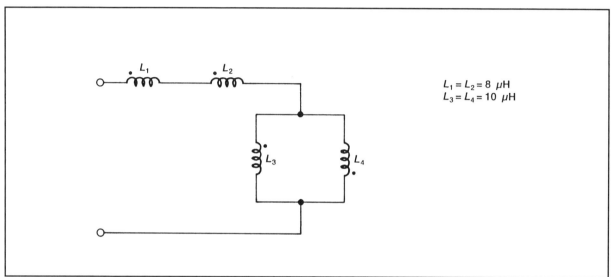

Figure 6.10 Mutual inductance

Solution:

$M_{12} = kL_1$

$M_{21} = kL_2$

$M_{12} = M_{21}$

$(M_{12})^2 = k^2 L_1 L_2$

$M_{12} = k(L_1 L_2)^{1/2} = 0.25 \times [8 \times 10^{-6} \times 8 \times 10^{-6}]^{1/2} = 2 \times 10^{-6}$ henries

$M_{34} = k(L_3 L_4)^{1/2} = 0.4 \times [10 \times 10^{-6} \times 10 \times 10^{-6}]^{1/2} = 4 \times 10^{-6}$ henries

$L_A = L_1 + M_{12} = 8 \times 10^{-6} + 2 \times 10^{-6} = 10 \times 10^{-6}$ henries

$L_B = L_2 + M_{21} = 8 \times 10^{-6} + 2 \times 10^{-6} = 10 \times 10^{-6}$ henries

$L_C = L_3 - M_{34} = 10 \times 10^{-6} - 4 \times 10^{-6} = 6 \times 10^{-6}$ henries

$L_D = L_4 - M_{43} = 10 \times 10^{-6} - 4 \times 10^{-6} = 6 \times 10^{-6}$ henries

$L = L_A + L_B + (L_C \parallel L_D) = 10 \times 10^{-6} + 10 \times 10^{-6} + (3 \times 10^{-6}) = 23 \times 10^{-6}$ henries

6.11 Transformer

An ideal 24 VA transformer is used to step down 120 to 12 volts as shown in Figure 6.11.

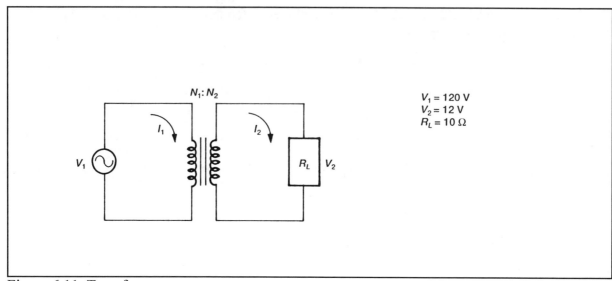

Figure 6.11 Transformer

1. The transformer turns ratio, $\alpha = N_1 / N_2$, is most nearly:

    a) 3   b) 10   c) 0.33   d) 0.1

2. The values of I_1 and I_2 are most nearly:

    a) 0.2 and 2   b) 1 and 0.1   c) 0.1 and 1   d) 2 and 0.2

3. The primary to secondary impedance ratio is most nearly:

    a) 10   b) 9   c) 6   d) 100

4. The reflected impedance is most nearly:

    a) 100   b) 600   c) 1,000   d) 660

6.11 Solution:

1. $\alpha = V_1 / V_2 = 120/12 = 10$

   The correct answer is (b).

2. $I_1 = VA / V_1 = 24/120 = 0.2$ amperes

   $I_2 = VA / V_2 = 24/12 = 2$ amperes

   $I_1 / I_2 = 0.2 / 2 = 1/10 = 1/\alpha$

   The correct answer is (a).

3. $Z_1 = V_1 / I_1 = 120/0.2 = 600$ ohms

   $Z_2 = V_2 / I_2 = 12/2 = 6$ ohms

   $Z_1 / Z_2 = 600/6 = 100 = \alpha^2$

   The correct answer is (d).

4. $R_{in} = \alpha^2 R_L = (10)^2 \times (10) = 1,000$ ohms

   The correct answer is (c).

6.12 Delta to Wye conversion

The impedance values of a delta connected network are shown in Figure 6.12a. Calculate the impedance values for the corresponding wye connected network of Figure 6.12b.

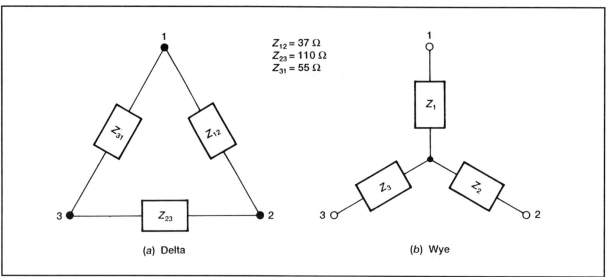

Figure 6.12 Delta to Wye conversion

Solution:

a) $Z_1 = \dfrac{Z_{31}Z_{12}}{Z_{12}+Z_{23}+Z_{31}} = \dfrac{55 \times 37}{37+110+55}$

$Z_1 = 2{,}035/202 = 10$ ohms

b) $Z_2 = \dfrac{Z_{12}Z_{23}}{Z_{12}+Z_{23}+Z_{31}} = \dfrac{37 \times 110}{37+110+55}$

$Z_2 = 4{,}070/202 = 20$ ohms

c) $Z_3 = \dfrac{Z_{23}Z_{31}}{Z_{12}+Z_{23}+Z_{31}} = \dfrac{110 \times 55}{37+110+55}$

$Z_3 = 6{,}050/202 = 30$ ohms

6.13 Wye to Delta conversion

The impedance values of a wye connected network are shown in Figure 6.13a. Calculate the impedance values for the corresponding delta connected network of Figure 6.13b.

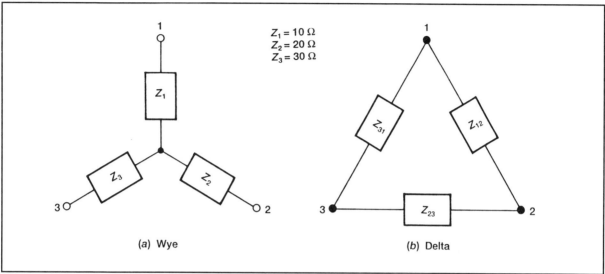

Figure 6.13 Wye to Delta conversion

Solution:

a) $Z_{12} = \dfrac{Z_1 Z_2 + Z_2 Z_3 + Z_1 Z_3}{Z_3} = \dfrac{200 + 600 + 300}{30}$

$Z_{12} = 1{,}100/30 = 37$ ohms

b) $Z_{23} = \dfrac{Z_1 Z_2 + Z_2 Z_3 + Z_1 Z_3}{Z_1} = \dfrac{200 + 600 + 300}{10}$

$Z_{23} = 1{,}100/10 = 110$ ohms

c) $Z_{31} = \dfrac{Z_1 Z_2 + Z_2 Z_3 + Z_1 Z_3}{Z_2} = \dfrac{200 + 600 + 300}{20}$

$Z_{31} = 1{,}100/20 = 55$ ohms

6.14 Four wire Wye:Wye

A three-phase, four-wire system provides power to a balanced resistive load as shown in Figure 6.14. Find the phase currents and the neutral current for a line-to-line voltage of 208 volts.

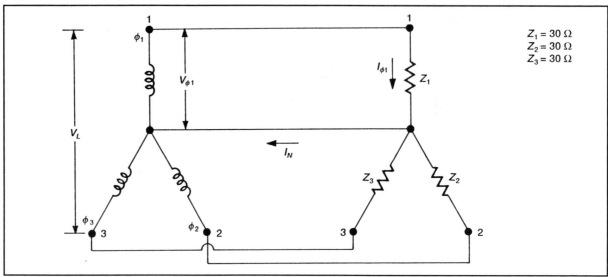

Figure 6.14 Four wire Wye:Wye

Solution:

$I_L = I_\phi$

$V_L = 208$ volts

$V_{12} = V_L \angle 0°$, $\quad V_{23} = V_L \angle -120°$, $\quad V_{31} = V_L \angle 120°$

$V_\phi = V_L / \sqrt{3} = 208 / \sqrt{3} = 120$ volts

$V_{\phi 1} = (V_L / \sqrt{3}) \angle -30°$

$V_{\phi 2} = (V_L / \sqrt{3}) \angle -150°$

$V_{\phi 3} = (V_L / \sqrt{3}) \angle 90°$

a) $I_{\phi 1} = V_{\phi 1} / Z_1 = (120/30) \angle -30° = 4 \angle -30° = 3.46 - j2.0$ amperes

b) $I_{\phi 2} = V_{\phi 2} / Z_2 = (120/30) \angle -150° = 4 \angle -150° = -3.46 - j2.0$ amperes

c) $I_{\phi 3} = V_{\phi 3} / Z_3 = (120/30) \angle 90° = 4 \angle 90° = 0.0 + j4.0$ amperes

d) $I_N = I_{\phi 1} + I_{\phi 2} + I_{\phi 3}$

$I_N = 3.46 - j2.0 - 3.46 - j2.0 + 0.0 + j4.0 = 0$ amperes

6.15 Three wire Wye:Wye

A three-phase, three-wire system provides power to a balanced reactive load as shown in Figure 6.15. Determine (a) the phase currents, (b) total power, (c) total VAR, and (d) total VA, for a line-to-line voltage of 208 volts.

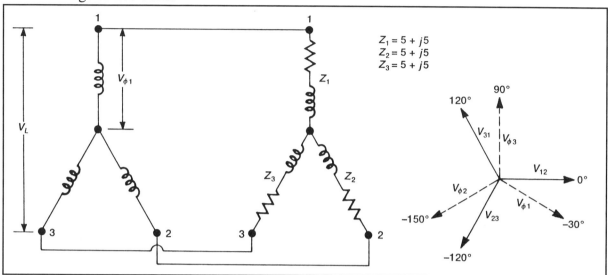

Figure 6.15 Three wire Wye:Wye

Solution:

$I_L = I_\phi$

$V_L = 208$ volts

$V_{12} = V_L\angle 0°$, $\quad V_{23} = V_L\angle -120°$, $\quad V_{31} = V_L\angle 120°$

$V_\phi = V_L/\sqrt{3} = 208/\sqrt{3} = 120$ volts

$V_{\phi 1} = (V_L/\sqrt{3})\angle -30°$, $\quad V_{\phi 2} = (V_L/\sqrt{3})\angle -150°$, $\quad V_{\phi 3} = (V_L/\sqrt{3})\angle 90°$

a) $I_{\phi 1} = V_{\phi 1}/Z_1 = [120\angle -30°]/[5+j5] = [120\angle -30°]/[7.07\angle 45°] = 16.96\angle -75°$ amperes

b) $I_{\phi 2} = V_{\phi 2}/Z_2 = [120\angle -150°]/[5+j5] = [120\angle -150°]/[7.07\angle 45°] = 16.96\angle -195°$ amperes

c) $I_{\phi 3} = V_{\phi 3}/Z_3 = [120\angle 90°]/[5+j5] = [120\angle 90°]/[7.07\angle 45°] = 16.96\angle 45°$ amperes

d) $P_\phi = V_\phi I_\phi \cos\theta \quad\quad (\theta$ of load $)$

$P_T = 3V_\phi I_\phi \cos\theta = 3(V_L/\sqrt{3})I_L\cos\theta = \sqrt{3}V_L I_L \cos\theta = 3\times 208\times 16.97\times \cos 45° = 4{,}323$ watts

e) $Q_T = \sqrt{3}V_L I_L \sin 45° = \sqrt{3}\times 208\times 16.97\times \sin 45° = 4{,}323$ VARs

f) $VA = \sqrt{3}V_L I_L = \sqrt{3}\times 208\times 16.97 = 6{,}113.72$

6.16 Three wire Wye:Delta

A three-phase wye connected system provides 624 watts of power to a delta connected balanced load that has a lagging power factor of 0.577 as shown in Figure 6.16. Find (a) the line current, I_L, (b) phase current, I_ϕ, and (c) load impedance, Z_L, for a line-to-line voltage of 208 volts.

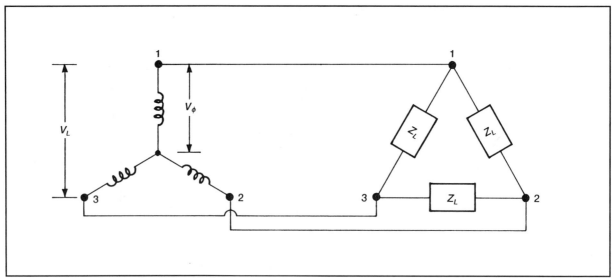

Figure 6.16   Three wire Wye:Delta

Solution:

For Wye source

$$P_T = 3V_\phi I_\phi \cos\theta = 3(V_L/\sqrt{3})I_L \cos\theta = \sqrt{3}V_L I_L \cos\theta$$

$$624 = \sqrt{3} \times 208 \times I_L \times 0.577$$

a) $I_L = (3\sqrt{3})/\sqrt{3} = 3$ amperes

For Delta load

b) $I_\phi = I_L/\sqrt{3} = 3/\sqrt{3} = \sqrt{3}$ amperes

c) $V_\phi = V_L = 208$ volts

$$Z_L = [V_\phi/I_\phi]\cos^{-1}(0.577) = [208/\sqrt{3}]\angle 54.76° = 120\angle 54.76° \text{ ohms}$$

6.17 Delta:Delta

A three-phase delta connected system provides power to a balanced resistive load as shown in Figure 6.17. Find (a) the line currents and (b) total power, P_T, for a line-to-line voltage of 240 volts.

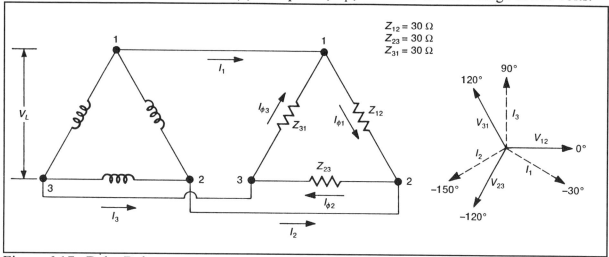

Figure 6.17   Delta:Delta

Solution:

$V_\phi = V_L = 240$ volts

$I_\phi = I_L / \sqrt{3}$ amperes

$I_{\phi 1} = V_{12} / Z_{12} = [240\angle 0°] / 30 = 8\angle 0° = 8 + j0$ amperes

$I_{\phi 2} = V_{23} / Z_{23} = [240\angle -120°] / 30 = 8\angle -120° = -4 - j6.93$ amperes

$I_{\phi 3} = V_{31} / Z_{31} = [240\angle 120°] / 30 = 8\angle 120° = -4 + j6.93$ amperes

a) $I_1 = I_{\phi 1} - I_{\phi 3} = (8 + j0) - (-4 + j6.93) = 12 - j6.93 = 13.86\angle -30°$

$I_2 = I_{\phi 2} - I_{\phi 1} = (-4 - j6.93) - (8 + j0) = -12 - j6.93 = 13.86\angle -150°$

$I_3 = I_{\phi 3} - I_{\phi 2} = (-4 + j6.93) - (-4 - j6.93) = 0 + j13.86 = 13.86\angle 90°$

$I_1 + I_2 + I_3 = 0$

b) $P_\phi = V_\phi I_\phi \cos\theta$ (θ of load)

$P_T = 3 V_\phi I_\phi \cos\theta$

$P_T = 3 V_L (I_L / \sqrt{3}) \cos\theta = \sqrt{3} V_L I_L \cos 0° = \sqrt{3} \times 240 \times 13.86 \times 1 = 5{,}761.49$ watts

6.18 Delta:Wye

A three-phase delta connected system provides 624 watts of power to a wye connected balanced load that has a lagging power factor of 0.577 as shown in Figure 6.18. Find (a) the line current, I_L, (b) phase current, I_ϕ, and (c) load impedance, Z_L, for a line-to-line voltage of 208 volts.

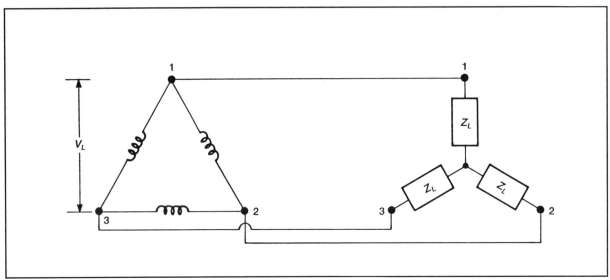

Figure 6.18   Delta:Wye

Solution:

For Delta source

$$P_T = \sqrt{3} V_L I_L \cos\theta$$

$$624 = \sqrt{3} \times 208 \times I_L \times (0.577)$$

a) $I_L = (3\sqrt{3})/\sqrt{3} = 3$ amperes

For Wye load

b) $I_\phi = I_L = 3$ amperes

c) $V_\phi = V_L/\sqrt{3} = 208/3 = 120$ volts

$Z_L = [V_\phi / I_\phi]\cos^{-1}(0.577) = [120/3]\angle 54.76° = 40\angle 54.76°$ ohms

6.19 Unbalanced Wye load

A three-phase wye connected system provides power to a wye unbalanced resistive load as shown in Figure 6.19. Find (a) the phase currents and (b) neutral current for a line-to-line voltage of 208 volts.

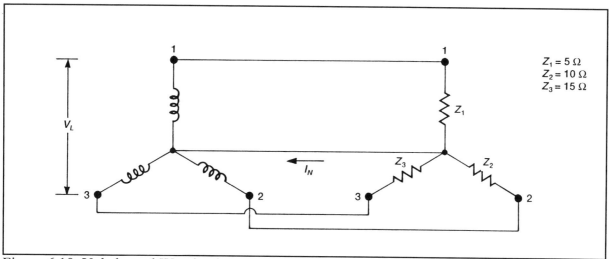

Figure 6.19 Unbalanced Wye load

Solution:

$I_L = I_\phi$

$V_L = 208$ volts

$V_\phi = V_L / \sqrt{3} = 208/\sqrt{3} = 120$ volts

$V_{12} = V_L \angle 0°$, $V_{23} = V_L \angle -120°$, $V_{31} = V_L \angle 120°$

$V_{\phi 1} = (V_L/\sqrt{3}) \angle -30°$, $V_{\phi 2} = (V_L/\sqrt{3}) \angle -150°$, $V_{\phi 3} = (V_L/\sqrt{3}) \angle 90°$

a) $I_{\phi 1} = V_{\phi 1} / Z_1 = [120 \angle -30°] / 5 = 24 \angle -30° = 20.78 - j12$ amperes

$I_{\phi 2} = V_{\phi 2} / Z_2 = [120 \angle -150°] / 10 = 12 \angle -150° = -10.39 - j6.0$ amperes

$I_{\phi 3} = V_{\phi 3} / Z_3 = [120 \angle 90°] / 15 = 8 \angle 90° = 0.0 + j8.0$ amperes

b) $I_N = I_{\phi 1} + I_{\phi 2} + I_{\phi 3} = 10.39 - j10.0$

$I_N = 14.48 \angle -43.9°$ amperes

6.20 Unbalanced Delta load

A three-phase delta connected system provides power to a delta unbalanced resistive load as shown in Figure 6.20. Find the resulting line currents for a line-to-line voltage of 240 volts.

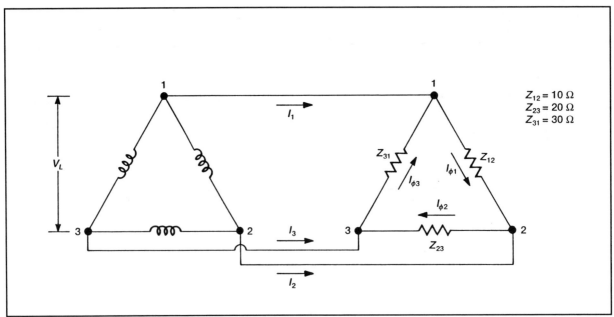

Figure 6.20 Unbalanced Delta load

Solution:

$$V_\phi = V_L = 240 \text{ volts}$$

$$I_\phi = I_L / \sqrt{3} \text{ amperes}$$

$$I_{\phi 1} = V_{12} / Z_{12} = [240\angle 0°] / 10 = 24\angle 0° = 24 + j0 \text{ amperes}$$

$$I_{\phi 2} = V_{23} / Z_{23} = [240\angle -120°] / 20 = 12\angle -120° = -6 - j10.4 \text{ amperes}$$

$$I_{\phi 3} = V_{31} / Z_{31} = [240\angle 120°] / 30 = 8\angle 120° = -4 + j6.93 \text{ amperes}$$

a) $I_1 = I_{\phi 1} - I_{\phi 3} = (24 + j0) - (-4 + j6.93) = 28.84\angle -13.51°$

b) $I_2 = I_{\phi 2} - I_{\phi 1} = (-6 - j10.4) - (24 + j0) = 31.75\angle -160.88°$

c) $I_3 = I_{\phi 3} - I_{\phi 2} = (-4 + j6.93) - (-6 - j10.4) = 17.44\angle 83.42°$

6.21 Generator effect

A 5 cm long conductor moves at a constant velocity of 0.2 meters/second between the poles of an electromagnet as shown in Figure 6.21. Determine the generated emf across the conductor if the magnetic flux density, **B**, equals 2.5 webers/meter^2, and the pole spacing, d, is 4 cm.

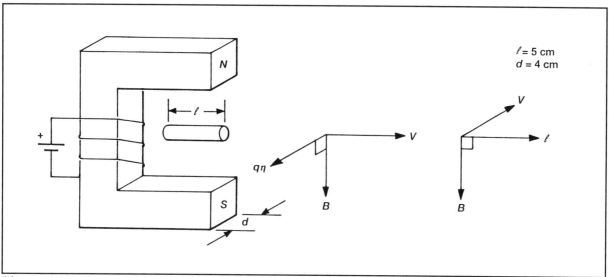

Figure 6.21 Generator effect

Solution:

$v = d/t = 0.2$ meter/second

$t = d/v = (4 \times 10^{-2}) / (2 \times 10^{-1}) = 0.2$ second

$A = dl = 4 \times 10^{-2} \times 5 \times 10^{-2} = 2 \times 10^{-3}$ meter^2

$F = q v \times B = qvB\sin\theta$

$V = (l/q)F$ (newton-meters/coulomb = joules/coulomb = volts)

$V = [l/q]qvB\sin\theta = (lvB)\sin\theta$

$V = l(d/t)B\sin\theta = [(Bdl)/t]\sin\theta = [(BA)/t]\sin\theta$

$$V = \left(\frac{BA}{t}\right)\sin\theta = \left(\frac{\Delta\Phi}{\Delta t}\right)\sin\theta = \left(\frac{d\Phi}{dt}\right)\sin\theta$$

$\theta = 90°$

$$V = \left(\frac{BA}{t}\right)\sin\theta = \left(\frac{2.5 \times 2 \times 10^{-3}}{2 \times 10^{-1}}\right) \times 1 = 25 \times 10^{-3} \quad \text{volts}$$

6.22 Motor effect

A 5 cm long current carrying conductor is placed between the poles of an electromagnet as shown in Figure 6.22. Determine the generated force on the conductor if its current is 0.125 amperes and the magnetic field, **H**, equals 2×10^6 amperes/meter.

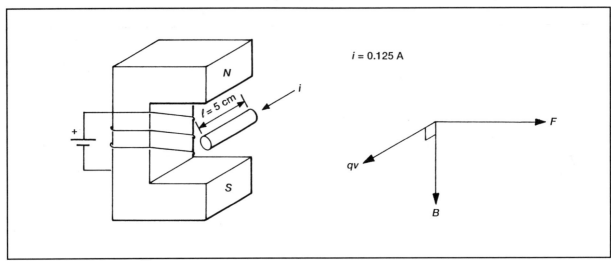

Figure 6.22 Motor effect

Solution:

$F = qv \times B = qvB\sin\theta$

$q = (It)$

$v = (l/t)$

$F = (It)(l/t)B\sin\theta = IlB\sin\theta$

$B = \mu_0 H$

$F = Il(\mu_0 H)\sin\theta$

$\theta = 90°$

$F = Il(\mu_0 H)\sin\theta = (0.125 \times 5 \times 10^{-2}) \times (4\pi \times 10^{-7} \times 2 \times 10^6) \times 1$

$F = 15 \times 10^{-3}$ newtons

6.23 Back EMF

The elementary ac motor shown in Figure 6.23 has a 100 turn stator winding which produces an airgap magnetic field, $\mathbf{H}_g$, equal to 2×10^6 amperes/meter. Its rotor has a radius of 2 cm and a length of 3 cm. Determine the generated back emf.

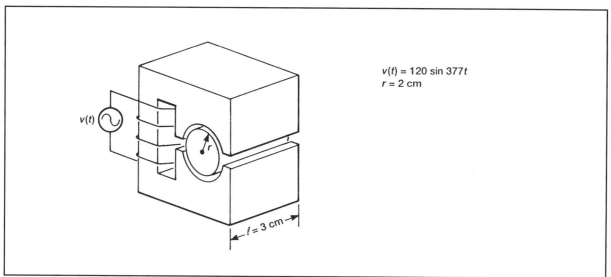

Figure 6.23 Back EMF

Solution:

$\varphi(t) = \Phi\cos\omega t = \Phi\cos 2\pi ft = \Phi\cos 377t$

$A = rl = 2\times10^{-2}\times3\times10^{-2} = 6\times10^{-4}$ meter^2

$\Phi = B_gA = \mu_0 H_g A = (4\pi\times10^{-7})\times(2\times10^6)\times(6\times10^{-4}) = 1.5\times10^{-3}$ webers

$N = 100$

$V_{emf} = -N(d\varphi/dt) = \omega N\Phi\sin\omega t = 2\pi fN\Phi\sin 377t$

$V_{emf} = (377\times100\times1.5\times10^{-3})\sin 377t = 56.55\sin 377t$

$V_{emf}(rms) = V_{emf}/\sqrt{2} = 40\sin 377t$ volts

6.24 Torque

The elementary ac motor shown in Figure 6.24 has a 100 turn stator winding which produces an airgap magnetic field, $\mathbf{H}_g$, equal to 2×10^6 amperes/meter. The airgap, d, is 2mm, the rotor has a radius of 2 cm and a length of 3 cm. Determine the generated torque on the rotor.

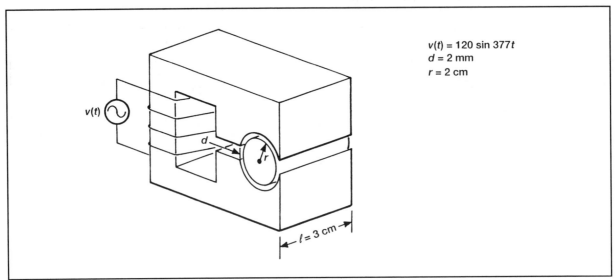

Figure 6.24 Torque

Solution:

$$F_g = \Phi_g^2 / (2\mu_0 A) = (\mathbf{B}_g A)^2 / (2\mu_0 A) = (\mu_0 \mathbf{H}_g A)^2 / (2\mu_0 A)$$

$$F_g = (\mu_0 \mathbf{H}_g^2 A)/2 \quad \text{newtons}$$

$$A = rl = 2\times10^{-2} \times 3\times10^{-2} = 6\times10^{-4} \quad \text{meters}^2$$

$$(\mu_0 \mathbf{H}_g A) = 4\pi\times10^{-7} \times 2\times10^6 \times 2\times10^{-4} = 1.5\times10^{-3} \quad \text{webers}$$

$$F_g = \frac{\mu_0 H_g^2 A}{2} = \left(\frac{H_g}{2}\right)\mu_0 H_g A = (1\times10^6)\times(1.5\times10^{-3}) = 1.5\times10^3 \quad \text{newtons}$$

$$T_g = F_g \times 2d = \mu_0 \mathbf{H}_g^2 A d$$

$$T_g = F_g \times 2d = 1.5\times10^3 \times 4\times10^{-3} = 6 \quad \text{newton-meters}$$

6.25 Separate winding DC generator

A separate winding dc generator produces 32 volts at 1,200 rpm with no load as shown in Figure 6.25. Determine (a) the load voltage for a load current of 10 amperes at 1,050 rpm, and (b) calculate the electromagnetic torque.

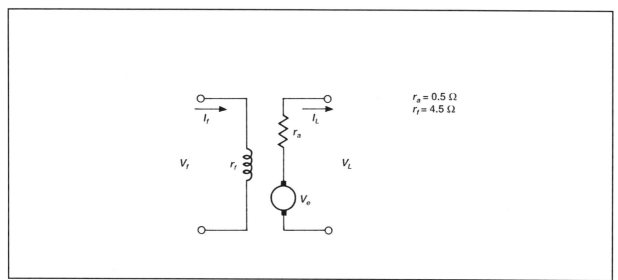

Figure 6.25 Separate winding DC generator

Solution:

a) $I_L = I_a$

$V_f = I_f r_f$

$V_e = V_{eo}(n/n_o) = 32 \times (1,050/1,200) = 28$ volts

$V_L = V_e - I_a r_a = 28 - 10(0.5) = 23$ volts

b) $T = (V_e I_a)/\omega$

$\omega = (2\pi n)/60$

$T = \dfrac{V_e I_a 60}{2\pi n} = \dfrac{28 \times 10 \times 60}{2\pi \times 1,050} = 2.54$ newton-meters

6.26 Series DC generator

A series winding dc generator produces 32 volts at 1,200 rpm with no load as shown in Figure 6.26. Determine (a) the load current for a load voltage of 25 volts at 1,125 rpm, and (b) calculate the output power.

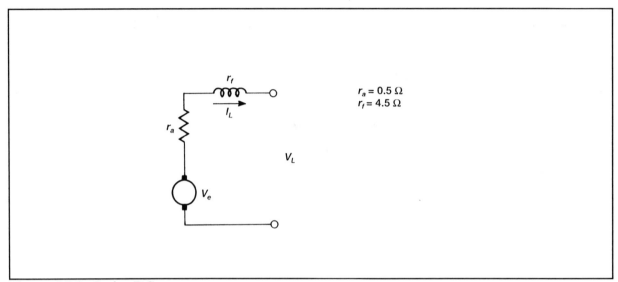

Figure 6.26 Series DC generator

Solution:

a) $I_L = I_a$

$V_e = V_{eo}(n/n_o) = 32 \times (1,125/1,200) = 30$ volts

$V_L = V_e - I_a(r_a + r_f)$

$25 = 30 - I_L(0.5 + 4.5)$

b) $I_L = (30 - 25)/5 = 1$ ampere

$P_L = V_L I_L = 25 \times 1 = 25$ watts

6.27 Shunt DC generator

A shunt winding dc generator produces 32 volts at 1,200 rpm with no load as shown in Figure 6.27. Determine the generator rpm for a load voltage of 22.5 volts and a load current of 6 amperes.

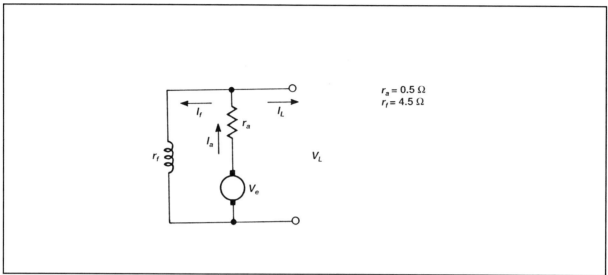

Figure 6.27 Shunt DC generator

Solution:

$V_L = V_e - I_a r_a$

$22.5 = V_e - I_a \times 0.5$

$I_a = I_L + I_f$

$I_a = 6 + V_L / r_f = 6 + 22.5/4.5 = 11$ amperes

$V_e = V_L + I_a r_a = 22.5 + 11 \times 0.5 = 28$ volts

$n = n_o (V_e / V_{eo}) = 1,200 \times (28/32) = 1,050$ rpm

6.28 Compound DC generator

A compound winding dc generator produces 32 volts at 1,200 rpm with no load as shown in Figure 6.28. Determine load current for a load voltage of 25 volts at 1,125 rpm.

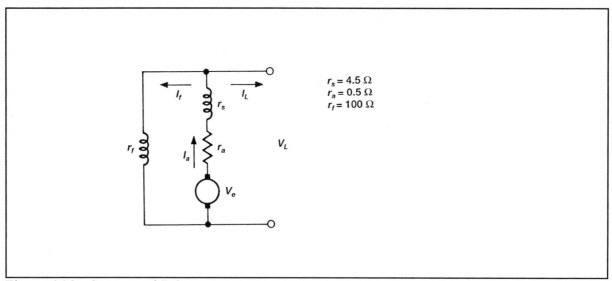

Figure 6.28 Compound DC generator

Solution:

$$V_L = V_e - I_a(r_a + r_s)$$

$$V_L = I_f r_f$$

$$I_a = I_L + I_f$$

$$I_L = I_a - I_f$$

$$I_L = \frac{V_e - V_L}{r_a + r_s} - \frac{V_L}{r_f}$$

$$V_e = V_{eo}(n/n_o) = 32 \times (1,125/1,200) = 30 \text{ volts}$$

$$I_L = \frac{30 - 25}{0.5 + 4.5} - \frac{25}{100} = 1 - 0.25 = 0.75 \text{ amperes}$$

6.29 Separate winding DC motor

A separate winding 28 volt dc motor is rotating at 1,200 rpm with no load as shown in Figure 6.29. Its field current is 0.25 amps with an armature voltage of 26 volts. Find the field current at 900 rpm.

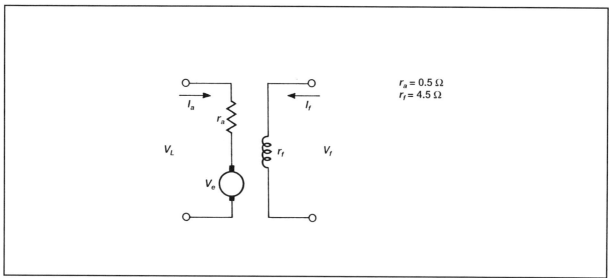

Figure 6.29 Separate winding DC motor

Solution:

$I_L = I_a$

$V_f = I_f r_f$

$V_L = V_e + I_a r_a = k I_f \omega_m + I_a r_a$

$28 = 26 + (0.5) I_a$

$I_a = \dfrac{28 - 26}{0.5} = 4$ amperes

$V_e = V_L - I_a r_a = k I_f \omega_m$

$\omega_m = (n_0 / 60) 2\pi = (1{,}200/60) \times 2\pi = 40\pi$

$k = \dfrac{V_L - I_a r_a}{I_f \omega_m} = \dfrac{28 - 4 \times 0.5}{0.25 \times 40\pi} = 0.828$

$\omega = (n/60) 2\pi = (900/60) \times 2\pi = 30\pi$

$I_L = \dfrac{V_L - I_a r_a}{k\omega} = \dfrac{28 - 4 \times 0.5}{0.828 \times 30\pi} = 0.333$ amperes

6.30 Series DC motor

A 120 volt series winding dc motor draws 10 amps at full load, and has an armature resistance, r_a, of 0.5 ohms as shown in Figure 6.30. Determine its back emf and horsepower for a field coil resistance, r_f, of 4.5 ohms.

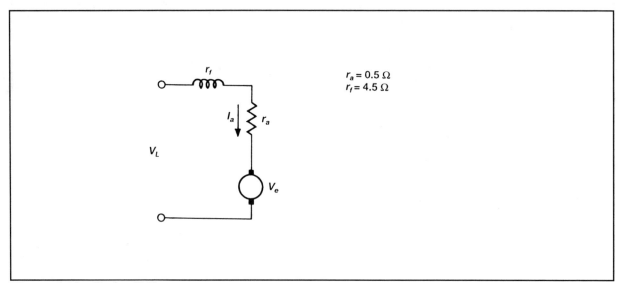

Figure 6.30 Series DC motor

Solution:

$$I_L = I_a$$

$$V_L = V_e + I_a(r_a + r_f) = k\varphi n + I_a(r_a + r_f)$$

$$V_e = V_L - I_a(r_a + r_f) = 120 - 10 \times (0.5 + 4.5) = 70 \text{ volts}$$

$$n = \frac{V_L - I_a(r_a + r_f)}{k\phi}$$

$$P_{in} = V_L I_L = 120 \times 10 = 1{,}200 \text{ watts}$$

$$P_{loss} = I_a^2(r_a + r_f) = (10)^2 \times 5 = 500 \text{ watts}$$

$$P_{mech} = P_{in} - P_{loss} = 1{,}200 - 500 = 700 \text{ watts}$$

$$hp = P_{mech} / 746 = 700/746 = 0.94$$

6.31 Shunt DC motor

A 28 volt shunt winding dc motor draws 2 amps with no load and rotates at 1,200 rpm as shown in Figure 6.31. When the load is connected it slows down to 1,100 rpm. Determine the armature load current for an armature resistance, r_a, of 0.5 ohms and a field coil resistance, r_f, of 4.5 ohms.

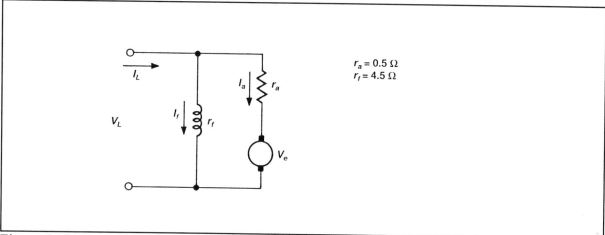

Figure 6.31 Shunt DC motor

Solution: $I_L = I_f + I_a$

No load

$$I_a = I_L \left[\frac{r_f}{r_a + r_f}\right] = 2\left[\frac{4.5}{0.5 + 4.5}\right] = 1.8 \quad \text{amperes}$$

$$I_f = I_L \left[\frac{r_a}{r_a + r_f}\right] = 2\left[\frac{0.5}{0.5 + 4.5}\right] = 0.2 \quad \text{amperes}$$

$$V_L = V_e + I_a r_a = k\varphi n + I_a r_a$$

$$k\varphi = \frac{V_L - I_a r_a}{n} = \frac{28 - 1.8 \times 0.5}{1,200} = \frac{27.1}{1,200} = 0.0225$$

$$n = (V_L - I_a r_a) / k\varphi$$

With load

$$I_a = \frac{V_L - k\varphi n}{r_a} = \frac{28 - 0.02254 \times 1,100}{0.5} = 6.5 \quad \text{amperes}$$

6.32 Compound DC motor

A 28 volt compound winding dc motor draws 5 amps with full load and rotates at 1,125 rpm as shown in Figure 6.32. Find the no load condition (a) armature voltage and (b) current at 1,200 rpm.

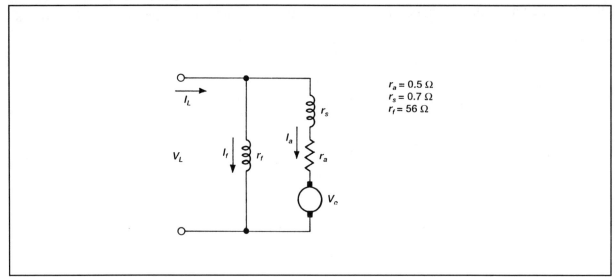

Figure 6.32 Compound DC motor

Solution:

$$I_L = I_f + I_a = V_L / r_f + I_a(r_a + r_s)$$

$$I_f = V_L / r_f = 28/56 = 0.5 \text{ ampere}$$

$$I_a = I_L - I_f = 5 - 0.5 = 4.5 \text{ amperes}$$

$$V_L = V_e + I_a(r_a + r_s)$$

a) $V_e = V_L - I_a(r_a + r_s) = 28 - 4.5 \times 1.2 = 22.6 \text{ volts}$

$$V_{eo} = (n_o / n)V_e = (1{,}200/1{,}125) \times 22.6 = 24.1 \text{ volts}$$

$$V_L = V_{eo} + I_{ao}(r_a + r_s)$$

b) $I_{a0} = \dfrac{V_L - V_{eo}}{r_a + r_s} = \dfrac{28 - 24.1}{0.5 + 0.7} = \dfrac{3.9}{1.2} = 3.25 \text{ amperes}$

6.33 AC Delta generator

A three-phase delta connected generator provides 50 kwatts at 208 volts, 60 Hz line-to-line voltage as shown in Figure 6.33. Determine the phase current for a power factor of 0.9.

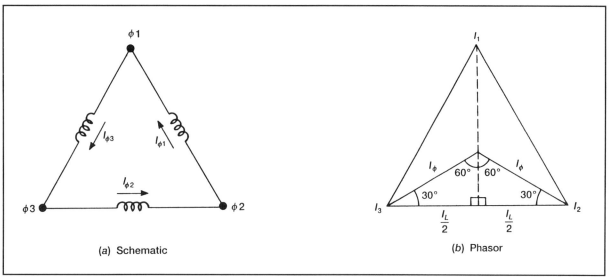

Figure 6.33   AC Delta generator

Solution:

$0.5 I_L = I_\phi \sin 60°$

$I_\phi = I_L / (2 \sin 60°) = I_L / \sqrt{3}$

$0.5 I_L = I_\phi \cos 30°$

$I_L = 2 \cos 30°$

$I_\phi = \sqrt{3} I_\phi$

For Delta $V_\phi = V_L = 208$ volts

$I_\phi = I_L / \sqrt{3}$ amperes

$P = 3 P_\phi = 3 V_\phi I_\phi \cos\theta = 3 V_L (I_L/\sqrt{3}) \cos\theta = \sqrt{3} V_L I_L \times 0.9$

$I_L = \dfrac{P}{\sqrt{3} V_L \cos\theta} = \dfrac{50 \times 10^3}{\sqrt{3} \times 208 \times 0.9} = 154.2$ amperes

$I_\phi = I_L / \sqrt{3} = 89$ amperes

6.34 AC Wye generator

A three-phase wye connected generator provides 50 kwatts at 208 volts, 60 Hz line-to-line voltage as shown in Figure 6.34. Determine the line current for a power factor of 0.9.

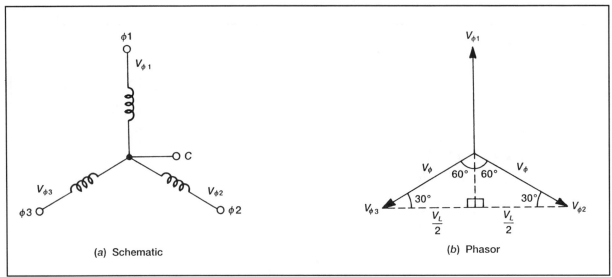

Figure 6.34 AC Wye generator

Solution:

$0.5V_L = V_\phi \cos 30°$

$V_\phi = V_L / (2\cos 30°) = V_L / \sqrt{3} = 208 / \sqrt{3} = 120$ volts

$0.5V_L = V_\phi \sin 60°$

$V_L = 2\sin 60°$

$V_\phi = \sqrt{3} V_\phi = 208$ volts

For Wye $\quad I_\phi = I_L$

$V_\phi = V_L / \sqrt{3}$ volts

$P = 3P_\phi = 3V_\phi I_\phi \cos\theta = 3(V_L/\sqrt{3})I_L \cos\theta = \sqrt{3} V_L I_L \times 0.9$

$I_L = \dfrac{P}{\sqrt{3} V_L \cos\theta} = \dfrac{50 \times 10^3}{\sqrt{3} \times 208 \times 0.9} = 154.2 \quad \text{amperes}$

6.35 AC Delta motor

A three-phase delta connected four-pole 10 hp induction motor is supplied with 208 volts, 60 Hz line-to-line voltage as shown in Figure 6.35. It has a full load speed of 1,764 rpm with a power factor of 0.9.

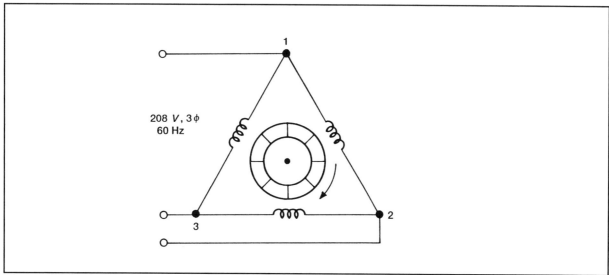

Figure 6.35 AC Delta motor

1. The motor no load rpm is most nearly:

   a) 2,000   b) 1,800   c) 1,900   d) 1,825

2. The motor slip (%) is most nearly:

   a) 11   b) 3.3   c) 2   d) 7

3. The motor input power is most nearly:

   a) 8,289   b) 9,000   c) 7,460   d) 10,000

4. The motor output torque is most nearly:

   a) 30   b) 22   c) 35   d) 40

6.35 Solution:

1. $f_s = (2f)/P = (2 \times 60)/4 = 30$ rps

   $\omega_s = 2\pi f_s = 2\pi \times 30 = 188.5$ radians/second

   rpm $= n_s = f_s \times 60 = 30 \times 60 = 1,800$

   The correct answer is (b).

2. slip $= \dfrac{n_s - n}{n_s} \times 100 = \dfrac{1,800 - 1,764}{1,800} \times 100 = 2\%$

   The correct answer is (c).

3. $P_{out} = 746 \times hp = 7,460$ watts

   $P_{in} = P_{out}/\cos\theta = 7,460/0.9 = 8,289$ watts

   The correct answer is (a).

4. $T = \dfrac{hp \times 33,000}{2\pi n} = \dfrac{10 \times 33,000}{2\pi \times 1,764} = 29.79$ ft-lbs

   $T = 29.79 \times 1.3558$ (N-m / ft-lb) $= 40.39$ newton-meters

   The correct answer is (d).

6.36 AC Wye motor

A three-phase wye connected four-pole induction motor is supplied with 208 volts, 60 Hz line-to-line voltage as shown in Figure 6.36. It has a no load rpm of 1,800, and a total power loss of 369 watts.

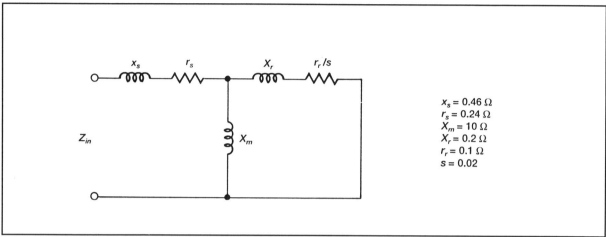

Figure 6.36 AC Wye motor

1. The per-phase impedance is most nearly:

   a) 4.5   b) 4.0   c) 3.8   d) 2.0

2. The output power is most nearly:

   a) 8.1×10^3 b) 7.9×10^3 c) 7.6×10^3 d) 8.3×10^3

3. The power factor is most nearly:

   a) 0.43   b) 0.6   c) 0.78   d) 0.89

4. The output torque is most nearly:

   a) 31   b) 41   c) 42   d) 44

6.36 Solution:

1. $Z_{in} = r_s + x_s + R_r + jX_r = 0.24 + j0.46 + \dfrac{jX_m[(r_r/s) + jX_r]}{(r_r/s) + j(X_m + X_r)}$

   $Z_{in} = 0.24 + j0.46 + \dfrac{j10[5 + j0.2]}{5 + j(10 + 0.2)} = 0.24 + j0.46 + 3.76 + j1.54$

   $Z_{in} = 4.0 + j2.0 = 4.47\angle 26.6°$ ohms

   The correct answer is (a).

2. $I_{in} = V_\phi / Z_{in} = (V_L / 3) / Z_{in} = 120 / [4.47\angle 26.6°] = 26.85\angle -26.6°$

   $P_{gap} = 3I_{in}^2 R_r = 3(26.85)^2 \times 3.76 = 8{,}132$ watts

   $P_{mech} = P_{gap}(1 - \text{slip}) = 8{,}132(1 - 0.02) = 7{,}969$ watts

   $P_{out} = P_{mech} - P_{loss} = 7{,}969 - 369 = 7{,}600$ watts

   The correct answer is (c).

3. $PF = \cos\theta = \cos 26.6° = 0.89$

   The correct answer is (d).

4. $\omega_s = (n_s / 60)2\pi = (1{,}800/60) \times 2\pi = 188.4$ radians/second

   $T_{out} = \dfrac{P_{out}}{\omega_r} = \dfrac{P_{out}}{\omega_s(1 - slip)} = \dfrac{7{,}600}{188.4 \times (1 - 0.02)} = 41.16$ newton-meters

   The correct answer is (b).

6.37 AC induction motor

The 120 volt, 60 Hz shaded pole motor shown in Figure 6.37 has four poles and a slip factor of 14 %. Determine its actual speed and horsepower if it develops a torque of 0.5 newton-meters.

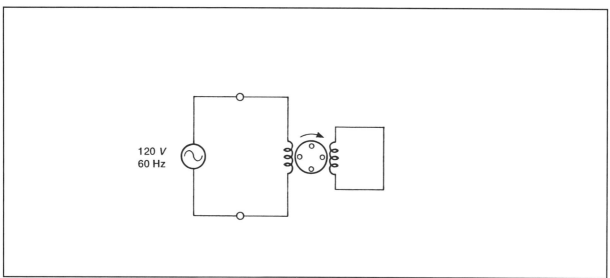

Figure 6.37 AC induction motor

Solution:

$$f_s = (2f)/P = (2 \times 60)/4 = 30 \text{ rps}$$

$$\text{speed} = n_s = f_s \times 60 = 30 \times 60 = 1{,}800 \text{ rpm}$$

$$\text{slip} = [(n_s - n)/n_s] \times 100$$

$$0.14 = (1{,}800 - n)/1{,}800$$

$$n = 1{,}800 - 1{,}800 \times 0.14 = 1{,}548 \text{ rpm}$$

$$\text{hp} = \frac{2\pi n T}{24{,}340} = \frac{2\pi \times 1{,}548 \times 0.5}{24{,}340}$$

$$\text{hp} = 0.2$$

6.38 AC synchronous motor

A three-phase synchronous motor is connected to 208 volts, 400 Hz line-to-line voltage and rotates at 4,000 rpm as shown in Figure 6.38. Determine (a) the number of poles and (b) its angular velocity, ω_s.

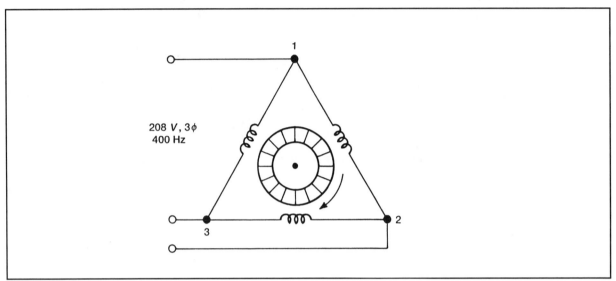

Figure 6.38 AC synchronous motor

Solution:

a) $f = (P/2)f_s = (P/2)(n_s/60)$

$P = (120f)/n_s = (120 \times 400)/4,000 = 12$

b) $\omega_s = \omega(2/P)$

$\omega = 2\pi f$

$\omega_s = 2\pi f(2/P)$

$\omega_s = (4\pi f)/12 = (400\pi)/3 = 418.9$ radians/second

6.39 Resolver

A resolver accepts an ac reference input voltage, $V_{ref} = 120\sin 377t$, at its rotor terminals and has a shaft angle, θ, of $60°$ as shown in Figure 6.39. Determine the two stator output signals using a coupling coefficient of 0.25 for both windings.

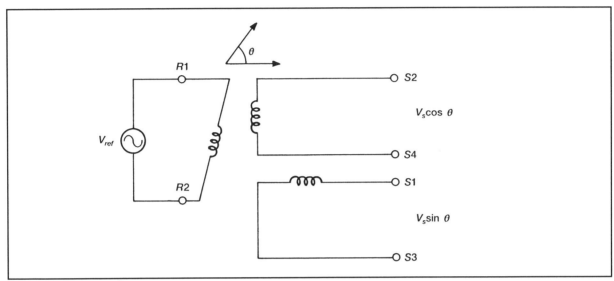

Figure 6.39 Resolver

Solution:

$V_{ref} = 120\sin 377t$

$V_s = 0.25 V_{ref} = 0.25 \times 120 = 30$ volts

$V_{13} = (V_s \sin\theta)\sin 377t$

$V_{13} = (30\sin 60°)\sin 377t = 26\sin 377t$

$V_{24} = (V_s \cos\theta)\sin 377t$

$V_{24} = (30\cos 60°)\sin 377t = 15\sin 377t$

6.40 Synchro

A synchro accepts an ac reference input voltage, $V_{ref} = 120\sin 377t$, at its rotor terminals and has a shaft angle, θ, of $30°$ as shown in Figure 6.40. Determine the three stator output signals using a coupling coefficient of 0.25 for all windings.

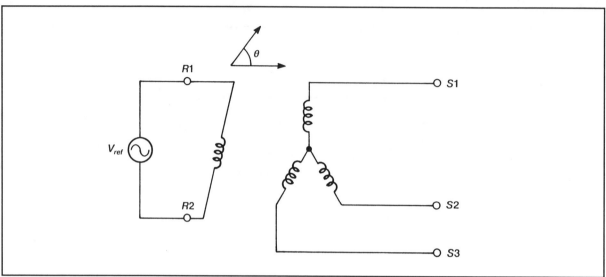

Figure 6.40 Synchro

Solution:

$V_{ref} = 120\sin 377t$

$V_s = 0.25 V_{ref} = 0.25 \times 120 = 30$ volts

$V_{12} = (V_s \sin\theta)\sin 377t$

$V_{12} = (30\sin 30°)\sin 377t = 15\sin 377t$

$V_{23} = [V_s \sin(\theta - 120°)]\sin 377t$

$V_{23} = [30\sin(30° - 120°)]\sin 377t = -30\sin 377t$

$V_{31} = [V_s \sin(\theta + 120°)]\sin 377t$

$V_{31} = [30\sin(30° + 120°)]\sin 377t = 15\sin 377t$

6.41 Servo

A servo tracking system uses two resolvers to form a feedback control loop as shown in Figure 6.41. Using an ac reference input voltage, $V_{ref} = 120\sin 377t$, and an input shaft angle, θ, of 72° determine the error voltage. Let the output shaft angle, ϕ, be 30° with a coupling coefficient of 0.25 for all windings.

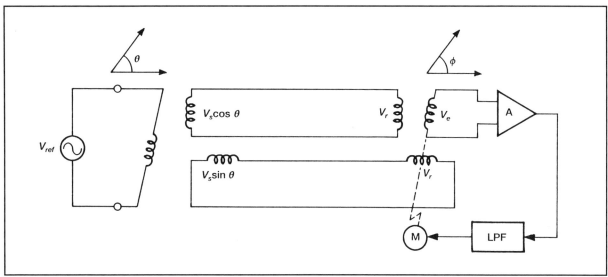

Figure 6.41 Servo

Solution:

$V_{ref} = 120\sin 377t$

$V_s = 0.25 V_{ref} = 0.25 \times 120 = 30$ volts

$V_r = 0.25 V_s = 0.25 \times 30 = 7.5$ volts

$V_e = [V_r \sin\theta \cos\phi][\sin 377t] - [V_r \cos\theta \sin\phi][\sin 377t]$

Using $\sin(A - B) = \sin A \cos B - \cos A \sin B$

$V_e = V_r \sin(\theta - \phi)[\sin 377t]$

$V_e = 7.5\sin(72° - 30°)[\sin 377t]$

$V_e = 5[\sin 377t]$

6.42 Stepper motor

A stepper motor design uses eight rotor teeth and four stator coils as shown in Figure 6.42. The coils have a uniform inductance of 1×10^{-3} henries and an internal resistance of 5 ohms.

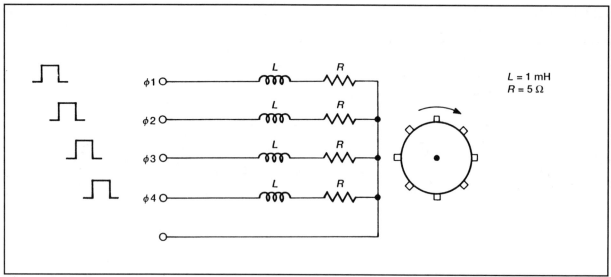

Figure 6.42 Stepper motor

1. The number of steps per revolution is most nearly:

   a) 64 b) 32 c) 16 d) 8

2. The minimum step size is most nearly:

   a) 11.2° b) 22.4° c) 5.6° d) 44.8°

3. The maximum input pulse rate is most nearly:

   a) 1.5×10^3 b) 1×10^3 c) 5×10^2 d) 2×10^3

4. The maximum angular velocity is most nearly:

   a) 196 b) 147 c) 49 d) 98

6.42 Solution:

1. N = 2×phases = 2×4 = 8 poles

   RT = 8 rotor teeth

   SPR = N×RT = 8×8 = 64 steps/revolution

   The correct answer is (a).

2. SS = 360° / SPR = 5.625° per step

   The correct answer is (c).

3. τ = L / R = 1×10^{-3} / 5 = 0.2×10^{-3} seconds

   T = 5τ = $5\times0.2\times10^{-3}$ = 1×10^{-3} seconds

   PPS = 1 / T = 1×10^{3} pulses/second

   The correct answer is (b).

4. ω = (2π×PPS×SS) / 360° = (2π×PPS) / SPR

   ω = (2π×1,000) / 64 = 98 radians /second

   The correct answer is (d).

6.43 Half wave rectifier

A half wave rectifier circuit accepts an input voltage, $v_{in} = 120\sin 377t$, and provides a dc output voltage as shown in Figure 6.43. Determine (a) the output dc current, I_{dc}, (b) rms current, I_{rms}, and (c) ripple factor, r_f, respectively.

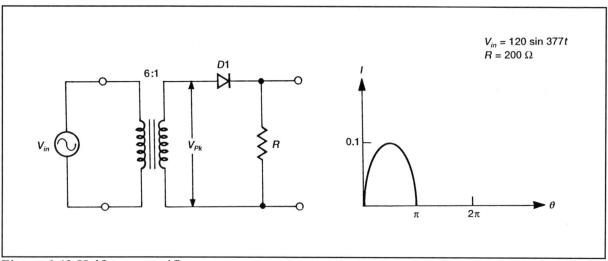

Figure 6.43 Half wave rectifier

Solution:

$$V_{pk} = 120/6 = 20 \text{ volts}$$

$$I_{pk} = V_{pk}/R = 20/200 = 100\times10^{-3} \text{ amperes}$$

a) $I_{dc} = \dfrac{1}{2\pi}\int_0^{\pi} I_{pk}\sin\theta\, d\theta = -\dfrac{I_{pk}}{2\pi}[\cos]_0^{\pi} = \dfrac{I_{pk}}{2\pi}[1+1] = \dfrac{I_{pk}}{\pi} = 0.318 I_{pk}$

$$I_{dc} = (0.318)\times 100\times 10^{-3} = 31.8\times 10^{-3} \text{ amperes}$$

b) $I_{rms} = \left(\dfrac{1}{2\pi}\int_0^{\pi}[I_{pk}\sin\theta]^2 d\theta\right)^{1/2} = \left(\dfrac{I_{pk}^2}{2\pi}\int_0^{\pi}\left[\dfrac{1}{2} - \dfrac{1}{2}\cos 2\theta\right] d\theta\right)^{1/2}$

$$I_{rms} = \left(\dfrac{I_{pk}^2}{2\pi}\left[\dfrac{1}{2}\theta - \dfrac{1}{2}\sin 2\theta\right]_0^{\pi}\right)^{1/2} = \left(\dfrac{I_{pk}^2}{2\pi}\times\dfrac{\pi}{2}\right)^{1/2} = \dfrac{I_{pk}}{2}$$

$$I_{rms} = 0.5 I_{pk} = 0.5\times 100\times 10^{-3} = 50\times 10^{-3} \text{ amperes}$$

c) $r_f = [(I_{rms}/I_{dc})^2 - 1]^{1/2} = [(1.57)^2 - 1]^{1/2} = 1.21$

6.44 Full wave rectifier

A full wave rectifier circuit accepts an input voltage, $v_{in} = 120\sin 377t$, and provides a dc output voltage as shown in Figure 6.44. Determine (a) the output dc voltage, V_{dc}, (b) rms voltage, V_{rms}, and (c) ripple factor, r_f, respectively.

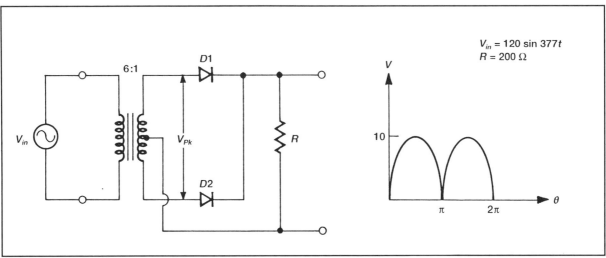

Figure 6.44 Full wave rectifier

Solution:

$$V_{pk} = 120/6 = 20 \text{ volts}$$

$$V'_{pk} = V_{pk}/2 = 20/2 = 10 \text{ volts}$$

a) $$V_{dc} = \frac{1}{2\pi}\int_0^{2\pi} V'_{pk}\sin\theta\, d\theta = -\frac{2V'_{pk}}{2\pi}[\cos]_0^\pi = \frac{V'_{pk}}{\pi}[1+1] = \frac{2V'_{pk}}{\pi} = 0.636 V'_{pk}$$

$$V_{dc} = 0.636 \times V'_{pk} = (0.636) \times 10 = 6.36 \text{ volts}$$

b) $$V_{rms} = \left(\frac{1}{2\pi}\int_0^{2\pi}[V'_{pk}\sin\theta]^2 d\theta\right)^{1/2} = \left(\frac{2V'^2_{pk}}{2\pi}\int_0^\pi[\frac{1}{2} - \frac{1}{2}\cos 2\theta]d\theta\right)^{1/2}$$

$$V_{rms} = \left(\frac{V'^2_{pk}}{\pi}\left[\frac{1}{2}\theta - \frac{1}{2}\sin 2\theta\right]_0^\pi\right)^{1/2} = \left(\frac{V'^2_{pk}}{\pi} \times \frac{\pi}{2}\right)^{1/2} = \frac{V'_{pk}}{\sqrt{2}}$$

$$V_{rms} = 0.707 V'_{pk} = 10 \times 0.707 = 7.07 \text{ volts}$$

c) $$r_f = [(V_{rms}/V_{dc})^2 - 1]^{1/2} = [(1.11)^2 - 1]^{1/2} = 0.482$$

6.45 Bridge rectifier

A bridge rectifier circuit accepts an input voltage, $v_{in} = 120\sin 377t$, and provides a dc output voltage as shown in Figure 6.45. Determine (a) the output dc voltage, V_{dc}, (b) the output rms voltage, V_{rms} And (c) find the capacitor value, C, needed to limit the ripple to 0.1 volts.

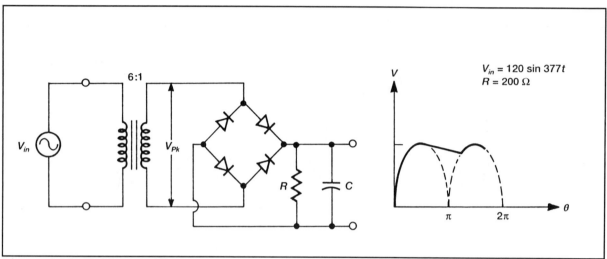

Figure 6.45 Bridge rectifier

Solution:

$V_{pk} = 120/6 = 20$ volts

(a) $V_{dc} = (2V_{pk})/\pi = 0.636 V_{pk}$

$V_{dc} = (0.636) \times 20 = 12.72$ volts

(b) $V_{rms} = V_{pk}/\sqrt{2} = 0.707 V_{pk}$

$V_{rms} = 20 \times 0.707 = 14.14$ volts

(c) $I_{pk} = V_{pk}/R = 20/200 = 100 \times 10^{-3}$ amperes

$T = 1/f = 1/60 = 16.66 \times 10^{-3}$ seconds

$C = Q/V = (IT)/V$

$C = \dfrac{I_{pk} T}{V_r} = \dfrac{100 \times 10^{-3} \times 16.66 \times 10^{-3}}{0.1} = 16.66 \times 10^{-6}$ farads

6.46 Voltage regulator

A series voltage regulator circuit accepts an input ranging between 10 and 15 volts dc, and maintains an output of 5 volts dc, at a maximum current of one ampere as shown in Figure 6.46. The transistor parameters are $V_{be} = 0.3$ volts, $r_{be} = 50$ ohms, and $\beta = 100$. Calculate (a) the zener diode voltage, V_z, (b) the value of R for an I_z (min) of 10 ma, and (c) find the transistor maximum power dissipation.

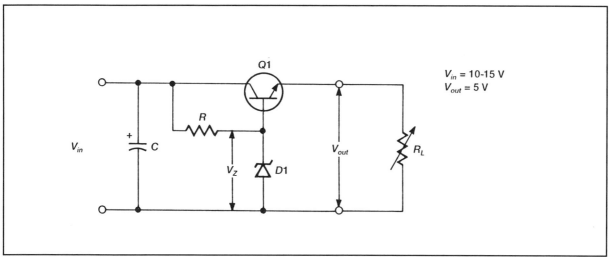

Figure 6.46 Voltage regulator

Solution:

$R_L = V_L / I_L = 5/1 = 5$ ohms

$V_L = (1 + \beta) I_b R_L$

$I_b = V_L / [(1 + \beta) R_L] = 5 / [101 \times 5] = 9.9 \times 10^{-3}$ amperes

a) $V_z = I_b [r_{be} + (1 + \beta) R_L] + V_{be} = 9.9 \times 10^{-3} \times [50 + 101 \times 5] + 0.3 = 5.8$ volts

b) $R = [V_{in}(min) - V_z] / I_z(min) = [10 - 5.8] / [10 \times 10^{-3}] = 420$ ohms

c) $V_{CE} = V_{in}(max) - V_L = 15 - 5 = 10$ volts

$I_C = I_L = 1$ ampere

$P_C(max) = V_{CE} I_C = 10 \times 1 = 10$ watts

6.47 SCR

A light dimmer control circuit uses a diode triggered, silicon controlled rectifier (SCR) as shown in Figure 6.47 with an input voltage, $v_{in} = 120\sin 377t$. Using the provided component values, determine the phase angle, ϕ, at which the SCR turns ON.

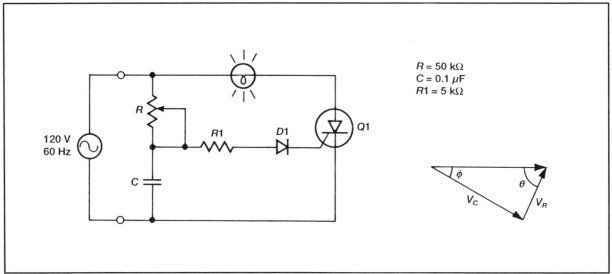

Figure 6.47 SCR

Solution:

$$X_C = \frac{1}{2\pi fC} = \frac{1}{2\pi \times 60 \times 0.1 \times 10^{-6}} = 25.54 \times 10^3 \text{ ohms}$$

$$Z = [(R)^2 + (X_C)^2]^{\frac{1}{2}} = [(50 \times 10^3)^2 + (25.54 \times 10^3)]^{\frac{1}{2}} = 56.6 \times 10^3 \text{ ohms}$$

$$I_{pk} = V_{pk}/Z = 120/(56.6 \times 10^3) = 2.12 \times 10^{-3} \text{ amperes}$$

$$V_C = I_{pk}X_C = 2.12 \times 10^{-3} \times 25.54 \times 10^3 = 54.145 \text{ volts}$$

$$\theta = \tan^{-1}(X_C/R) = \tan^{-1}(25.54 \times 10^3 / 50 \times 10^3) = 27.96°$$

$$V_{ON} = V_C \sin\varphi \approx 0.7 \text{ volts (diode forward drop)}$$

$$\varphi = \sin^{-1}(V_{ON}/V_C) = \sin^{-1}(0.7/54.145) = 0.74°$$

$$\phi \text{ (ON)} = 90° - \theta + \varphi = 90° - 27.96° + 0.74° = 62.78°$$

6.48 TRIAC

A light dimmer control circuit uses a diac triggered, triac as shown in Figure 6.48 with an input voltage, $v_{in} = 120\sin 377t$. Using the provided component values and a diac breakdown voltage, V_{bo} of 10 volts. Determine the phase angle, ϕ, at which the triac turns ON.

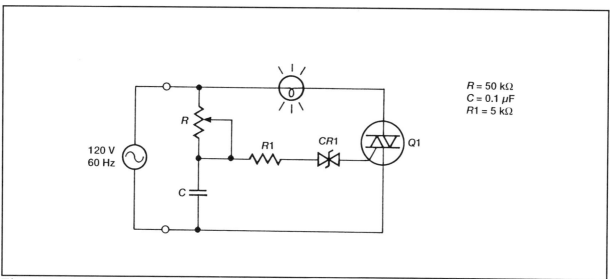

Figure 6.48 TRIAC

Solution:

$$X_C = \frac{1}{2\pi f C} = \frac{1}{2\pi \times 60 \times 0.1 \times 10^{-6}} = 25.54 \times 10^3 \text{ ohms}$$

$$Z = [(R)^2 + (X_C)^2]^{1/2} = [(50\times10^3)^2 + (25.54\times10^3)^2]^{1/2} = 56.6\times10^3 \text{ ohms}$$

$$I_{pk} = V_{pk}/Z = 120/(56.6\times10^3) = 2.12\times10^{-3} \text{ amperes}$$

$$V_C = I_{pk} X_C = 2.12\times10^{-3} \times 25.54\times10^3 = 54.145 \text{ volts}$$

$$\theta = \tan^{-1}(X_C/R) = \tan^{-1}[(25.54\times10^3)/(50\times10^3)] = 27.96°$$

$$V_{bo} = V_C \sin\varphi = 10 \text{ volts}$$

$$\varphi = \sin^{-1}(V_{bo}/V_C) = \sin^{-1}(10/54.145) = 10.64°$$

$$\phi(\text{ON}) = 90° - \theta + \varphi = 90° - 27.96 + 10.64° = 72.68°$$

6.49 Batteries in series

Two identical 12 volt batteries are connected in series to drive a 20 ohm resistive load as shown in Figure 6.49. Calculate the resulting circuit current using a battery internal resistance, r, of 2 ohms.

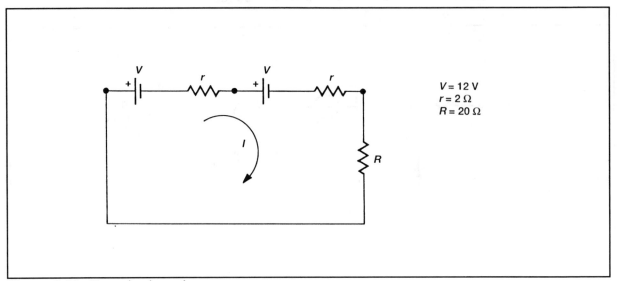

Figure 6.49 Batteries in series

Solution:

$$V + V - Ir - Ir - IR = 0$$

$$V + V = (r + r + R)I$$

$$I = \frac{2V}{2r + R} = \frac{2 \times 12}{2 \times 2 + 20} = \frac{24}{24}$$

$$I = 1 \text{ ampere}$$

For n batteries in series,

$$I = \frac{nV}{nr + R}$$

6.50 Batteries in parallel

Two identical 12 volt batteries are connected in parallel to drive a 20 ohm resistive load as shown in Figure 6.50a. Calculate the resulting load current using a battery internal resistance, r, of 2 ohms.

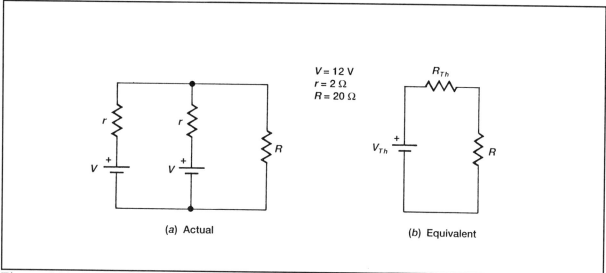

Figure 6.50 Batteries in parallel

Solution:

$$V_{Th} = \frac{I}{Y} = \frac{\frac{V}{r} + \frac{V}{r}}{\frac{1}{r} + \frac{1}{r}} = \frac{\frac{2V}{r}}{\frac{2}{r}} = V = 12 \quad \text{volts}$$

$$R_{Th} = \frac{1}{\frac{1}{r} + \frac{1}{r}} = \frac{1}{\frac{2}{r}} = \frac{r}{2} = \frac{2}{2} = 1 \quad \text{ohm}$$

$$I = \frac{V_{Th}}{R_{Th} + R} = \frac{12}{1 + 20} = \frac{12}{21} = 0.57 \quad \text{ampere}$$

For n batteries in parallel,

$$I = \frac{V}{\frac{r}{n} + R}$$

6.51 Per unit notation

A single phase generator unit provides power to a load via a transmission line as shown in Figure 6.51. The generator is rated at 4.8 kVA, and supplies 400 volts, with a short circuit capacity of 10 kVA. The generator kVA rating is used a base.

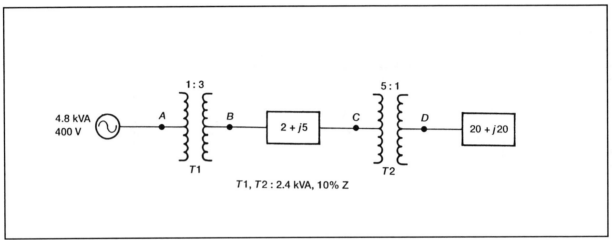

Figure 6.51 Per unit notation

1. The generator per unit impedance is most nearly:

   a) 2.0   b) 0.48   c) 0.1   d) 0.8

2. The T1 transformer per unit impedance is most nearly:

   a) 0.6   b) 0.1   c) 1.0   d) 0.2

3. The base impedance at point B is most nearly:

   a) 300   b) 150   c) 200   d) 100

4. The per unit impedance at point D is most nearly:

   a) 3 + j3   b) 12 + j20   c) 1.6 + j1.6   d) 4 + j4

5. The reflected per unit impedance at point A is most nearly:

   a) 5.3   b) 3.6   c) 4.2   d) 2.66

6.51 Solution:

1. $Z_g = Z_{rated}(kVA_{base}/kVA_{sc}) = 1.0 \times (4.8/10) = 0.48$ pu

   The correct answer is (b).

2. $Z_{T1} = Z_{rated}(kVA_{base}/kVA_{rated}) = 0.1 \times (4.8/2.4) = 0.2$ pu

   The correct answer is (d).

3. $\alpha_1 = 1/3$

   $Z_{base}(B) = V^2/kVA_{base} = [V(A)/\alpha_1]^2/kVA_{base} = [400 \times 3]^2/[4,800] = 300$ ohms

   The correct answer is (a).

4. $\alpha_2 = 5/1$

   $Z_{base}(D) = V^2/kVA_{base} = [V(A)/(\alpha_1\alpha_2)]^2/kVA_{base} = [400 \times 3/5]^2/[4,800] = 12$ ohms

   $Z_{load} = Z_{rated}/Z_{base}(D) = (20+j20)/12 = 1.66+j1.66$ ohms

   The correct answer is (c).

5. $Z_{line} = Z_{rated}/Z_{base}(B) = (2+j5)/300 = 0.0066+j0.0166$ ohms

   $Z(A) = Z_{T1} + Z_{line} + Z_{T2} + Z_{load}$

   $Z(A) = j0.2 + (0.0066+j0.0166) + j0.2 + (1.66+j1.66)$

   $Z(A) = 1.66+j2.08 = 2.66\angle 51°$

   The correct answer is (d).

6.52 Ground faults

A three-phase generator unit provides power to a synchronous motor via a transmission line as shown in Figure 6.52. The generator reactances are $X'' = 12\%$, $X_2 = 12\%$, $X_0 = 6\%$. The motor reactances are $X_m'' = 20\%$, $X_{m2} = 20\%$, $X_{m0} = 12\%$. The generator ratings are used as a base.

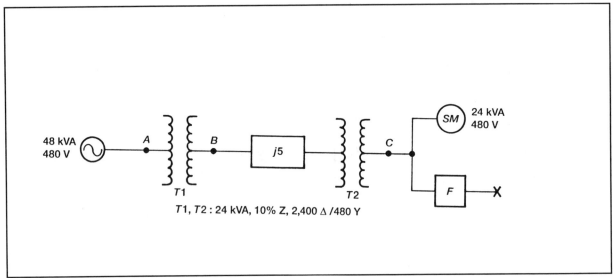

Figure 6.52 Ground faults

1. The motor base current and per unit impedance at point C is most nearly:

   a) 29 and 0.4   b) 100 and 0.2   c) 58 and 0.4   d) 100 and 0.4

2. The per unit transformer and line impedances are most nearly:

   a) 0.2 and 0.004   b) 2.5 and 0.2   c) 0.4 and 5   d) 0.2 and 2.5

3. The current due to a three-phase fault at point C is most nearly:

   a) 104   b) 145   c) 173   d) 250

4. The current due to a line-to-ground fault at point C is most nearly:

   a) 142   b) 82   c) 242   d) 103

5. The value of R_n at T2 to limit the line-to-ground fault to 15 amps is most nearly:

   a) 32   b) 9   c) 18.5   d) 16

6.52 Solution:

1. $I_{base} = kVA_{base} / (\sqrt{3}V_{base}) = 48\times10^3 / (\sqrt{3}\times480) = 57.7$ amps

   $Z_{m2} = Z_{rated} (kVA_{base} / kVA_{rated}) = 0.2\times(48/24) = 0.4$ pu

   The correct answer is (c).

2. $Z_T = Z_{rated} (kVA_{base} / kVA_{rated}) = 0.1\times(48/24) = 0.2$ pu

   $Z_{base}(B) = kV^2 / kVA_{base} = (2{,}400)^2 / (48\times10^3) = 120$ ohms

   $Z_{line} = Z_{rated} / Z_{base} = j5/120 = j0.004$ ohms

   The correct answer is (a).

3. $Z_g = Z_{rated} (kVA_{base} / kVA_{rated}) = 0.2\times(48/48) = 0.12$ pu

   $Z_1 = Z_g + Z_T + Z_{line} + Z_T = 0.12 + 0.2 + 0.004 + 0.2 = 0.56$ pu

   $Z_2 = Z_{m2} = 0.4$ pu

   $Z_{eq} = Z_1 \| Z_2 = 0.233$ pu

   $I_{fault} = 1/Z_{eq} = 4.3$ pu

   $I = I_{fault} \times I_{base} = 4.3\times57.7 = 250$ amps

   The correct answer is (d).

4. Since $X'' = X_2$, $Z_1 = Z_{eq}$ and $Z_2 = Z_{eq}$

   $Z_{m0} = Z_{rated} (kVA_{base} / kVA_{rated}) = 0.12\times(48/24) = 0.24$ pu

   $Z_0 = Z_T \| Z_{m0} = 0.1$

   $I_{fault} = 1/(Z_1 + Z_2 + Z_0) = 1.42$ pu

   $I = I_{fault}\times I_{base} = 1.42\times57.7 = 82$ amps

   The correct answer is (b).

5. $R_n = V_L/\sqrt{3}I_n = 480/(\sqrt{3}\times15) = 18.5$ ohms

   The correct answer is (c).

6.53 Breaker ratings

A three-phase generator unit provides power to a synchronous and an induction motor via a transmission line as shown in Figure 6.53. The synchronous motor reactances are $X''_{sm} = 10\%$, $X'_{sm} = 18\%$. The induction motor reactance, X_{im}, is 19%. The generator short circuit capacity is 72,000 kVA, and its ratings are used as a base.

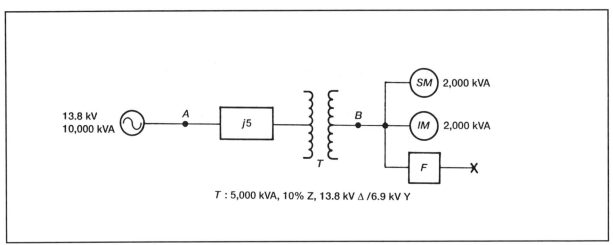

Figure 6.53 Breaker ratings

1. The current base at point B is most nearly:

   a) 483   b) 724   c) 418   d) 839

2. The per unit line impedance at point A is most nearly:

   a) 0.26   b) 1.06   c) 3.8   d) 0.94

3. The per unit generator side Z_{th} at point B is most nearly:

   a) 1.4   b) 0.38   c) 0.6   d) 0.45

4. The momentary duty rating of breaker F for a three-phase fault at point X is most nearly:

   a) 1,398   b) 3,776   c) 1,678   d) 883

5. The interrupting duty rating of breaker F for a three-phase fault at point X is most nearly:

   a) 1,398   b) 776   c) 932   d) 2,330

6.53 Solution:

1. $I_{base} = kVA_{base} / (\sqrt{3}V_{base}) = (10 \times 10^3) / (\sqrt{3} \times 6.9) = 839$ amps

   The correct answer is (d).

2. $Z_{base} = kV^2 / kVA_{base} = (13.8)^2 / 10 = 19$ ohms

   $Z_{line} = Z_{rated} / Z_{base} = 5/19 = 0.26$ pu

   The correct answer is (a).

3. $Z_g = Z_{rated} (kVA_{base} / kVA_{sc}) = 1.0 \times (10/72) = 0.14$ pu

   $Z_T = Z_{rated} (kVA_{base} / kVA_{rated}) = 0.1 \times (10/5) = 0.2$ pu

   $Z_{th} = Z_g + Z_{line} + Z_T = 0.14 + 0.26 + 0.2 = 0.6$ pu

   The correct answer is (c).

4. $Z"_{sm} = Z_{rated} (kVA_{base} / kVA_{rated}) = 0.1 \times (10/2) = 0.5$ pu

   $Z_{im} = Z_{rated} (kVA_{base} / kVA_{rated}) = 0.19 \times (10/2) = 0.95$ pu

   $Z_{eq} = Z_{th} || Z"_{sm} || Z_{im} = 0.6 || 0.5 || 0.95 = 0.22$ pu

   $I_{fault} = 1/Z_{eq} = 4.5$ pu

   $I = I_{fault} \times I_{base} = 4.5 \times 839 = 3{,}776$ amps

   The correct answer is (b).

5. $Z'_{sm} = Z_{rated} (kVA_{base} / kVA_{rated}) = 0.18 \times (10/2) = 0.9$ pu

   $Z_{eq} = Z_{th} || Z'_{sm} = 0.6 || 0.9 = 0.36$ pu

   $I_{fault} = 1/Z_{eq} = 2.77$ pu

   $I = I_{fault} \times I_{base} = 2.77 \times 839 = 2{,}330$ amps

   The correct answer is (d).

6.54 Power factor correction

A 60 Hz, three-phase generator unit provides power, via a step down transformer, to a synchronous and an induction motor as shown in Figure 6.54.

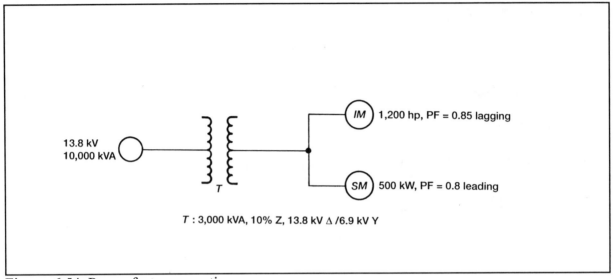

Figure 6.54 Power factor correction

1. The real and reactive power supplied to the induction motor is most nearly:

    a) 1,020 and 1,412   b) 1,053 and 760   c) 895 and 555   d) 760 and 1,053

2. The reactive and total power supplied to the synchronous motor is most nearly:

    a) 375 and 625   b) 300 and 400   c) 625 and 833   d) 400 and 500

3. The input kVA and PF supplied to both motors is most nearly:

    a) 1,407 and 0.13   b) 1,395 and 0.68   c) 1,676 and 0.83   d) 1,407 and 0.99

4. The per phase current supplied by the transformer is most nearly:

    a) 118   b) 68   c) 40   d) 204

5. The per phase capacitance needed to change the power factor to 1.0 is most nearly:

    a) 20.1×10^{-6} b) 10×10^{-6} c) 3.34×10^{-6} d) 6.7×10^{-6}

6.54 Solution:

1. S_{im} = (hp×746) / (1,000×PF) = 1,053 kVA

   PF = $\cos\theta$ = 0.85; θ = $\cos^{-1}(0.85)$ = 31.8°

   P_{im} = $S_{im}\cos\theta$ = 1,053×0.85 = 895 kwatts

   Q_{im} = $S_{im}\sin\theta$ = 1,053×0.53 = 555 kVARs

   The correct answer is (c).

2. PF = $\cos\theta$ = 0.80; θ = $\cos^{-1}(0.80)$ = 36.9°

   S_{sm} = P_{sm} / $\cos\theta$ = 500/0.8 = 625 kVA

   Q_{sm} = $S_{sm}\sin\theta$ = 625×0.6 = 375 kVARs

   The correct answer is (a).

3. P_T = P_{im} + P_{sm} = 895 + 500 = 1,395 kwatts

   Q_T = Q_{im} - Q_{sm} = 555 - 375 = 180 kVARs

   S_T = $[(P_T)^2 + (Q_T)^2]^{1/2}$ = 1,407 kVA

   θ = $\tan^{-1}(Q_T / P_T)$ = 7.35°;   PF = $\cos\theta$ = 0.99

   The correct answer is (d).

4. I = kVA_{total} / ($\sqrt{3}kV_{rated}$) = 1,407/ ($\sqrt{3}$×6.9) = 118 amps

   I_Φ = I / $\sqrt{3}$ = 68 amps per phase

   The correct answer is (b).

5. When Q_T = Q_C,   PF = 1.0

   Q_T = Q_C = 180 kVARs

   Q_{phase} = Q_C / 3 = 180/3 = 60 kVARs

   C_{phase} = Q_{phase} = $Q_C / [\omega V^2]$ = $[60\times10^3] / [377\times(6.9\times10^3)^2]$ = 3.34×10^{-6} farads

   The correct answer is (c).

6.55 Motor rating

A 100 hp induction motor is connected to a 480 volt, 60 Hz, three-phase power source as shown in Figure 6.55. The rotor has four poles, the slip factor is 0.02, the power factor equals 0.85 lagging, and the efficiency is 80%.

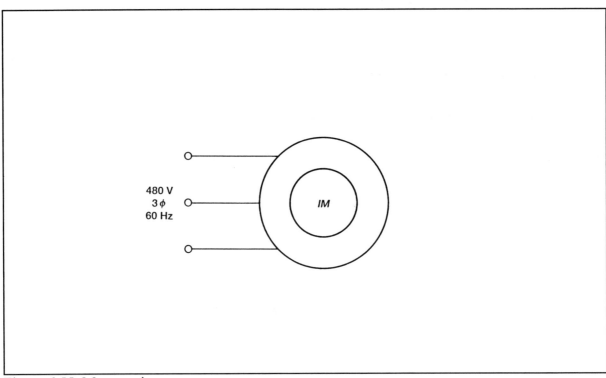

Figure 6.55 Motor rating

1. The real and imaginary power supplied to the motor is most nearly:

   a) 57.8 and 93.4   b) 70.2 and 87.8   c) 93.4 and 57.8   d) 87.8 and 70.2

2. The motor full load rpm is most nearly:

   a) 1,800   b) 1,836   c) 1,782   d) 1,764

3. The motor output torque is most nearly:

   a) 404   b) 495   c) 395   d) 505

4. The motor input current is most nearly:

   a) 194   b) 65   c) 112   d) 140

5. The per phase capacitance needed to correct the power factor to 1.0 is most nearly:

   a) 128×10^{-6} b) 222×10^{-6} c) 666×10^{-6} d) 139×10^{-5}

6.55 Solution:

1. S_{im} = (hp×746) / (PF×Eff) = (100×746) / (0.85×0.8) = 109.7 kVA

   PF = $\cos\theta$ = 0.85; θ = $\cos^{-1}(0.85)$ = 31.8°

   P_{im} = $S_{im}\cos\theta$ = 109.7×0.85 = 93.3 kwatts

   Q_{im} = $S_{im}\sin\theta$ = 109.7×0.53 = 57.8 kVARs

   The correct answer is (c).

2. n_s = 120f / P = (120×60)/4 = 1,800 rpm

   n = $n_s(1 - S)$ = 1,800×(1 - 0.02) = 1,764 rpm

   The correct answer is (d).

3. ω_s = $(2\pi n_s) / 60$ = $(2\pi \times 1,800)/60$ = 188.5 rads/sec

   T_{out} = $P_{out} / (1 - S)\omega_s$ = $(93.3 \times 10^3 \times 0.8) / (0.98 \times 188.5)$ = 404 N-m

   The correct answer is (a).

4. I = $P_{in} / \sqrt{3}V$ = $(93.3 \times 10^3 \times 0.8) / (0.98 \times 188.5)$ = 112 amps

   The correct answer is (c).

5. When Q_{im} = Q_C = 57.8 kVARs, PF = 1.0

   Q_{phase} = $Q_C / 3$ = 57.8/3 = 19.26 kVARs

   C_{phase} = Q_{phase} = $Q_C/(\omega V^2)$ = $[19.26 \times 10^3] / [377 \times (480)^2]$ = 222×10^{-6} farads

   The correct answer is (b).

6-66

NOTES:

7.0 DIGITAL
7.1 Minterms

The following truth table defines the input-output relationship of the logic block diagram shown in Figure 7.1. Derive the corresponding minterm canonical form.

A B C D	f
0 0 0 0	1
0 0 0 1	0
0 0 1 0	1
0 0 1 1	0
0 1 0 0	0
0 1 0 1	1
0 1 1 0	0
0 1 1 1	1
1 0 0 0	0
1 0 0 1	1
1 0 1 0	0
1 0 1 1	1
1 1 0 0	1
1 1 0 1	0
1 1 1 0	1
1 1 1 1	0

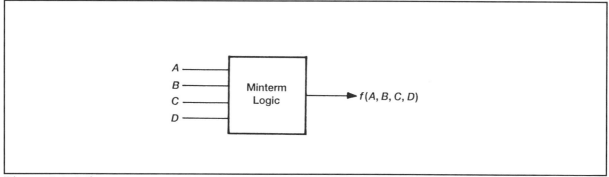

Figure 7.1 Minterms

Solution:

Use a "1" to represent an uncomplemented literal and a "0" to represent a complemented literal.

$$f(A, B, C, D) = \sum (m_0, m_2, m_5, m_7, m_9, m_{11}, m_{12}, m_{14})$$

$$f(A,B,C,D) = \overline{A}\,\overline{B}\,\overline{C}\,\overline{D} + \overline{A}\,\overline{B}\,C\,\overline{D} + \overline{A}\,B\,\overline{C}\,D + \overline{A}\,B\,C\,D + A\,\overline{B}\,\overline{C}\,D + A\,\overline{B}\,C\,D + A\,B\,\overline{C}\,\overline{D} + A\,B\,C\,\overline{D}$$

7.2 Maxterms

The following truth table defines the input-output relationship of the logic block diagram in Figure 7.2. Derive the corresponding maxterm canonical form.

A B C D	g
0 0 0 0	1
0 0 0 1	0
0 0 1 0	1
0 0 1 1	0
0 1 0 0	0
0 1 0 1	1
0 1 1 0	0
0 1 1 1	1
1 0 0 0	0
1 0 0 1	1
1 0 1 0	0
1 0 1 1	1
1 1 0 0	1
1 1 0 1	0
1 1 1 0	1
1 1 1 1	0

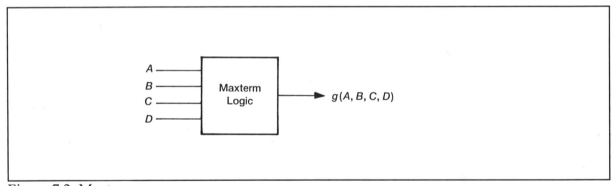

Figure 7.2 Maxterms

Solution:

Using a "0" to represent an uncomplemented literal and a "1" to represent a complemented literal.

$$g(A, B, C, D) = \prod (M_1, M_3, M_4, M_6, M_8, M_{10}, M_{13}, M_{15})$$

$$g(A,B,C,D) = (A + B + C + \bar{D})(A + B + \bar{C} + \bar{D})(A + \bar{B} + C + D)(A + \bar{B} + \bar{C} + D) \cdot$$
$$(\bar{A} + B + C + D)(\bar{A} + B + \bar{C} + D)(\bar{A} + \bar{B} + C + \bar{D})(\bar{A} + \bar{B} + \bar{C} + \bar{D})$$

7.3 DeMorgan's Theorem 1

A Boolean function of four variables is expressed as a logical sum of eight minterms by $f(A, B, C, D) = \sum (m_0, m_2, m_5, m_7, m_9, m_{11}, m_{12}, m_{14})$. The logic block diagram shown in Figure 7.3 implements the function and its complement. Express the two functions in their canonical forms.

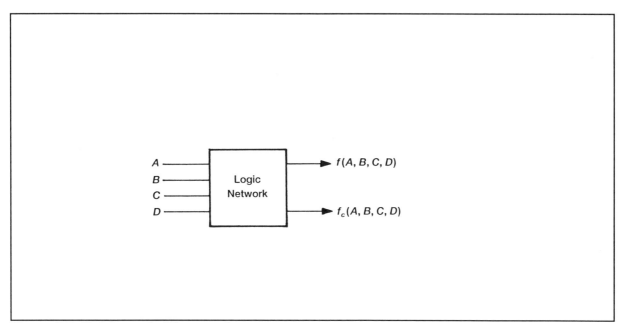

Figure 7.3 DeMorgan's Theorem 1

Solution:

$$f(A, B, C, D) = \sum (m_0, m_2, m_5, m_7, m_9, m_{11}, m_{12}, m_{14})$$

$$f(A,B,C,D) = \overline{A}\,\overline{B}\,\overline{C}\,\overline{D} + \overline{A}\,\overline{B}C\overline{D} + \overline{A}B\overline{C}D + \overline{A}BCD + A\overline{B}\overline{C}D + A\overline{B}CD + AB\overline{C}\,\overline{D} + ABC\overline{D}$$

Using DeMorgan's Theorem 1: $\overline{AB} = \overline{A} + \overline{B}$

$$f_c(A, B, C, D) = \prod (M_0, M_2, M_5, M_7, M_9, M_{11}, M_{12}, M_{14})$$

$$f_c(A,B,C,D) = (A + B + C + D)(A + B + \overline{C} + D)(A + \overline{B} + C + \overline{D})(A + \overline{B} + \overline{C} + \overline{D}) \cdot$$
$$(\overline{A} + B + C + \overline{D})(\overline{A} + B + \overline{C} + \overline{D})(\overline{A} + \overline{B} + C + D)(\overline{A} + \overline{B} + \overline{C} + D)$$

7.4 DeMorgan's Theorem 2

A Boolean function of four variables is expressed as a logical sum of eight maxterms by $g(A, B, C, D) = \prod(M_1, M_3, M_4, M_6, M_8, M_{10}, M_{13}, M_{15})$. The logic block diagram shown in Figure 7.4 implements the function and its complement. Express the two functions in their canonical forms.

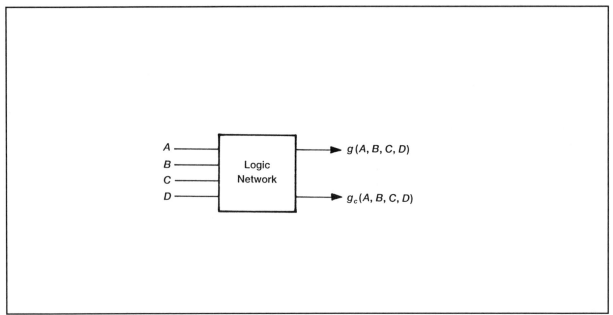

Figure 7.4 DeMorgan's Theorem 2

Solution:

$$g(A, B, C, D) = \prod(M_1, M_3, M_4, M_6, M_8, M_{10}, M_{13}, M_{15})$$

$$g(A,B,C,D) = (A + B + C + \bar{D})(A + B + \bar{C} + \bar{D})(A + \bar{B} + C + D)(A + \bar{B} + \bar{C} + D) \bullet$$
$$(\bar{A} + B + C + D)(\bar{A} + B + \bar{C} + D)(\bar{A} + \bar{B} + C + \bar{D})(\bar{A} + \bar{B} + \bar{C} + \bar{D})$$

Using DeMorgan's Theorem 2: $\overline{A + B} = \bar{A}\bar{B}$

$$g_c(A, B, C, D) = \sum(m_1, m_3, m_4, m_6, m_8, m_{10}, m_{13}, m_{15})$$

$$g_c(A,B,C,D) = \bar{A}\bar{B}\bar{C}D + \bar{A}\bar{B}CD + \bar{A}B\bar{C}\bar{D} + \bar{A}BC\bar{D} + A\bar{B}\bar{C}\bar{D} + A\bar{B}C\bar{D} + AB\bar{C}D + ABCD$$

7.5 Logic analysis

A combinational logic diagram of two Boolean functions is shown in Figure 7.5. Derive the output logic equations.

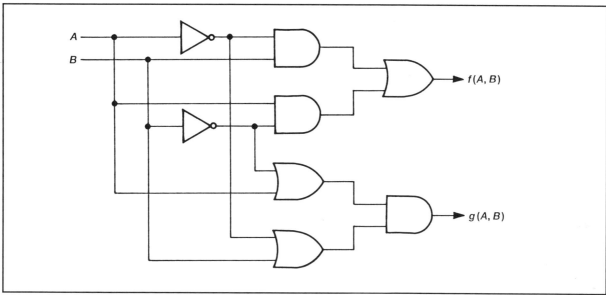

Figure 7.5 Logic analysis

Solution:

$$f(A,B) = \overline{A}B + A\overline{B} = A \oplus B$$

Using DeMorgan's Theorem 1: $\overline{AB} = \overline{A} + \overline{B}$

$$\overline{f}(A,B) = (A + \overline{B})(\overline{A} + B)$$

$$\overline{f}(A,B) = A\overline{A} + \overline{A}\,\overline{B} + AB + B\overline{B} = AB + \overline{A}\,\overline{B}$$

$$g(A,B) = (\overline{A} + B)(A + \overline{B})$$

$$g(A,B) = \overline{A}A + \overline{A}\,\overline{B} + AB + B\overline{B} = AB + \overline{A}\,\overline{B}$$

Hence, $g(A,B) = \overline{f}(A,B) = \overline{A \oplus B}$

7.6 NAND logic

Two combinational logic diagrams are shown in Figure 7.6. Determine their corresponding output logic equations.

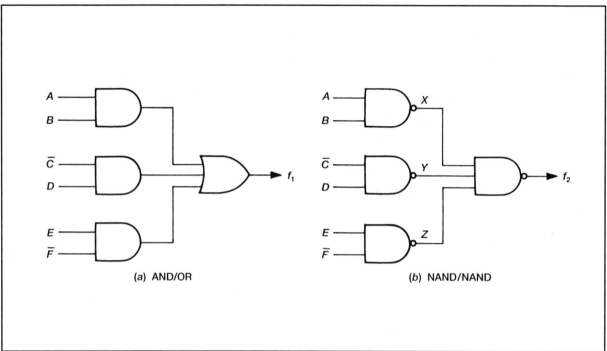

Figure 7.6 SOP Logic Equivalence

Solution:

$$f_1 = AB + \overline{C}D + E\overline{F}$$

Let

$$X = \overline{AB} = (\overline{A} + \overline{B}), \qquad Y = \overline{\overline{C}D} = (C + \overline{D}), \qquad Z = \overline{E\overline{F}} = (\overline{E} + F)$$

$$f_2 = \overline{XYZ} = \overline{(\overline{A} + \overline{B})(C + \overline{D})(\overline{E} + F)}$$

$$f_2 = AB + \overline{C}D + E\overline{F}$$

Hence $f_1 = f_2$

7.7 NOR logic

Two combinational logic diagrams are shown in Figure 7.7. Determine their corresponding output logic equations.

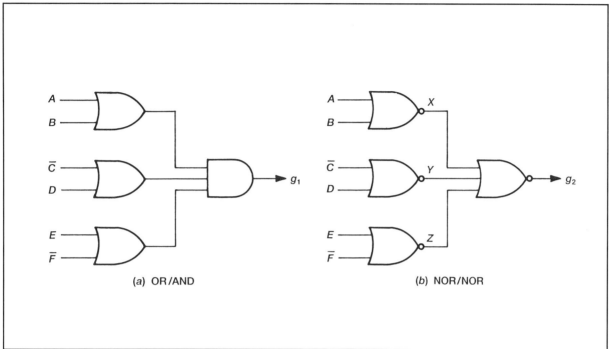

Figure 7.7 POS Logic Equivalence

Solution:

$$g_1 = (A + B)(\bar{C} + D)(E + \bar{F})$$

Let

$$X = \overline{(A + B)} = \bar{A}\bar{B}, \qquad Y = \overline{(\bar{C} + D)} = C\bar{D}, \qquad Z = \overline{(E + \bar{F})} = \bar{E}F$$

$$g_2 = \overline{X + Y + Z} = \overline{\bar{A}\bar{B} + C\bar{D} + \bar{E}F}$$

$$g_2 = (A + B)(\bar{C} + D)(E + \bar{F})$$

Hence $g_1 = g_2$

7.8 SOP Standard form

A Boolean function of four variables is expressed as a logical sum of eight minterms by
$f(A, B, C, D) = \sum(m_0, m_2, m_5, m_7, m_9, m_{11}, m_{12}, m_{14})$. Use the Karnaugh map to derive its Sum-Of-Products (SOP) standard form.

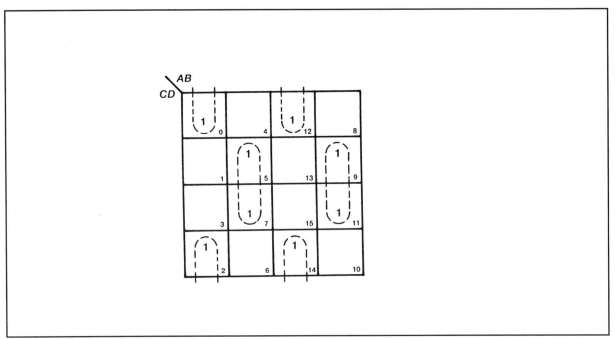

Figure 7.8 SOP Minimization

Solution:

Placing the minterms on their corresponding Karnaugh map locations, Figure 7.8, results in four adjacent pairs.

$$f(A, B, C, D) = (m_0, m_2) + (m_5, m_7) + (m_9, m_{11}) + (m_{12}, m_{14})$$

$$f(A,B,C,D) = (\overline{A}\overline{B}\overline{C}\overline{D} + \overline{A}\overline{B}C\overline{D}) + (\overline{A}B\overline{C}D + \overline{A}BCD) + (A\overline{B}\overline{C}D + A\overline{B}CD) + (AB\overline{C}\overline{D} + ABC\overline{D})$$

Using $AB + A\overline{B} = A(B + \overline{B}) = A \cdot 1 = A$ repeatedly

$$f(A,B,C,D) = \overline{A}\overline{B}\overline{D} + \overline{A}BD + A\overline{B}D + AB\overline{D}$$

7.9 POS Standard form

A Boolean function of four variables is expressed as a logical sum of eight maxterms by
$g(A, B, C, D) = \prod (M_1, M_3, M_4, M_6, M_8, M_{10}, M_{13}, M_{15})$. Use the Karnaugh map to derive its Product-Of-Sums (POS) standard form.

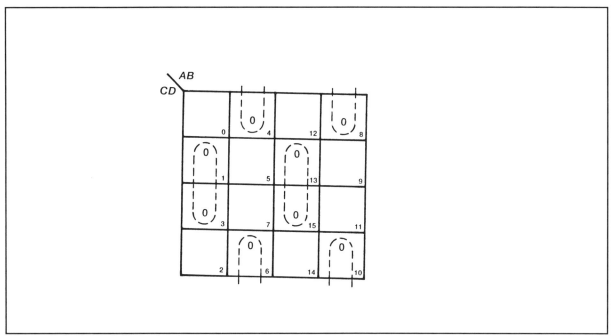

Figure 7.9 POS Minimization

Solution:

Placing the maxterms on their corresponding Karnaugh map locations, Figure 7.9, results in four adjacent pairs

$g(A, B, C, D) = (M_1, M_3)(M_4, M_6)(M_8, M_{10})(M_{13}, M_{15})$

$g(A,B,C,D) = (A + B + C + \bar{D})(A + B + \bar{C} + \bar{D})(A + \bar{B} + C + D)(A + \bar{B} + \bar{C} + D) \cdot$
$(\bar{A} + B + C + D)(\bar{A} + B + \bar{C} + D)(\bar{A} + \bar{B} + C + \bar{D})(\bar{A} + \bar{B} + \bar{C} + \bar{D})$

Using $(A + B)(A + \bar{B}) = AA + A\bar{B} + AB + B\bar{B} = A + AB + A\bar{B} + 0 =$
$A + A(B + \bar{B}) = A + A \cdot 1 = A$ repeatedly

$g(A,B,C,D) = (A + B + \bar{D})(A + \bar{B} + D)(\bar{A} + B + D)(\bar{A} + \bar{B} + \bar{D})$

7.10 Multiplexer logic

A Boolean function of four variables is expressed as a logical sum of eight minterms by
$f(A, B, C, D) = \sum (m_0, m_2, m_5, m_7, m_9, m_{11}, m_{12}, m_{14})$. Use an 8:1 multiplexer module to implement the function.

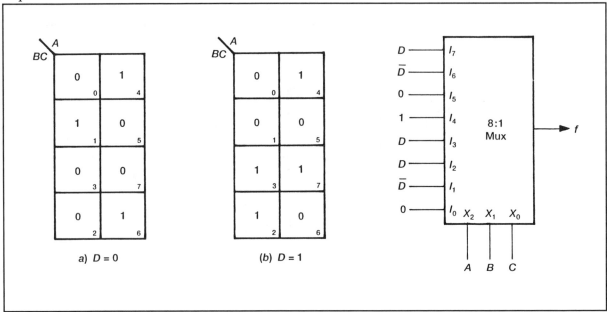

Figure 7.10 Multiplexer logic

Solution:
1. Split $f(A, B, C, D)$ into two parts, one for $D = 0$ and one for $D = 1$, then place them on the Karnaugh map.
2. Compare the corresponding values of $f(A, B, C, D)$ on the two maps and determine the value for each multiplexer input, I_i, using the following table.

	D = 0	D = 1	I_i
f	0	0	0
f	0	1	D
f	1	0	$\bar{D}$
f	1	1	1

And the resulting multiplexer inputs, for Figure 7.10, become

$I_0 = 0$, $I_1 = \bar{D}$, $I_2 = D$, $I_3 = D$, $I_4 = 1$, $I_5 = 0$, $I_6 = \bar{D}$, $I_7 = D$

7.11 Sequencer

A sequential machine is designed to produce an output, Z, when two or three consecutive zero inputs are followed by an one input, and no output otherwise. Determine the generalized next state and output equations.

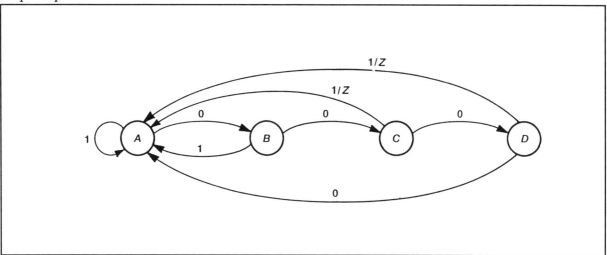

Figure 7.11 State diagram

Solution:

The state diagram is derived from the problem statement as shown in Figure 7.11.

Let $X_1 = 0$ and $X_2 = 1$

P / S	X_1	X_2
A	B	A
B	C	A
C	D	A/Z
D	A	A/Z

The generalized next state equations become

$A = AX_2 + BX_2 + CX_2 + DX_2$

$B = AX_1$

$C = BX_1$

$D = CX_1$

The generalized output equation is
$Z = CX_2 + DX_2$

7.12 Moore machine

A Moore type sequential machine logic block diagram is shown in Figure 7.12, and it is defined by the following state table. A given next state, Y_i, is defined as $Y_i = f_1(X,Y)$ and a given output, Z_i, is defined as $Z_i = f_2(Y)$. Prepare the PROM programming table using encoded states and inputs.

P/S	X_1 X_2 X_3 X_4	Z_4 Z_3 Z_2 Z_1
A	B C D A	0 0 0 1
B	C D A B	0 0 1 0
C	B A D C	0 1 0 0
D	C B A D	1 0 0 0

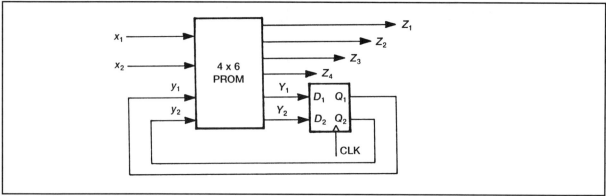

Figure 7.12 Moore machine

Solution:

Let $X_1 = \bar{x}_2\bar{x}_1 = 00$, $X_2 = \bar{x}_2 x_1 = 01$, $X_3 = x_2\bar{x}_1 = 10$, $X_4 = x_2 x_1 = 11$

Let $A = \bar{y}_2\bar{y}_1 = 00$, $B = \bar{y}_2 y_1 = 01$, $C = y_2\bar{y}_1 = 10$, $D = y_2 y_1 = 11$

Inputs	Next State	Outputs
y_2 y_1 x_2 x_1	Y_2 Y_1	Z_4 Z_3 Z_2 Z_1
0 0 0 0	0 1	0 0 0 1
0 0 0 1	1 0	0 0 0 1
0 0 1 0	1 1	0 0 0 1
0 0 1 1	0 0	0 0 0 1
0 1 0 0	1 0	0 0 1 0
0 1 0 1	1 1	0 0 1 0
0 1 1 0	0 0	0 0 1 0
0 1 1 1	0 1	0 0 1 0
1 0 0 0	0 1	0 1 0 0
1 0 0 1	0 0	0 1 0 0
1 0 1 0	1 1	0 1 0 0
1 0 1 1	1 0	0 1 0 0
1 1 0 0	1 0	1 0 0 0
1 1 0 1	0 1	1 0 0 0
1 1 1 0	0 0	1 0 0 0
1 1 1 1	1 1	1 0 0 0

7.13 Mealy machine

A Mealy type sequential machine logic block diagram is shown in Figure 7.13, and it is defined by the following state table. A given next state, Y_i, is defined as $Y_i = f_1(X,Y)$ and a given output, Z_i, is defined as $Z_i = f_2(X,Y)$. Prepare the PROM programming table using encoded states and inputs.

P/S	X_1	X_2	X_3	X_4
A	B	C	D	A/Z_1
B	C	D	A/Z_2	B
C	B	A/Z_3	D	C
D	C/Z_4	B	A	D

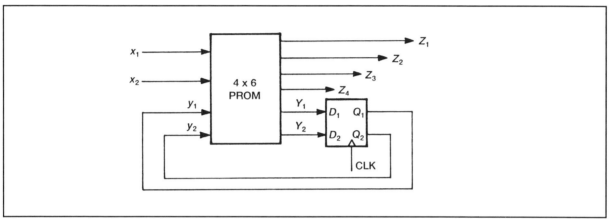

Figure 7.13 Mealy machine

Solution:

Let $X_1 = \bar{x}_2\bar{x}_1 = 00$, $\quad X_2 = \bar{x}_2 x_1 = 01$, $\quad X_3 = x_2\bar{x}_1 = 10$, $\quad X_4 = x_2 x_1 = 11$

Let $A = \bar{y}_2\bar{y}_1 = 00$, $\quad B = \bar{y}_2 y_1 = 01$, $\quad C = y_2\bar{y}_1 = 10$, $\quad D = y_2 y_1 = 11$

Inputs	Next State	Outputs
$y_2\ y_1\ x_2\ x_1$	$Y_2\ Y_1$	$Z_4\ Z_3\ Z_2\ Z_1$
0 0 0 0	0 1	0 0 0 0
0 0 0 1	1 0	0 0 0 0
0 0 1 0	1 1	0 0 0 0
0 0 1 1	0 0	0 0 0 1
0 1 0 0	1 0	0 0 0 0
0 1 0 1	1 1	0 0 0 0
0 1 1 0	0 0	0 0 1 0
0 1 1 1	0 1	0 0 0 0
1 0 0 0	0 1	0 0 0 0
1 0 0 1	0 0	0 1 0 0
1 0 1 0	1 1	0 0 0 0
1 0 1 1	1 0	0 0 0 0
1 1 0 0	1 0	1 0 0 0
1 1 0 1	0 1	0 0 0 0
1 1 1 0	0 0	0 0 0 0
1 1 1 1	1 1	0 0 0 0

7.14 Decoded states

A sequential machine logic block diagram is shown in Figure 7.14, and it is defined by the following state table. Prepare the PAL programming table using decoded states and inputs.

P / S	X_1	X_2	X_3	X_4
A	D	A	B	C
B	B	C	D	A
C	C	A	B	D
D	A	B	C	D

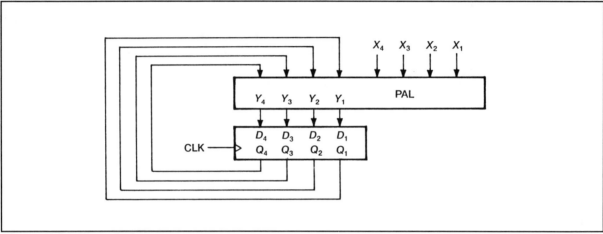

Figure 7.14 Decoded states

Solution: Let $Q_1 = A$, $Q_2 = B$, $Q_3 = C$, $Q_4 = D$

$Y_1 = Q_1X_2 + Q_2X_4 + Q_3X_2 + Q_4X_1$; $Y_2 = Q_1X_3 + Q_2X_1 + Q_3X_3 + Q_4X_2$

$Y_3 = Q_1X_4 + Q_2X_2 + Q_3X_1 + Q_4X_3$; $Y_4 = Q_1X_1 + Q_2X_3 + Q_3X_4 + Q_4X_4$

P-term #	P-term Equation	Present State Q_4 Q_3 Q_2 Q_1	Inputs X_4 X_3 X_2 X_1	Next State Y_4 Y_3 Y_2 Y_1
1	Q_1X_2	- - - 1	- - 1 -	- - - 1
2	Q_2X_4	- - 1 -	1 - - -	- - - 1
3	Q_3X_2	- 1 - -	- - 1 -	- - - 1
4	Q_4X_1	1 - - -	- - - 1	- - - 1
5	Q_1X_3	- - - 1	- 1 - -	- - 1 -
6	Q_2X_1	- - 1 -	- - - 1	- - 1 -
7	Q_3X_3	- 1 - -	- 1 - -	- - 1 -
8	Q_4X_2	1 - - -	- - 1 -	- - 1 -
9	Q_1X_4	- - - 1	1 - - -	- 1 - -
10	Q_2X_2	- - 1 -	- - 1 -	- 1 - -
11	Q_3X_1	- 1 - -	- - - 1	- 1 - -
12	Q_4X_3	1 - - -	- 1 - -	- 1 - -
13	Q_1X_1	- - - 1	- - - 1	1 - - -
14	Q_2X_3	- - 1 -	- 1 - -	1 - - -
15	Q_3X_4	- 1 - -	1 - - -	1 - - -
16	Q_4X_4	1 - - -	1 - - -	1 - - -

7.15 Encoded states

A sequential machine logic block diagram is shown in Figure 7.15, and it is defined by the following state table. Prepare the PLA programming table using encoded states and decoded inputs.

P/S	X_1	X_2	X_3	X_4
A	D	A	B	C
B	B	C	D	A
C	C	A	B	D
D	A	B	C	D

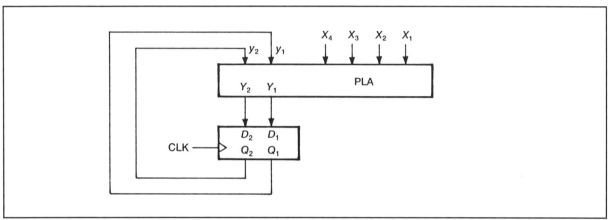

Figure 7.15 Encoded states

Solution:

Let $A = \bar{y}_2\bar{y}_1 = 00$, $\quad B = \bar{y}_2 y_1 = 01$, $\quad C = y_2\bar{y}_1 = 10$, $\quad D = y_2 y_1 = 11$

The resulting P-terms become

P/S	X_1	X_2	X_3	X_4	X_1	X_2	X_3	X_4
01	10	01	11	00	P_1	P_4	P_6	-
11	11	00	10	01	P_2	-	P_7	P_8
00	00	01	11	10	-	P_4	P_6	P_9
10	01	11	00	10	P_3	P_5	-	P_9

$Y_1 = P_1 + P_2 + P_5 + P_6 + P_7 + P_9$; $\quad Y_2 = P_2 + P_3 + P_4 + P_5 + P_6 + P_8$

P-term #	P-term Equation	Present State $y_2\ y_1$	Inputs $X_4\ X_3\ X_2\ X_1$	Next State $Y_2\ Y_1$
1	$\bar{y}_2\ y_1\ X_1$	0 1	- - - 1	1 -
2	$y_2\ y_1\ X_1$	1 1	- - - 1	1 1
3	$y_2\ \bar{y}_1\ X_1$	1 0	- - - 1	- 1
4	$\bar{y}_2\ X_2$	0 -	- - 1 -	- 1
5	$y_2\ y_1\ X_2$	1 0	- - 1 -	1 1
6	$\bar{y}_2\ X_3$	0 -	- 1 - -	1 1
7	$y_2\ y_1\ X_3$	1 1	- 1 - -	1 -
8	$y_2\ y_1\ X_4$	1 1	1 - - -	- 1
9	$\bar{y}_1\ X_4$	- 0	1 - - -	1 -

7.16 Flip-flop logic

A JK flip-flop, shown in Figure 7.16, is known as the universal type because it can be used to implement the T, D, and RS types respectively. The JK flip-flop setup time, T_{su}, equals 6 nsec, the hold time, T_h, equals 3 nsec, and the clock-to-output propagation time, T_p, equals 6.5 nsec.

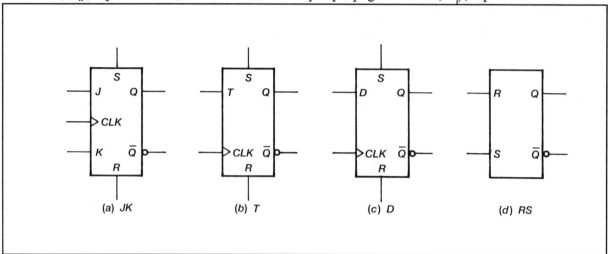

Figure 7.16 Flip-flop logic

1. The JK flip-flop state equation is:

   a) $Q_{n+1} = \bar{J}Q_n + KQ_n$ b) $Q_{n+1} = JQ_n + K\bar{Q}_n$
   c) $Q_{n+1} = J\bar{Q}_n + \bar{K}Q_n$ d) $Q_{n+1} = J\bar{Q}_n + KQ_n$

2. The T flip-flop state equation is:

   a) $Q_{n+1} = \overline{T \oplus Q_n}$ b) $Q_{n+1} = T \oplus \bar{Q}_n$ c) $Q_{n+1} = \bar{T} \oplus Q_n$ d) $Q_{n+1} = T \oplus Q_n$

3. The D flip-flop state equation is:

   a) $Q_{n+1} = D$ b) $Q_{n+1} = DQ_n$ c) $Q_{n+1} = D\bar{Q}_n$ d) $Q_{n+1} = \bar{D}$

4. The RS flip-flop state equation is:

   a) $Q_{n+1} = R + \bar{S}Q_n$ b) $Q_{n+1} = R + SQ_n$ c) $Q_{n+1} = S + \bar{R}Q_n$ d) $Q_{n+1} = S + RQ_n$

5. The maximum JK flip-flop clocking frequency is most nearly:

   a) 32×10^6 b) 80×10^6 c) 64×10^6 d) 40×10^6

7.16 Solution:

J K Q_n	Q_{n+1}
0 0 0	0
0 0 1	1
0 1 0	0
0 1 1	0
1 0 0	1
1 0 1	1
1 1 0	1
1 1 1	0

1. $Q_{n+1} = \sum(m_1, m_4, m_5, m_6) = (m_1, m_5) + (m_4, m_6)$

   $Q_{n+1} = J\overline{Q}_n + \overline{K}Q_n$

   The correct answer is (c).

2. Let $J = K = T$, $\quad Q_{n+1} = J\overline{Q}_n + \overline{K}Q_n = T\overline{Q}_n + \overline{T}Q_n = T \oplus Q_n$

   The correct answer is (d).

3. Let $J = D$ and $K = \overline{D}$, $\quad Q_{n+1} = J\overline{Q}_n + \overline{K}Q_n = D\overline{Q}_n + DQ_n = D(\overline{Q}_n + Q_n) = D$

   The correct answer is (a).

S R Q_n	Q_{n+1}
0 0 0	0
0 0 1	1
0 1 0	0
0 1 1	0
1 0 0	1
1 0 1	1
1 1 0	d
1 1 1	d

4. $Q_{n+1} = \sum(m_1, m_4, m_5, d_6, d_7) = (m_1, m_5) + (m_4, m_5, d_6, d_7)$

   $Q_{n+1} = S + \overline{R}Q_n$

   The correct answer is (c).

5. $T = T_p + T_{su} = 6.5 + 6 = 12.5$ nsec; $\quad f_{max} = 1/T = 80 \times 10^6$ Hz

   The correct answer is (b).

7.17 D Flip-flop counter

A synchronous counter using D type flip-flops is shown in Figure 7.17. The input clock has a 50% duty cycle. The flip-flop setup time, T_{su}, equals 5 nsec, the hold time, T_h, equals 3 nsec, and the clock-to-output propagation time, T_p, equals 6.5 nsec. The gate input-to-output propagation delay, T_{pg}, equals 4.5 nsec.

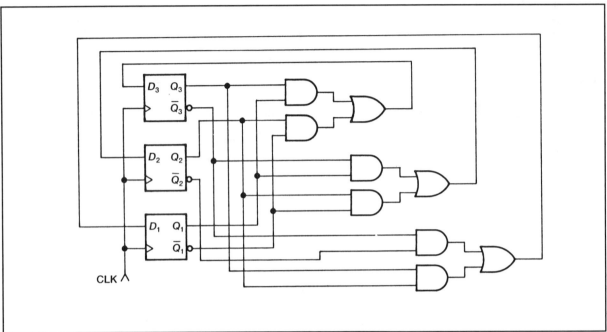

Figure 7.17 D Flip-flop counter

1. The maximum clocking frequency is most nearly:

   a) 68.9×10^6 b) 71.4×10^6 c) 62.5×10^6 d) 52.6×10^6

2. The logic equation for the D_3 input is:

   a) $D_3 = Q_3 Q_1 + Q_2 \overline{Q}_1$ b) $D_3 = \overline{Q}_3 Q_1 + Q_2 \overline{Q}_1$ c) $D_3 = \overline{Q}_3 \overline{Q}_2 + Q_3 Q_2$ d) $D_3 = Q_3 \overline{Q}_2 + \overline{Q}_3 Q_2$

3. The canonical form logic equation for the D_2 input is:

   a) $D_2 = m_2 + m_5 + m_6 + m_7$ b) $D_2 = m_1 + m_2 + m_3 + m_6$

   c) $D_2 = m_0 + m_1 + m_6 + m_7$ d) $D_2 = m_2 + m_3 + m_4 + m_5$

4. The skew between the input clock and D_3 is most nearly:

   a) 10.5×10^{-9} b) 12.5×10^{-9} c) 14.5×10^{-9} d) 15.5×10^{-9}

5. The number of valid counter states is:

   a) 4   b) 8   c) 5   d) 6

7.17 Solution:

1. $T = T_p + T_{su} + T_{pg} = 6.5 + 5 + 4.5 = 16$ nsec

   $f_{max} = 1/T = 62.5 \times 10^6$ Hz

   The correct answer is (c).

2. By inspection

   $D_3 = Q_3 Q_1 + Q_2 \overline{Q_1}$

   The correct answer is (a).

3. By inspection

   $D_2 = \overline{Q_3} Q_1 + Q_2 \overline{Q_1} = \overline{Q_3} Q_1 (Q_2 + \overline{Q_2}) + Q_2 \overline{Q_1}(Q_3 + \overline{Q_3})$
   $D_2 = \overline{Q_3}\overline{Q_2} Q_1 + \overline{Q_3} Q_2 \overline{Q_1} + \overline{Q_3} Q_2 Q_1 + Q_3 Q_2 \overline{Q_1} = m_1 + m_2 + m_3 + m_6$

   The correct answer is (b).

4. Skew $= T_p + 2T_{pg} = 6.5 + 2 \times 4.5 = 15.5$ nsec

   The correct answer is (d).

$Q_3\ Q_2\ Q_1$	$D_3\ D_2\ D_1$
0 0 0	0 0 1
0 0 1	0 1 1
0 1 1	0 1 0
0 1 0	1 1 0
1 1 0	1 1 1
1 1 1	1 0 1
1 0 1	1 0 0
1 0 0	0 0 0

5. $D_3 = Q_3 Q_1 + Q_2 \overline{Q_1}$ $D_2 = \overline{Q_3} Q_1 + Q_2 \overline{Q_1}$ $D_1 = \overline{Q_3}\overline{Q_2} + Q_3 Q_2$

   The correct answer is (b).

7.18 JK Flip-flop counter

A synchronous counter using JK type flip-flops is shown in Figure 7.18. The input clock has a 50% duty cycle.

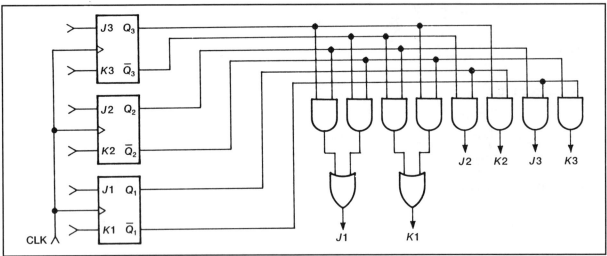

Figure 7.18 JK Flip-flop counter

1. The logic equation for the J_1 input is :

   a) $J_1 = Q_2 \oplus Q_3$ b) $J_1 = \overline{Q_1 \oplus Q_3}$ c) $J_1 = Q_1 \oplus Q_3$ d) $J_1 = \overline{Q_2 \oplus Q_3}$

2. The logic equation for the K_1 input is :

   a) $K_1 = Q_2 \oplus Q_3$ b) $K_1 = \overline{Q_2 \oplus Q_3}$ c) $K_1 = Q_1 \oplus Q_3$ d) $K_1 = \overline{Q_1 \oplus Q_3}$

3. The canonical form logic equation for the K_1 input is :

   a) $K_1 = m_0 + m_1 + m_6 + m_7$ b) $K_1 = m_0 + m_1 + m_2 + m_3$

   c) $K_1 = m_2 + m_3 + m_4 + m_5$ d) $K_1 = m_4 + m_5 + m_6 + m_7$

4. The state equation for the $Q_3(t+1)$ output is :

   a) $Q_3(t+1) = Q_2(t)\overline{Q_1}(t) + Q_3(t)Q_1(t)$ b) $Q_3(t+1) = Q_2(t)Q_3(t) + Q_3(t)Q_1(t)$

   c) $Q_3(t+1) = Q_2(t)\overline{Q_1}(t) + Q_2(t)Q_3(t)$ d) $Q_3(t+1) = Q_2(t)Q_1(t) + \overline{Q_2}(t)Q_1(t)$

5. For a present counter state $Q_3Q_2Q_1 = 010$, the next state is :

   a) $Q_3Q_2Q_1 = 011$ b) $Q_3Q_2Q_1 = 001$ c) $Q_3Q_2Q_1 = 100$ d) $Q_3Q_2Q_1 = 110$

7.18 Solution:

1. $J_1 = \bar{Q}_3\bar{Q}_2 + Q_3Q_2 = \overline{Q_2 \oplus Q_3}$

   The correct answer is (d).

2. $K_1 = \bar{Q}_3Q_2 + Q_3\bar{Q}_2 = Q_2 \oplus Q_3$

   The correct answer is (a).

3. $K_1 = \bar{Q}_3Q_2 + Q_3\bar{Q}_2 = \bar{Q}_3Q_2(Q_1 + \bar{Q}_1) + Q_3\bar{Q}_2(Q_1 + \bar{Q}_1)$
   $K_1 = \bar{Q}_3Q_2\bar{Q}_1 + \bar{Q}_3Q_2Q_1 + Q_3\bar{Q}_2\bar{Q}_1 + Q_3\bar{Q}_2Q_1$

   $K_1 = m_2 + m_3 + m_4 + m_5$

   The correct answer is (c).

4. The state equation for the $Q_3(t+1)$ output is:

   $Q_{n+1} = J\bar{Q}_n + \bar{K}Q_n$

   $J_3 = Q_2\bar{Q}_1,$ $K_3 = \bar{Q}_2\bar{Q}_1$

   $Q_3(t+1) = (Q_2\bar{Q}_1)\bar{Q}_3 + \overline{(\bar{Q}_2\bar{Q}_1)}Q_3 = \bar{Q}_3Q_2\bar{Q}_1 + Q_3Q_2 + Q_3Q_1$
   $Q_3(t+1) = (Q_3 + \bar{Q}_3)(Q_2\bar{Q}_1 + Q_2) + Q_3Q_1 = Q_2\bar{Q}_1 + Q_3Q_1$
   $Q_3(t+1) = Q_2(t)\bar{Q}_1(t) + Q_3(t)Q_1(t)$

   The correct answer is (a).

5. $Q_3Q_2Q_1(t) = 010$

   $K_3 = Q_2\bar{Q}_1 = 1$ $K_2 = \bar{Q}_3Q_1 = 0$ $K_1 = \overline{Q_2 \oplus Q_3} = 0$
   $J_3 = \bar{Q}_2\bar{Q}_1 = 0$ $J_2 = Q_3Q_1 = 0$ $J_1 = Q_2 \oplus Q_3 = 1$

   Q_3 is Set Q_2 no change Q_1 is Reset

   $Q_3Q_2Q_1(t+1) = 110$

   The correct answer is (d).

7.19 Shift register logic

The four-stage shift register shown in Figure 7.19 uses feedback logic to generate a maximum length sequence. The input clock frequency is 3×10^3 Hz.

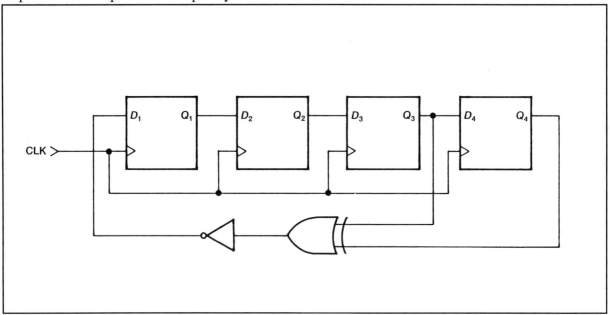

Figure 7.19 Shift register logic

1. The maximum sequence length is:

    a) 16   b) 8   c) 15   d) 4

2. The maximum length sequence repetition period is most nearly:

    a) 3.33×10^{-4} b) 1×10^{-2} c) 6.66×10^{-4} d) 5×10^{-3}

3. The maximum (-3dB) frequency is most nearly:

    a) 1,350   b) 1,500   c) 1,485   d) 1,770

4. For a present state of $Q_1Q_2Q_3Q_4 = 1001$, the next state is:

    a) $Q_1Q_2Q_3Q_4 = 0011$ b) $Q_1Q_2Q_3Q_4 = 1010$

    c) $Q_1Q_2Q_3Q_4 = 0110$ d) $Q_1Q_2Q_3Q_4 = 0100$

5. The shift register forbidden state is:

    a) $Q_1Q_2Q_3Q_4 = 1111$ b) $Q_1Q_2Q_3Q_4 = 1000$

    c) $Q_1Q_2Q_3Q_4 = 0001$ d) $Q_1Q_2Q_3Q_4 = 0000$

7.19 Solution:

1. $N = 2^n - 1 = 2^4 - 1 = 15$

Q_n	1, 0, 8, 12, 14, 7, 11, 13, 6, 3, 9, 4, 10, 5, 2
Q_{n+1}	0, 8, 12, 14, 7, 11, 13, 6, 3, 9, 4, 10, 5, 2, 1

   The correct answer is (c).

2. $f_{min} = f_c / N = 3 \times 10^3 / 15 = 200$ Hz

   $T = 1 / f_{min} = 1/200 = 5 \times 10^{-3}$ seconds

   The correct answer is (d).

3. f_{max} (-3dB) $\doteq 0.45 f_c = 0.45 \times 3 \times 10^3 = 1,350$ Hz

   The correct answer is (a).

4. $Q_1Q_2Q_3Q_4(t) = 1001$

   $f = \overline{Q_3 \oplus Q_4} = \overline{0 \oplus 1} = 0$

   $Q_1(t+1) = f = 0$

   $Q_2(t+1) = Q_1(t) = 1$

   $Q_3(t+1) = Q_2(t) = 0$

   $Q_4(t+1) = Q_3(t) = 0$

   $Q_1Q_2Q_3Q_4(t+1) = 0100$

   The correct answer is (d)

5. When $Q_1Q_2Q_3Q_4(t) = 1111$

   $f = \overline{Q_3 \oplus Q_4} = \overline{1 \oplus 1} = 1$

   $Q_1Q_2Q_3Q_4(t+1) = 1111$ the shift register locks up

   The correct answer is (a).

7.20 DDS logic

The direct digital synthesizer block diagram shown in Figure 7.20 is used to generate precision sinusoidal waveforms using digital logic. The input clock frequency, f_c, is 4.096×10^3 Hz, the number of tuning bits, k, equals 15, the number input bits, n, of the sine table is 10, and the number of output bits, m, of the sine table is 11.

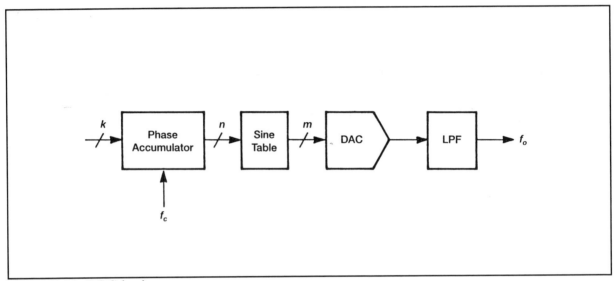

Figure 7.20 DDS logic

1. For an input tuning value of k = 15, the output frequency is:

   a) 30 b) 120 c) 68 d) 60

2. The minimum output frequency is:

   a) 8 b) 2 c) 4 d) 0.25

3. The maximum output frequency is:

   a) 2.048×10^3 b) 1.024×10^3 c) 500 d) 4.096×10^3

4. The output signal-to-noise ratio (dB) due to amplitude quantization is most nearly:

   a) 60 b) 68 c) 64 d) 66

5. For an input value of $x = 128_{10}$ to the sine table, the output angle is most nearly:

   a) 90° b) 52° c) 45° d) 38°

7.20 Solution:

1. $f_o = (f_c/N)k = (f_c/2^n)k = [(4.096 \times 10^3)/(1{,}024)] \times 15 = 60$ Hz

   The correct answer is (d).

2. $f_{min} = f_c/N = (4.096 \times 10^3)/(1{,}024) = 4$ Hz

   The correct answer is (c).

3. $f_{max} = f_c/2 = (4.096 \times 10^3)/2 = 2.048 \times 10^3$ Hz

   The correct answer is (a).

4. For a sinewave signal

   $S/N = (2^m \sqrt{12})/(2\sqrt{2})$

   $S/N(dB) = 6.02m + 1.76 = 6.02 \times 11 + 1.76 \approx 68$ dB

   The correct answer is (b).

5. $y = \sin\theta = \sin[2\pi(x/N)] = \sin[2\pi(128/1{,}024)] = 0.785$ radians

   $\theta = 0.785(360°/2\pi) = 45°$

   The correct answer is (c).

NOTES:

BIBLIOGRAPHY

1. Angelo, E. J., "Electronics: BJTs, FETs, and Microcircuits," McGraw-Hill, New York, NY, 1969.
2. Brogan, W. L., "Modern Control Theory," 3rd Edition, Prentice-Hall, Englewood Cliffs, NJ, 1991.
3. Carlson, B. A., "Communication Systems," 3rd Edition, McGraw-Hill, New York, NY, 1986.
4. Davenport, W. B., "Probability and Random Processes", McGraw-Hill, New York, NY,1970.
5. DeRusso, P. M., R. J. Roy, C. Close, "State Variables for Engineers," 2nd Edition, John Wiley & Sons Inc., New York, NY, 1997.
6. Dorf, R. C., R. H. Bishop, "Modern Control Systems," 8th Edition, Addison-Wesley, Reading, MA, 1997.
7. Fitzgerald, A.E., C. Kingsley, S. D. Umans, "Electric Machinery," 5th Edition, McGraw-Hill, New York, NY, 1990.
8. Hayt, W. H., J. Kemmerly, "Engineering Circuit Analysis," 5th Edition, McGraw-Hill, New York, NY, 1993.
9. Houpis, C. H., G. B. Lamont, "Digital Control Systems," 2nd Edition, McGraw-Hill, New York, NY, 1991.
10. Hayt, W. H., "Engineering Electromagnetics," 5th Edition, McGraw-Hill, New York, NY, 1989.
11. Karalis, E., "Digital Design Principles and Computer Architecture," Prentice-Hall, Upper Saddle River, NJ, 1996.
12. Korneff, T., "Introduction to Electronics," Academic Press, New York, NY, 1966.
13. Lance, A. L., "Introduction to Microwave Theory and Measurements," McGraw-Hill, NY, 1964.
14. Lorrain, P., D. R. Corson, "Electromagnetic Fields and Waves," 3rd Edition, W. H. Freeman and Co., New York, NY, 1988.
15. Lurch, E. N., "Fundamentals of Electronics," 3rd Edition, John Wiley & Sons Inc., New York, NY, 1981.
16. Millman, J., C. C. Halkias, "Integrated Electronics: Analog and Digital Circuits and Systems," McGraw-Hill, New York, NY, 1972.
17. Murdoch, J. B., "Network Theory,", McGraw-Hill, New York, NY, 1970.
18. Ogata, K., "Modern Control Engineering," 3rd Edition, Prentice-Hall, Englewood Cliffs, NJ, 1986.
19. Pettit, J. M., M. McWhorter, "Electronic Switching, Timing, and Pulse Circuits," McGraw-Hill, New York, NY, 1970.
20. Ramo, S., J. R. Whinnery, T. Van Duzer, "Fields & Waves in Communication Electronics," John Wiley & Sons Inc., 3rd Edition, New York, NY, 1994.
21. Sanmugam, K. S., "Digital and Analog Communication Systems," John Wiley & Sons Inc., New York, NY, 1979.
22. Savant, C. J., M. S. Roden, G. L. Carpenter, "Electronic Design: Circuits and Systems," 2nd Edition, Benjamin/Cummings, CA, 1987.
23. Schilling, D. L., C. Belove, "Electronic Circuits: Discrete and Integrated," 3rd Edition, McGraw-Hill, New York, NY, 1989.
24. Schwartz, R. J., "Information Transmission, Modulation and Noise," 4th Edition, McGraw-Hill, New York, NY, 1990.
25. Skilling, H. H., "Electrical Engineering Circuits," 2nd Edition, John Wiley & Sons, New York, NY, 1965.
26. Grainger, J. J., W. D. Stevenson, "Power System Analysis," McGraw-Hill, New York, NY, 1994.
27. Strauss, L., "Wave Generation and Shaping," 2nd Edition, McGraw-Hill, New York, NY, 1970.
28. Takahashi, Y., R. J. Rabins, D. M. Auslander, "Control and Dynamic Systems," Addison-Wesley, Reading, MA, 1970.
29. Taub, H., D. L. Schilling, "Principles of Communication Systems," 2nd Edition, McGraw-Hill, NY, 1986.
30. Van Valkenberg, M. E., M. Elwyn, "Network Analysis," 3nd Edition, Prentice-Hall, Englewood Cliffs, NJ, 1974.
31. Veinott, C. G., J. E. Martin,"Fractional and Subfractional Horsepower Motors," McGraw-Hill, New York, NY, 1986.
32. Wakerly, J. F., "Digital Design: Principles and Practices," Prentice-Hall, Englewood Cliffs, NJ, 1994.
33. Zadeh, L. A., C. A. Desoer, "Linear Systems Theory, The State Space Approach," Krieger, New York, NY, 1979.

Electrical Engineering PE Review Materials

Electrical Engineering License Review
Lincoln D. Jones, M.S., P.E., Electrical Engineer
A Completely New Book. Learn from the Professor's success in training thousands of electrical engineers. A very practical review book with numerous special test taking tips. Over 100 problems in Circuit Analysis; Electromagnetic Fields; Machinery, Power Distribution; Electronics; Control Systems; Digital Computers; and Engineering Economics. Sample Examination. 30% Text. 70% Problems but no Solutions. (Order the Problems and Solution book below). 8½ x 11 ISBN 1-57645-032-5
Order No. 325 $51.50   8th Ed 1999 Paperback   400 Pages

**Electrical Engineering License
Problems And Solutions**   Includes a Sample Exam !
Lincoln D. Jones, M.S., P.E., Electrical Engineer
Companion book to *Electrical Engineering License Review*. Here the end-of-chapter problems have been repeated and detailed Step-by-Step solutions are provided. Also included is a sample exam (same as 35X below), with detailed step-by-step solutions.100% Problems and Solutions.8½ x 11   ISBN 1-57645-033-3
Order No. 333 $30.50   8th Ed 1999   300 Pgs Paperback

A Programmed Review for Electrical Engineering Professional Engineer's Exam
James Bentley, P.E.
Covers 111 problems, solutions, and explanations for the topics on the exam. Easy-to-use tables, charts, graphs and formulas provide background needed to solve problems. The chapters cover Fundamental Concepts of Electrical Engr.; Basic Circuits; Power; Machinery; Control theory; Electronics; Logic; and Communications. 30% Text. 70% Problems. 8½ x 11. ISBN 1-57645-034-1.
Order No. 341 $43.50   3rd Ed 1999   238 Pages Paperback

Electrical Engineering Sample Exam
James Bentley, P.E.
A complete 8 hour 24 problem exam with Step-by-Step solutions. This is the sample exam included in *Electrical Engineering License Problems And Solutions*.
8 1/2 x11 paperback. ISBN 1-57645-035-X
Order No. 35X $19.50   1999   90 Pages

EE Rapid Problem Index   Essential !
Our editors have carefully classified by key word each of the problems found in three of our most popular EE review books: *Electrical Engineering License Review; Electrial Engineering License Problems and Solutions;* and *A Programmed Review for Electrical Engineering Professional Engineering Exam.* This complete index will save you valuable time seaching for the right problem as you attempt the principles and practice exam.ISBN 1-57645-026-0
Order No. 260 $4.95   1998   32 Pages

Order By Phone
TOLL FREE: 1 (800) 800-1651
  Hours: 9 AM - 5 PM EST   Mon - Fri
      (6 AM - 2PM PST)
Only credit card orders accepted by phone.
Visa, MasterCard, Discover, Am Express.

Order By Mail Please confirm current prices before ordering. See our website or call 800-800-1651. Complete the order form and send with your check including shipping charges to:

Engineering Press
P.O. Box 200129
Austin, TX 78720-0129

Order By Fax
Fax your order along with your credit card number and expiration date.
Our toll-free fax number:
      (800)700-1651
Only credit card orders accepted by fax.

INDEX

AC Delta motor, 6-37
AC induction motor, 6-41
AC synchronous motor, 6-42
AC voltage divider, 1-40
AC Wye generator, 6-36
AC Wye motor, 6-39
AC Delta generator, 6-35
Adaptive control, 3-47
ADC, 4-60
Admittance, 1-26
AM receiver, 4-50
AM/FM modulation, 4-35
AM modulation, 4-30
Ampere's law, 5-22
Antenna radiation, 5-32
Astable multivibrator, 2-53
Attenuator, 1-64
Back EMF, 6-25
Balanced modulator, 4-45
Bandpass filter, 4-25
Bandreject filter, 4-26
Batteries in series, 6-54
Batteries in parallel, 6-55
Bipolar junction transistor, 2-3
Bistable multivibrator, 2-55
Block diagrams, 3-3
Bode plot, 3-13
Breaker ratings, 6-60
Bridge rectifier, 6-50
Buffer amplifier, 2-66
Capacitance, 1-5, 5-3
Cascode amplifier, 2-62
CDF function, 4-11
Channel capacity, 4-56
Circular wave guide, 5-45
Clamper, 2-42
Clipper, 2-41
Closed-loop control, 3-2
Coaxial cable, 5-39
Collpitts oscillator, 2-49
Common collector amplifier, 2-9
Common drain amplifier, 2-12
Common emitter amplifier, 2-7
Common gate amplifier, 2-11
Common source amplifier, 2-10
Common base amplifier, 2-8

Comparator, 2-40
Complex plane, 1-41, 3-22
Compound DC motor, 6-34

Compound DC generator, 6-30
Conditional probability, 4-13
Control loop ratios, 3-51
Controllability, 3-35
Convolution, 4-3
Correlation, 4-4
Coupled circuit, 1-62
Cross spectral density, 4-7
Cross correlation, 4-6
Current amplifier, 2-26
Current divider, 1-8
Current sweep generator, 2-46
Current, 1-2
Cylindrical cavity, 5-50
Cylindrical capacitor, 5-6
D Flip-flop counter, 7-18
DAC, 4-58
Darlington amplifier, 2-60
DDS logic, 7-24
Decoded states, 7-14
Delay line LPF, 4-27
Delay line HPF, 4-28
Delta:Delta, 6-19
Delta:Wye, 6-20
Delta to Wye conversion, 6-14
DeMorgan's Theorem 2, 7-4
DeMorgan's Theorem 1, 7-3
Describing function, 3-41
Diagonal matrix, 3-33
Differential amplifier, 2-32
Differentiating op amp, 2-38
Diode, 2-1
Dipole antenna, 5-36
Direct digital control, 3-50
Discrete-time control, 3-49
Duality theorem, 1-18
Dynamic range, 4-20
Eigenvalues and eigenvectors, 3-32
Electric dipole, 5-28
Electric energy, 5-12
Electric field, 5-8
Electric field strength, 5-9

Electric flux density, 5-11
Electric force, 5-7
Electric interference, 5-55
Electric potential, 5-10
Electric quadrupole, 5-29
Electric deflection, 5-13
Encoded states, 7-15
Envelope detector, 4-46
Far field, 5-34
Faraday's law, 5-23
Feedback amplifier, 5-30
Field effect transistor, 2-4
First order loop, 3-6
Flip-flop logic, 7-16
FM receiver, 4-51
FM/AM modulation, 4-36
FM modulation, 4-31
Four wire Wye:Wye, 6-16
Fourier series, 4-1
Fourier integral, 4-2
Frequency domain, 3-12
Frequency response, 1-51
Frequency analysis, 3-57
FSK modulation, 4-43
Full wave rectifier, 6-49
Gain and phase margins, 3-17
Gauss's law, 5-24
Generator effect, 6-23
Ground faults, 6-58
Half wave rectifier, 6-47
Hall effect, 2-59
Hartley oscillator, 2-48
High frequency model, 2-16
Highpass filter, 4-24
Hybrid parameters, 1-60
Hybrid amplifier, 2-64
Impedance matching, 5-53
Impedance, 1-25
Impulse response, 1-49
Impulse function, 1-44
Inductance, 1-4, 5-2
Insertion loss, 1-54
Integrating op amp, 2-37
Intercept point, 4-22
Intermodulation distortion, 4-21
Inverting op amp, 2-34
JK Flip-flop counter, 7-20
Joint probability, 4-12
Kalman filter, 3-46
Kirchhoff's voltage law, 1-13

Kirchhoff's current law, 1-14
Lag compensation, 3-19
Laplace transform, 1-43
Laplace's equation, 5-21
Large signal amplifier, 2-22
Lead compensation, 3-18
Lead-lag compensation, 3-20
Limit cycles, 3-40
Limiter, 2-43
Line inductance, 6-2
Line parameters, 5-51
Line resistance, 6-1
Line capacitance, 6-3
Log attenuator, 2-57
Logic analysis, 7-5
Long line, 6-6
Loop antenna, 5-37
Low frequency model, 2-15
Lowpass filter, 4-23
Lyaponov criterion, 3-37
Magnetic dipole, 5-30
Magnetic energy, 5-19
Magnetic field, 5-15
Magnetic field strength, 5-16
Magnetic flux density, 5-18
Magnetic force, 5-14
Magnetic interference, 5-57
Magnetic potential, 5-17
Magnetic quadrupole, 5-31
Magnetic deflection, 5-20
Mason's rule, 3-5
Matched filter, 4-29
Maximum principle, 3-45
Maximum power transfer, 6-8
Maxterms, 7-2
Maxwell's law, 5-25
Mealy machine, 7-13
Mean squared value, 4-15
Mean, 4-14
Medium line, 6-5
Miller's theorem, 2-44
Millman's theorem, 1-17
Minterms, 7-1
Monostable multivibrator, 2-54
Moore machine, 7-12
MOSFET enhancement mode, 2-6
MOSFET depletion mode, 2-5
Motor rating, 6-64
Motor effect, 6-24
Multiplexer logic, 7-10

Multiplier, 2-58
Mutual inductance, 6-11
NAND logic, 7-6
Near field, 5-33
nFM modulation, 4-33
Nichols chart, 3-14
Noise figure, 4-18
Noise Power, 4-54
Noninverting op amp, 2-35
Nonlinear control, 3-38
NOR logic, 7-7
Norton's theorem, 1-16
Nyquist plot, 3-15
Observability, 3-36
Ohms's Law, 1-6
One port network, 1-52
Op amp filter, 2-68
Open-loop control, 3-1
Operational amplifier, 2-33
Optimal control, 3-44
PAM modulation, 4-39
Parallel conductors, 5-40
Parallel plate capacitor, 5-4
Parallel resonance, 1-31
Parallel to series, 1-28
Partial fractions, 3-23
PCM modulation, 4-42
PD controller, 3-27
PDF function, 4-10
PDM modulation, 4-40
Per unit notation, 6-56
Phase plane, 3-39
Phase shift oscillator, 4-47
Phasor, 1-50
PI controller, 3-28
PID controller, 3-29
Pierce oscillator, 2-50
PM modulation, 4-32
POS Standard form, 7-9
Power amplifier, 2-28
Power factor, 6-7
Power factor correction, 6-62
PPM modulation, 4-41
Probability, 4-8
PSK modulation, 4-44
QAM modulation, 4-38
Quarter wave stub, 5-41
Radar, 5-59
Ramp function, 1-46
Ratio detector, 4-49

RC charge circuit, 1-9
RC coupled amplifier, 2-13
RC decay circuit, 1-10
RC forced response, 1-34
RC natural response, 1-33
Reactance, 1-24
Reciprocity theorem, 1-20
Rectangular cavity, 5-49
Rectangular wave guide, 5-44
Reflections (R<Z), 5-42
Reflections (R>Z), 5-43
Resistance, 1-3, 5-1
Resolver, 6-43
RL charge circuit, 1-11
RL decay circuit, 1-12
RL forced response, 1-36
RL natural response, 1-35
RLC forced response, 1-38
RLC natural response, 1-37
Root locus, 3-26
Ruth-Hurwitz, 3-25
S-plane analysis, 3-55
Sample and Hold, 4-52
Schmitt trigger, 2-56
SCR, 6-52
Second order loop, 3-7
Self inductance, 6-10
Sensitivity, 3-11, 4-19
Separate winding DC generator, 6-27
Separate winding DC motor, 6-31
Sequencer, 7-11
Series DC generator, 6-28
Series DC motor, 6-32
Series resonance, 1-29
Series to parallel, 1-27
Servo, 6-45
Shift register logic, 7-22
Short line, 6-4
Shunt DC generator, 6-29
Shunt DC motor, 6-33
Signal flow graphs, 3-4
Sinusoidal function, 1-22
Small signal amplifier, 2-21
Solenoid, 5-27
SOP Standard form, 7-8
Spectral density, 4-5
Spherical capacitor, 5-5
Square law detector, 4-47
Square wave, 1-47
SSB modulation, 4-37

Stability, 3-16
Stability criterion, 3-42
Standard deviation, 4-17
State space model, 3-30
State variable controller, 3-31
Step Function, 1-45
Stepper motor, 6-46
Stochastic control, 3-43
Substitution theorem, 1-21
Summing op amp, 2-36
Superposition theorem, 1-19
Switched capacitor network, 1-55
Synchro, 6-44
Synchronous detector, 4-48
TE wave, 5-47
TEM wave, 5-48
Thevenin's theorem, 1-15
Three wire Wye:Delta, 6-18
Three wire Wye:Wye, 6-17
Time analysis, 3-53
Time domain, 3-21
TM wave, 5-46
Toroid, 5-26
Torque, 6-26
Transfer function, 3-24
Transfer function analysis, 3-59
Transfer function op amp, 2-39
Transformer, 6-12
Transition matrix, 3-34
Transmission constants, 1-58
Transmission line, 5-38
TRIAC, 6-53
Tuned amplifier, 2-17
Two port network, 1-53
Type 0 error, 3-8
Type 1 error, 3-9
Type 2 error, 3-10
UJT oscillator, 2-52
Unbalanced Delta load, 6-22
Unbalanced Wye load, 6-21
Variance, 4-16
Vertical antenna, 5-35
Voltage, 1-1
Voltage amplifier, 2-24
Voltage divider, 1-7
Voltage regulator, 6-51
Voltage sweep generator, 2-45
Waveform analysis, 1-56
wFM modulation, 4-34
Wideband amplifier, 2-19

Wye to Delta conversion, 6-15
z Transform, 3-48
Zener diode, 2-2